STUDY GUIDE

CHERYL V. R

SECOND E

BEGINNING ALGEBRA

K. ELAYN MARTIN-GAY

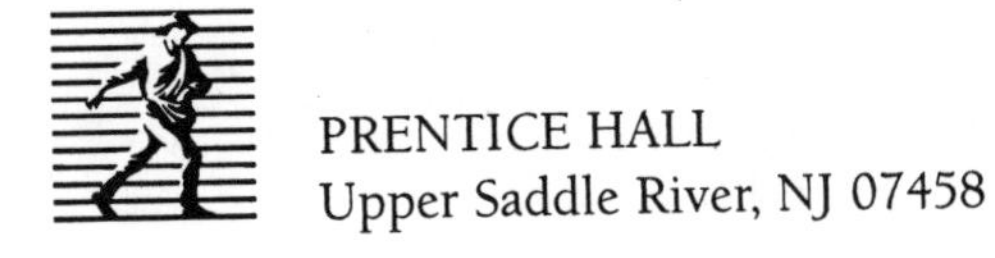

PRENTICE HALL
Upper Saddle River, NJ 07458

Acquisitions Editor: ***April Thrower***
Production Editor: ***James Buckley***
Production Supervisor: ***Joan Eurell***
Production Coordinator: ***Alan Fischer***
Art Director: ***Amy Rosen***

Simon & Schuster/A Viacom Company
Upper Saddle River, NJ 07458

Printed in the United States of America

10 9 8 7 6 5 4 3 2 1

ISBN 0-13-568403-X

Prentice-Hall International (UK) Limited, *London*
Prentice-Hall of Australia Pty. Limited, *Sydney*
Prentice-Hall Canada, Inc., *Toronto*
Prentice-Hall Hispanoamericana, S.A., *Mexico*
Prentice-Hall of India Private Limited, *New Delhi*
Prentice-Hall of Japan, Inc., *Tokyo*
Simon & Schuster Asia Pte. Ltd., *Singapore*
Editora Prentice-Hall do Brasil, Ltda., *Rio de Janeiro*

TABLE OF CONTENTS

CHAPTER 1 REVIEW OF REAL NUMBERS

1.1 Symbols and Sets of Numbers	1
1.2 Fractions	2
1.3 Exponents and Order of Operations	3
1.4 Introduction to Variable Expressions and Equations	4
1.5 Adding Real Numbers	6
1.6 Subtracting Real Numbers	6
1.7 Multiplying and Dividing Real Numbers	8
1.8 Properties and Real Numbers	9
1.9 Reading Graphs	10
Practice Test	14

CHAPTER 2 EQUATION, INEQUALITIES, AND PROBLEM SOLVING

2.1 Simplifying Algebraic Expressions	16
2.2 The Addition Property of Equality	17
2.3 The Multiplication Property of Equality	18
2.4 Solving Linear Equations	19
2.5 An Introduction to Problem Solving	21
2.6 Formulas and Problem Solving	24
2.7 Percent and Problem Solving	25
2.8 Further Problem Solving	27
2.9 Solving Linear Inequalities	30
Practice Test	33

CHAPTER 3 GRAPHING

3.1 The Rectangular Coordinate System	34
3.2 Graphing Linear Equations	36
3.3 Intercepts	38
3.4 Slope	41
3.5 Graphing Linear Inequalities	43
Practice Test	45

CHAPTER 4 EXPONENTS AND POLYNOMIALS
4.1 Exponents 46
4.2 Adding and Subtracting Polynomials 47
4.3 Multiplying Polynomials 48
4.4 Special Products 48
4.5 Negative Exponents and Scientific Notation 49
4.6 Division of Polynomials 50
Practice Test 52

CHAPTER 5 FACTORING POLYNOMIALS
5.1 The Greatest Common Factor and Factoring by Grouping 54
5.2 Factoring Trinomials of the Form 54
5.3 Factoring Trinomials of the Form 55
5.4 Factoring Binomials 57
5.5 Choosing a Factoring Strategy 57
5.6 Solving Quadratic Equations by Factoring 58
5.7 Quadratic Equations and Problem Solving 60
Practice Test 63

CHAPTER 6 RATIONAL EXPRESSIONS
6.1 Simplifying Rational Expressions 64
6.2 Multiplying and Dividing Rational Expressions 65
6.3 Adding and Subtracting Rational Expressions Common and Least Common Denominator 67
6.4 Adding and Subtracting Rational Expressions with Unlike Denominators 69
6.5 Simplifying Complex Fractions 71
6.6 Solving Equations Containing Rational Expressions 72
6.7 Ratio and Proportion 74
6.8 Rational Equations and Problem Solving 76
Practice Test 80

CHAPTER 7 FURTHER GRAPHING
7.1 The Slope-Intercept Form 82
7.2 The Point-Slope Form 83
7.3 Graphing Nonlinear Equations 85
7.4 An Introduction to Functions 88
Practice Test 90

CHAPTER 8 SOLVING SYSTEMS OF LINEAR EQUATIONS
8.1 Solving Systems of Linear Equations by Graphing 92
8.2 Solving Systems of Linear Equations by Substitution 94
8.3 Solving Systems of Linear Equations by Addition 96
8.4 Systems of Linear Equations and Problem Solving 97
8.5 Systems of Linear Inequalities 101
Practice Test 104

CHAPTER 9 ROOTS AND RADICALS
9.1 Introduction to Radicals 105
9.2 Simplifying Radicals 106
9.3 Adding and Subtracting Radical Expressions 107
9.4 Multiplying and Dividing Radical Expressions 108
9.5 Solving Equations Containing Radicals 110
9.6 Radical Equations and Problem Solving 112
9.7 Rational Exponents 115
Practice Test 117

CHAPTER 10 SOLVING QUADRATIC EQUATIONS
10.1 Solving Quadratic Equations by the Square Root Method 118
10.2 Solving Quadratic Equations by Completing the Square 119
10.3 Solving Quadratic Equations by the Quadratic Formula 120
10.4 Summary of Methods for Solving Quadratic Equations 123
10.5 Complex Solutions to Quadratic Equations 125
10.6 Graphing Quadratic Equations 127
Practice Test 130

PRACTICE FINAL EXAMINATION 131

SOLUTIONS 136

Study for Success

Congratulations! You have made a responsible decision in owning this Study Guide that supplements your textbook Beginning Algebra, 2e. Elayn Martin-Gay. This Study Guide provides you with additional resources to help you be successful in your mathematics course.

In the first part of the Study Guide there are examples and exercises to supplement each section of the textbook. Also there is a practice test to accompany each chapter and a practice final examination. All of the examples, exercises, and tests are similar to those in the textbook.

At the beginning of each section the examples with their solutions are presented. These are followed by an exercise set of 15-30 problems. At the end of each chapter a practice test is presented. At the end of all of the chapters a 100 question practice final examination is featured.

In the second part of the Study Guide complete step-by-step solutions are worked out for all exercises, tests, and examinations. The methods of solving the problems are just like those presented in the textbook.

Consider the following tips for becoming more successful in your mathematics course.

- Attend class everyday. If you have to miss a class, borrow the notes from another student and make a copy. Call and get the homework assignment so that you do not get behind.
- In the classroom be sure you have a good view of the board and that you can hear.
- Come to class prepared. Have a notebook just for this class, preferably consisting of a section for notes and a section for homework. Bring writing utensils.
- Since a lecture can move rapidly you are forced to take notes quickly. Set up your own abbreviation system. Take notes as thoroughly as possible and after class you can re-write your notes more legibly and fill in the gaps.
- Be sure you know what is assigned for homework. Do your homework as soon after lecture as possible. This way the information is still fresh in your mind.
- When reading your textbook, don't just read the examples. The entire section is important in your understanding of the material.
- As you do homework problems and read the textbook, write down your questions. Don't rely on your memory, because when you get to class you will most likely have forgotten some of your questions.
- Don't be afraid to ask questions. If something is not clear ask for further explanation. Usually there are other students with the same questions.

- When doing homework or taking a test give yourself plenty of paper. Show every step of a problem in an organized manner. This will help reduce the number of careless errors.

- It is important to study your mathematics as many days as you can possibly fit into your schedule. Even if it is just for 10 or 15 minutes, this is beneficial. Do not wait until just before a test to do a crash study session. The material needs to be understood and learned, not memorized. Most likely you will have another mathematics course after this one. You are gaining tools in this class that you will use in the next class. **Retention is the key word.**

  Of course you can't remember everything, that's why it's important to have good notes. Hopefully in the next class even if you can't remember exactly how to do a problem, you will remember that you have done it before and you can find it in your notes. Then you refresh your memory more quickly.

- When doing your homework try each type of problem that was assigned. It is easy to sit in a lecture and a problem appear very simple as the instructor does the work. However, when you are on your own it may all of a sudden become more difficult.

- Practice is what improves your skills in every area including mathematics. Practice is what will make you become comfortable with problems. If there are certain assigned problems that you are having difficulty with, do more of this type until you can do them easily. Remember this Study Guide provides you with lots of additional problems. Once you have done them you can then check them with the completely worked out solutions that are also provided.

- Have assignments in on time. Do not lose points needlessly.

- Take advantage of all extra-credit opportunities if they exist.

- Once more let me stress the importance of preparing daily for a test not just the day before a test.

- Many times students can do their homework assignments, but then "freeze-up" on a test. Their minds go blank, they panic and hence they do not do well on their test. If this happens to you, it can be overcome by practicing taking tests. When studying for a test actually make yourself a test. Then find a quiet spot and pretend you are in class. Take this test just as if you were in class. The more you practice taking tests, the more comfortable you will be in class during the actual event. Don't forget the Practice Tests provided in this Study Guide. When you use them you will not only be able to check your answers, but also your work since complete solutions are provided.

- When your instructor hands back a graded test, be sure to make corrections to any problems you missed. It is important to clear up these mistakes now because you will generally be using these concepts again. Also when exam time rolls around, your old tests are a great study source. Remember there is a Practice Final Examination along with complete solutions provided in this Study Guide. Use it to prepare for your exam.

- If you have taken advantage of the previous tips, but you are still having a lot of difficulty, look for other sources of help. See your instructor during office hours, work with a fellow class-mate, form a study group, go to the math lab if one exists at your school, or get a tutor. Don't let things get out of hand, get help early. Be a responsible student.

Good luck in your mathematics course! Hopefully this Study Guide and these helpful hints will contribute to your success in this course. However, remember it is up to you to be a responsible student and take advantage of all resources provided and take all of the steps necessary to being successful!

Cheryl V. Roberts
Northern Virginia Community College

1.1 SYMBOLS AND SETS OF NUMBERS

Example 1: Insert <, >, or = between 10 and 14 to make a true statement.

Solution: $10 < 14$ since 10 is to the left of 14 on the number line.

Example 2: Tell whether the statement $6 \geq 6$ is true or false.

Solution: True, since $6 = 6$ is true.

Example 3: Translate the following sentence into a mathematical statement.

Seven is greater than or equal to four.

Solution:

Seven	is greater than or equal to	four
7	$\geq$	4

Example 4: Find the following absolute value: $|-6|$

Solution: $|-6| = 6$ since -6 is 6 units from 0 on the number line.

1.1 EXERCISES

Insert <, >, or = in the space between the paired numbers to make each statement true.

1. 12   21
2. 8.13   8.13
3. -18 -19
4. $|-22|$ $|-23|$
5. -48 $|-48|$
6. $\left|-\frac{4}{3}\right|$ $\frac{4}{3}$

Write each sentence as a mathematical statement.

7. Seven is less than 10.
8. Negative twelve is greater than or equal to negative twenty.
9. Nine is not equal to negative nine.

Are the following statements true or false?

10. $30 \geq 30$
11. 0 is a positive number.
12. $4(6) < 4 + 6$
13. Every natural number is an integer.
14. $|-11| > |-10|$
15. $|0| \geq 0$

1.2 FRACTIONS

Example 1: Write the fraction $\frac{87}{116}$ in lowest terms.

Solution: $\frac{87}{116} = \frac{3 \cdot 29}{2 \cdot 2 \cdot 29}$ Write the numerator and denominator as products of primes.

$= \frac{3}{4}$ Apply the Fundamental Principle of Fractions.

Example 2: Find the product of $\frac{3}{16}$ and $\frac{4}{7}$. Write the product in lowest terms.

Solution: $\frac{3}{16} \cdot \frac{4}{7} = \frac{3 \cdot 4}{16 \cdot 7}$ Multiply numerators. Multiply denominators.

$= \frac{3 \cdot 2 \cdot 2}{2 \cdot 2 \cdot 2 \cdot 2 \cdot 7}$ Write the numerator and denominator as products of primes.

$= \frac{3}{28}$ Apply the Fundamental Principle of Fractions.

Example 3: Find the quotient: $\frac{3}{5} \div 15$ Write the answer in lowest terms.

Solution: $\frac{3}{5} \div 15 = \frac{3}{5} \div \frac{15}{1}$ Write 15 as $\frac{15}{1}$.

$= \frac{3}{5} \cdot \frac{1}{15}$ Multiply by the reciprocal of 15.

$= \frac{3 \cdot 1}{5 \cdot 15}$

$= \frac{3}{5 \cdot 3 \cdot 5}$ Factor.

$= \frac{1}{25}$ Reduce.

Example 4: Add $\frac{7}{9}$ and $\frac{5}{6}$. Write the answer in lowest terms.

Solution: $\frac{7}{9} = \frac{7 \cdot 2}{9 \cdot 2} = \frac{14}{18}$ Write each fraction with the least common denominator, 18.

$\frac{5}{6} = \frac{5 \cdot 3}{6 \cdot 3} = \frac{15}{18}$

$$\frac{7}{9} + \frac{5}{6} = \frac{14}{18} + \frac{15}{18}$$

$$= \frac{14 + 15}{18} \quad \text{Add the numerators.}$$

$$= \frac{29}{18}$$

1.2 EXERCISES

Write each of the following numbers as a product of primes.

1. 180
2. 175
3. 48
4. 882

Write the following fractions in lowest terms.

5. $\frac{20}{45}$
6. $\frac{38}{57}$
7. $\frac{34}{85}$
8. $\frac{138}{161}$

Multiply or divide as indicated. Write the answer in lowest terms.

9. $\frac{4}{5} \cdot \frac{20}{14}$
10. $\frac{3}{8} \cdot 2\frac{4}{9}$
11. $\frac{5}{6} \div \frac{15}{24}$
12. $8 \div \frac{4}{3}$

Add or subtract as indicated. Write the answer in lowest terms.

13. $\frac{19}{26} - \frac{7}{26}$
14. $\frac{10}{9} + \frac{2}{27}$
15. $9 - \frac{2}{3}$
16. $5\frac{3}{7} + 2\frac{1}{2}$

Perform the following operations. Write answers in lowest terms.

17. $\frac{3}{4} + \frac{8}{5} - \frac{1}{3}$
18. $6\frac{2}{3} \div \frac{5}{3}$
19. $\frac{8}{9} + \frac{3}{4}$
20. $3\frac{1}{8} + 1\frac{2}{7} + \frac{3}{5}$

1.3 EXPONENTS AND ORDER OF OPERATIONS

Example 1: Simplify $2 \cdot 9 - 6 \div 6$

Solution: $2 \cdot 9 - 6 \div 6 = 18 - 1$ Multiply and divide left to right.

$= 17$ Subtract.

Example 2: Simplify $\frac{5 + |7 - 2| + 3^2}{9 - 4}$

Solution: $\frac{5 + |7 - 2| + 3^2}{9 - 4} = \frac{5 + |5| + 3^2}{9 - 4}$ Simplify inside the absolute value bars.

$$= \frac{5 + 5 + 3^2}{5}$$ Find the absolute value and simplify the denominator.

$$= \frac{5 + 5 + 9}{5}$$ Evaluate the exponential expression.

$$= \frac{19}{5}$$ Simplify the numerator.

Example 3: Simplify $6[5(4 + 3) - 8]$

Solution: $6[5(4 + 3) - 8] = 6[5(7) - 8]$ Simplify the expression in parentheses.
$= 6(35 - 8)$ Multiply 5 and 7.
$= 6(27)$ Subtract inside parentheses.
$= 162$ Multiply.

1.3 EXERCISES

Evaluate.

1. 7^3
2. $\left(\frac{3}{2}\right)^4$
3. $(0.03)^2$
4. 0^4
5. 1^3
6. 4^3

Simplify each expression.

7. $9 + 8 \cdot 3$
8. $11 \cdot 4 - 2 \cdot 7$
9. $6 \cdot 2^3$
10. $12(7 - 4)$
11. $\frac{5 - 2}{15 - 4}$
12. $\frac{2}{5} \cdot \frac{10}{3} - \frac{5}{6}$
13. $6[4 + 3(9 - 5)]$
14. $\frac{4 + 4(6 + 3)}{4^2 + 2}$
15. $\frac{21 + |18 - 4| + 5^2}{23 - 3}$

Translate each word statement to symbols.

16. The sum of eight and twelve is greater than fifteen.
17. The difference of twenty and four is less than nineteen.
18. The product of two and six is greater than ten.

1.4 INTRODUCTION TO VARIABLE EXPRESSIONS AND EQUATIONS

Example 1: Evaluate $x^2 + 2y$ if $x = 4$ and $y = 3$.

Solution: $x^2 + 2y = 4^2 + 2(3)$ Replace x with 4 and y with 3.
$= 16 + 6$
$= 22$

Example 2: Decide whether 3 is a solution of $5x + 9 = 8x$.

Solution:

$$5x + 9 = 8x$$
$$5(3) + 9 = 8(3) \quad \text{Replace } x \text{ with 3.}$$
$$15 + 9 = 24 \quad \text{Simplify each side.}$$
$$24 = 24 \quad \text{True}$$

Since we arrived at a true statement, 3 is a solution of the equation.

Example 3: Write the following sentence as an equation. Let x represent the unknown number.

Six less than a number is 15.

Solution:

Six less than a number

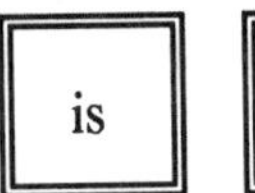

15

$x - 6 \quad = \quad 15$

1.4 EXERCISES

Evaluate each expression if $x = 2$, $y = 5$ and $z = 6$.

1. $4x - z$
2. $xz - 2y$
3. $3z^2$
4. $|z - 4| + 8x$
5. $\frac{z}{x} + y$
6. $\frac{x^2 + y}{y^2 - z}$

Write each of the following sentences as an equation. Use x to represent the unknown number.

7. The sum of a number and 4 is 18.
8. The quotient of 14 and a number is $\frac{1}{7}$.
9. Twice a number subtracted from 30 is 11.
10. The sum of triple a number and 5 is 84.
11. Eight minus four times a number is 31.
12. The product of $\frac{1}{5}$ and the sum of a number and 12 equals 7.
13. The sum of twice a number and one is 60.
14. The quotient of 15 and a number is $\frac{2}{3}$.

Decide whether the given number is a solution of the given equation.

15. $4x + 8 = 28$; 5
16. $3x - 2 = 4x$; 2
17. $7x + 10 = 9x - 6$; 8
18. $5x + 1 = 5x + 1$; 50
19. $9 = 8 - x$; 1
20. $8 - 3x = 14$; 2

1.5 ADDING REAL NUMBERS

Example 1: Find the sum: $8 + (-5) + (-9)$

Solution: $8 + (-5) + (-9)$

$= 3 + (-9)$ Adding numbers with different signs.

$= -6$ Adding numbers with different signs.

Example 2: Simplify $-|-8|$

Solution: $-|-8| = -(8)$ $|-8| = 8$

$= -8$ The opposite of 8 is -8.

1.5 EXERCISES

Find the sums.

1. $-12 + (-4)$
2. $11 + (-7)$
3. $8 + (+18)$
4. $-13 + 5$
5. $-\frac{8}{15} + \frac{3}{5}$
6. $\frac{5}{2} + \left(-\frac{1}{4}\right)$
7. $10 + (-30) + 25$
8. $-131 + (-64) + 17$
9. $|4 + (-19)|$
10. $[-2 + 6] + (-12)$
11. $-27 + (-4) + 13$
12. $|8 + (-14)| + |-10|$

Find the additive inverse or the opposite.

13. -9
14. $-\frac{2}{3}$
15. $|-17|$
16. The low temperature in Anoka Ramsey, Minnesota was -18 degrees last night. During the day it rose only 11 degrees. Find the high temperature for the day.

1.6 SUBTRACTING REAL NUMBERS

Example 1: Simplify: $-13 + [(-6 - 1) - 4]$

Solution: $-13 + [(-6 - 1) - 4] = -13 + [(-6 + (-1)) - 4]$ Rewrite $-6 - 1$ as a sum.

$= -13 + [-7 - 4]$ Add: $-6 + (-1)$

$= -13 + [-7 + (-4)]$ Rewrite $-7 - 4$ as a sum.

$= -13 + (-11)$ Add: $-7 + (-4)$

$= -24$ Add.

Example 2: If $x = -3$ and $y = 5$, evaluate $\frac{x - y}{10 + x}$.

Solution: $\frac{x - y}{10 + x} = \frac{-3 - 5}{10 + (-3)}$ Replace x with -3 and y with 5.

$= \frac{-3 + (-5)}{10 + (-3)}$ Rewrite $-3 - 5$ as a sum.

$= \frac{-8}{7}$ Add.

Example 3: Your stock posted a loss of $1\frac{3}{8}$ points yesterday. If it drops another $\frac{1}{4}$ point today, find its overall change for the two days.

Solution: loss yesterday plus loss today.

$-1\frac{3}{8} + -\frac{1}{4}$

$= -1\frac{3}{8} + \left(-\frac{1}{4}\right)$

$= -1\frac{3}{8} + \left(-\frac{2}{8}\right)$ $\quad -\frac{1}{4} = \frac{-1(2)}{4(2)}$

$= -1\frac{5}{8}$

1.6 EXERCISES

Find each difference.

1. $-8 - 13$
2. $7 - 12$
3. $-22 - (-11)$
4. $\frac{3}{4} - \frac{7}{2}$

Simplify each expression.

5. $-9 - (3 - 12)$
6. $1 - 4(7 - 5)$
7. $|-3| + 7^2 + (-5 - 4)$
8. $3\frac{5}{9} - \left(-2\frac{1}{3}\right)$
9. $7 - \{2[6 - (-10)] - 13\}$
10. $4 - (2 \cdot 8)^2$
11. $-17 + [(2 - 9) - (-14) - 20]$
12. $5 - 3 \cdot 2^3$

If $x = -5$, $y = -4$, and $t = 3$, evaluate each expression.

13. $3t - x$
14. $\frac{x - y}{t}$
15. $y - t^3$

16. A jet liner hits an air pocket and drops 220 feet. After climbing 180 feet, it drops another 195 feet. What is its overall vertical change?

1.7 MULTIPLYING AND DIVIDING REAL NUMBERS

Example 1: Find the product: $(-4)(7)(-2)$

Solution: $(-4)(7)(-2) = (-28)(-2)$ Multiply two factors at a time from left to right.

$= 56$

Example 2: Find the quotient: $\dfrac{-36}{-3}$

Solution: $\dfrac{-36}{-3} = -36 \cdot -\dfrac{1}{3}$ Definition of quotient of two numbers.

$= 12$ Multiply.

Example 3: Simplify $\dfrac{(-8)(-2) + 5}{-9 + 6}$.

Solution: $\dfrac{(-8)(-2) + 5}{-9 + 6} = \dfrac{16 + 5}{-3}$ Simplify the numerator and denominator separately.

$= \dfrac{21}{-3}$ Add in the numerator.

$= -7$ Divide.

Example 4: If $x = -1$ and $y = -7$, evaluate $4x - y^2$.

Solution: $4x - y^2 = 4(-1) - (-7)^2$ Replace x with -1 and y with -7.

$= -4 - (49)$
$= -4 + (-49)$
$= -53$

1.7 EXERCISES

Perform indicated operations.

1. $(-12)(10)$
2. $(-8)(5)(0)$
3. $\left(\dfrac{3}{2}\right)\left(-\dfrac{8}{9}\right)$
4. $(-6)^2$
5. $\dfrac{32}{-16}$
6. $\dfrac{0}{-8}$
7. $\dfrac{-5^2 + 7}{-9}$
8. $\dfrac{12 - (-3)^2}{11 - 14}$
9. $\dfrac{10 - 5(-4)}{7 - 2(-9)}$
10. $-8(4 - 15)$
11. $-6[(2 - 7) - (3 - 14)]$
12. $\left(-3\dfrac{2}{5}\right)\left(-\dfrac{25}{8}\right)$

If $x = -6$ and $y = -2$, evaluate each expression.

13. $\dfrac{5 - y}{7x}$

14. $3x^2 - 4y^3$

Write each of the following as an expression and evaluate.

15. The sum of −8 and the quotient of −20 and 10.

16. Subtract 4 from twice the difference of 10 and 17.

1.8 PROPERTIES OF REAL NUMBERS

Example 1: If $a = -6$ and $b = 3$, show that $a + b = b + a$.

Solution: Replace a with −6 and b with 3.

$$\begin{aligned} a + b &= b + a \\ -6 + 3 &= 3 + (-6) \\ -3 &= -3 \end{aligned}$$

True

Example: 2 Use the distributive property to write the expression without parentheses.

$-4(x + 2y - 3z)$

Solution:
$$\begin{aligned} -4(x + 2y - 3z) &= -4(x) + (-4)(2y) + (-4)(-3z) \\ &= -4x + (-8y) + 12z \\ &= -4x - 8y + 12z \end{aligned}$$

Example 3: Use the associative property of addition and the expression $(4 + 3) + 9$ to write a true statement.

Solution: $(4 + 3) + 9 = 4 + (3 + 9)$ Regroup.

1.8 EXERCISES

Name the property illustrated by each statement.

1. $7(2 + 1) = 7 \cdot 2 + 7 \cdot 1$

2. $\dfrac{1}{3} \cdot \dfrac{2}{5} = \dfrac{2}{5} \cdot \dfrac{1}{3}$

3. $(3 \cdot 5) \cdot 2 = (5 \cdot 3) \cdot 2$

4. $0 + (-7) = -7$

5. $4 \cdot \dfrac{1}{4} = 1$

6. $-8 \cdot 1 = -8$

Use the distributive property to write each expression without parentheses.

7. $-7(6x + 3y)$

8. $-(-a - b + 2c)$

9. $3(-5x + 4)$

Find the additive inverse or opposite of each of the following numbers.

10. $-\frac{10}{13}$

11. $-|-12|$

Find the multiplicative inverse or reciprocal of each of the following numbers.

12. $-2\frac{3}{7}$

13. $|-6|$

Use the indicated property and the given expression to write a true statement.

14. $3(a + b)$; commutative property of addition.

15. $5 \cdot (3 \cdot 8)$; associative property of multiplication.

16. $-7 + 0$; identity property of addition.

1.9 READING GRAPHS

Example 1: The following bar graph shows the number of banks that the FDIC closed during different intervals of years.

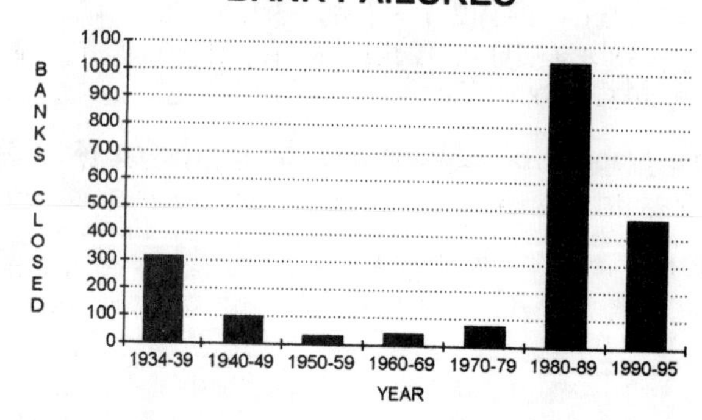

Data taken from USA Today

a. Which years had the most bank failures?

b. Approximate the number of bank failures during 1950–59.

Solution:

a. The tallest bar corresponds to the years with the most bank failures. 1980–89 had the greatest number of closings.

b. To approximate the number of bank failures during 1950–59, go to the top of the bar that corresponds to these years. From the T of the bar, move horizontally to the left until the vertical axis is reached. The height of the bar is approximately one-third of the way between 0 and 100. Therefore we conclude there were approximately 33 banks closed.

Example 2: The line graph below shows the number of Americans (in millions) traveling during the Fourth of July holiday for the years 1985 to 1995.

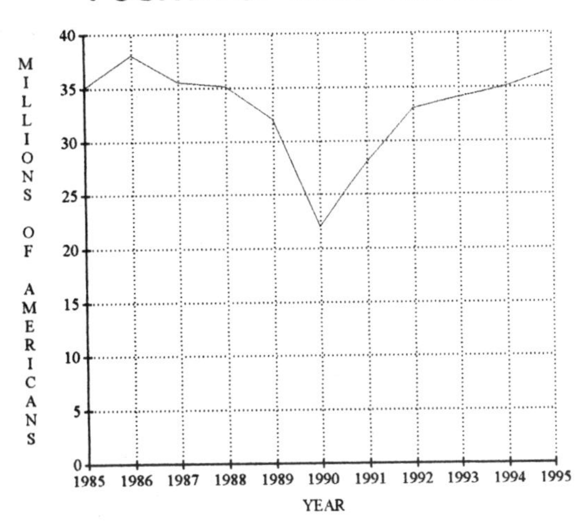

Data taken from USA Today

a. Which year had the least number of travelers?

b. Approximate the number of Americans traveling in 1986.

Solution: a. The lowest point on the graph indicates the year that had the least number of travelers. That year is 1990.

b. To approximate the number of Americans traveling in 1986, go to the point on the graph corresponding to that year. From this point, move horizontally to the left until the vertical axis is reached. This point is a little more than half way between 35 and 40. Therefore we conclude it is about 37 and it is measured in millions. Hence, in 1986 approximately 37 million Americans were traveling during the Fourth of July holiday.

1.9 EXERCISES

Use the bar graph in Example 1 *to answer the following.*

1. Approximate the number of bank failures during 1970–79.
2. Which interval of years had the greatest increase in bank closings?
3. Which years had the least number of banks fail?

Use the line graph in Example 2 to answer the following.

4. Which year had the greatest number of travelers?
5. Approximate the number of Americans traveling over the Fourth of July holiday in 1994.
6. Between what two years was there the least amount of change in the number of travelers?

The following pie chart shows the percentage of adult shoppers who prefer to pay for purchases in various ways.

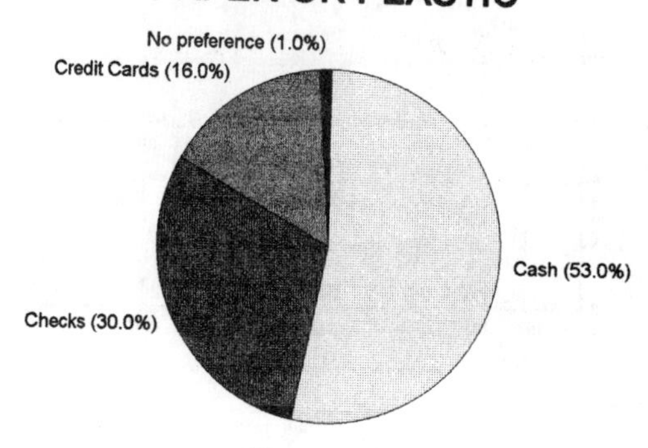

Data taken from USA Today

7. By what means do most adult shoppers prefer to pay?
8. What percentage of adult shoppers prefer to use checks or credit cards?
9. What percentage of adult shoppers do not prefer checks?

The following line chart shows the total number of traffic fatalities and the number of alcohol-related fatalities during the years 1982–1994.

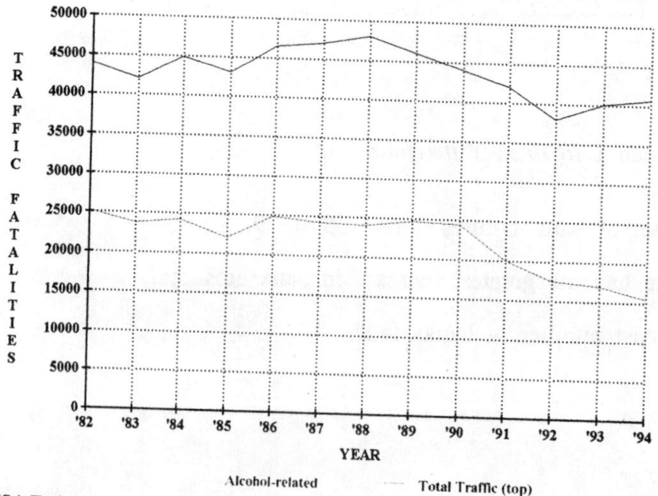

Data taken from USA Today

10. Which year had the greatest number of total traffic fatalities?

11. Which year had the least number of alcohol-related fatalities?

12. Approximate the number of alcohol-related fatalities in 1991.

13. During what years did the total traffic fatalities increase while the alcohol-related decreased?

The following 3D line chart shows the number of Americans over 65. Note that the number is projected for the years 1996-2030.

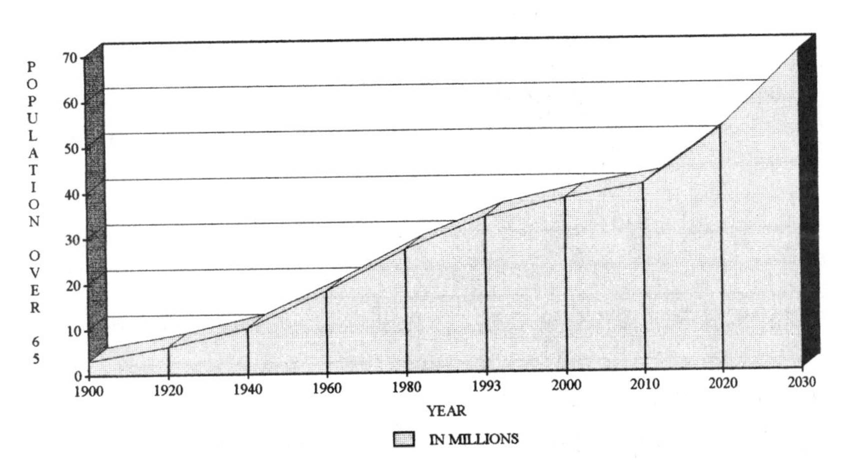

Data taken from USA Today

14. Approximate the population over the age of 65 in 1960.

15. Between what years does the population increase the least?

16. What is the difference between the population in 2020 and in 1980?

CHAPTER 1 PRACTICE TEST

Translate the statement into symbols.

1. The absolute value of negative nine is greater than four.
2. The sum of eight and three is greater than or equal to seven.

Simplify the expression.

3. $-12 + 5$
4. $-18 - (-4)$
5. $7 \cdot 2 - 9 \cdot 6$
6. $(15)(-4)$
7. $(-4)(-3)$
8. $\dfrac{|-24|}{-6}$
9. $\dfrac{-10}{0}$
10. $\dfrac{|-9| + 3}{4 - 5}$
11. $\dfrac{1}{3} - \dfrac{5}{12}$
12. $-2\dfrac{1}{4} + 6\dfrac{3}{8}$
13. $-\dfrac{2}{7} + \dfrac{13}{6}$
14. $4(-3)^2 - 60$
15. $7[4 + 3(2 - 6) - 5]$
16. $\dfrac{-25 + 2 \cdot 10}{5}$
17. $\dfrac{(-6)(-1)(0)}{-8}$

Insert <, >, or = in the appropriate space to make each of the following statements true.

18. $-5 \quad -8$
19. $6 \quad -9$
20. $|-7| \quad 6$
21. $|-3| \quad -2 - (-5)$
22. Given $\{-7, -2, \frac{1}{3}, 0, 2, 8, 12.3, \sqrt{11}, 5\pi\}$,
    list the numbers in this set that also belong to the set of:
    a. Natural numbers
    b. Whole numbers
    c. Integers
    d. Rational numbers
    e. Irrational numbers
    f. Real numbers

If $x = 3$, $y = -4$, and $z = -7$, evaluate each expression in exercises 23 - 26.

23. $x^2 - y^2$
24. $xy + z$
25. $5 - 2x + 3y$
26. $\dfrac{4x + y - 2}{x}$

Identify the property illustrated by each expression.

27. $7 \cdot (3 \cdot 4) = (7 \cdot 3) \cdot 4$
28. $5 + 9 = 9 + 5$
29. $-3(5 + 7) = (-3) \cdot 5 + (-3) \cdot 7$
30. $8 + (-8) = 0$
31. Find the opposite of -12.
32. Find the reciprocal of $-\dfrac{1}{8}$.

The Washington Redskins were 18 yards from the goal when the following series of gains and losses occurred.

	Gains and Losses in Yards
First down	4
Second down	-9
Third down	-1
Fourth down	24

33. During which down did the greatest loss of yardage occur?

34. Was a touchdown scored?

35. On Valentine's Day the temperature was 6 degrees below zero in the morning, but by noon it had risen 22 degrees. What was the temperature at noon?

36. Bob Suarez decided to sell 340 shares of stock, which decreased in value by $1.80 per share yesterday. How much money did he lose?

2.1 SIMPLIFYING ALGEBRAIC EXPRESSIONS

Example 1: Simplify by combining like terms:

$4x^2 - 6x + 2x$

Solution:
$$\begin{aligned} 4x^2 - 6x + 2x &= 4x^2 + (-6 + 2)x \\ &= 4x^2 + (-4)x \\ &= 4x^2 - 4x \end{aligned}$$

Example 2: Simplify: $12 - (8x + 1) - 4x$

Solution:
$$\begin{aligned} 12 - (8x + 1) - 4x &= 12 - 8x - 1 - 4x && \text{Apply the distributive property.} \\ &= (12 - 1) + (-8 - 4)x && \text{Combine like terms.} \\ &= 11 - 12x \end{aligned}$$

Example 3: Subtract $5x - 7$ from $4x + 11$.

Solution:
$$\begin{aligned} (4x + 11) - (5x - 7) &= 4x + 11 - 5x + 7 && \text{Apply the distributive property.} \\ &= (4 - 5)x + (11 + 7) && \text{Combine like terms.} \\ &= -x + 18 \end{aligned}$$

2.1 EXERCISES

Simplify each expression.

1. $5a - 6b + 9b + 8a$
2. $12 - 5c - 10c + 30$
3. $-2(4a - 3b) + 10a$
4. $-(x + 2y - z)$
5. $20x^2 + 6x - 8 + 4x - x^2$
6. $1 + 3y - 5y^2 + 9y - 10$
7. $4.6m - 5.3n + 2.1m + 7.8n$
8. $-17 - 4(r - 5)$
9. $18y - 12 - 13(y + 5)$
10. $-3(2x + 7) - 8x + 1$
11. $-19d - (5 - 3d)$
12. $(21 - 3x) - (7 + 6x) + 1$

Write the following phrases as algebraic expressions. Let x represent the unknown number.

13. Twice a number added to the product of 9 and the number.
14. The sum of a number and 6, divided by 10.
15. Four subtracted from the sum of a number and 7.

2.2 THE ADDITION PROPERTY OF EQUALITY

Example 1: Solve $x - 9 = 15$ for x.

Solution:

$x - 9 = 15$

$x - 9 + 9 = 15 + 9$ Add 9 to both sides.

$x = 24$ Simplify.

Solution set: $\{24\}$

Example 2: Solve $4x + 6x - 8 + 5 = 10x + 2 - x + 1$ for x.

Solution:

$4x + 6x - 8 + 5 = 10x + 2 - x + 1$

$10x - 3 = 9x + 3$ Combine like terms on each side of the equation.

$10x - 3 - 9x = 9x + 3 - 9x$ Subtract $9x$ from both sides.

$x - 3 = 3$ Combine like terms.

$x - 3 + 3 = 3 + 3$ Add 3 to both sides.

$x = 6$ Combine like terms.

Solution set: $\{6\}$

Example 3: Solve $-7(3y - 2) - (-8y + 5) = -14y$ for y.

Solution:

$-7(3y - 2) - (-8y + 5) = -14y$

$-21y + 14 + 8y - 5 = -14y$ Apply the distributive property.

$-13y + 9 = -14y$ Combine like terms.

$-13y + 9 + 13y = -14y + 13y$ Add $13y$ to both sides.

$9 = -y$ Combine like terms.

$-9 = y$ The opposite of y is 9, hence y is -9.

Solution set: $\{-9\}$

2.2 EXERCISES

Solve each equation.

1. $x + 17 = -6$
2. $4 + y = 18$
3. $3x = 2x - 8$
4. $5a - 3 = 4a + 9$
5. $17b - 1 = 16b + 31$
6. $9d + 18 - d = 7d - 15$
7. $x + 0.9 = 10.7$
8. $-6(y - 2) = 5 - 7y$
9. $10 - 5(x + 1) = 10x + 3 - 16x$
10. $2b - (b - 7) = 3b - 8 - b$
11. $-20m + 6(3m + 2) = 8 - 3(1 + m)$
12. $0.7x + 0.3(0.6 + x) = 3.61$
13. $-14(c - 3) - 5(8 - 3c) = -9$
14. Two numbers have a sum of 15. If one number is x, express the other number in terms of x.
15. A 25 foot board is cut into two pieces. If one piece is x feet long, express the other length in terms of x.

2.3 THE MULTIPLICATION PROPERTY OF EQUALITY

Example 1: Solve for x: $\frac{3}{5}x = 18$

Solution:

$$\frac{3}{5}x = 18$$

$$\frac{5}{3} \cdot \frac{3}{5}x = \frac{5}{3} \cdot 18 \quad \text{Multiply both sides by } \frac{5}{3}.$$

$$\left(\frac{5}{3} \cdot \frac{3}{5}\right)x = \frac{5}{3} \cdot \frac{18}{1} \quad \text{Apply the associative property.}$$

$$1x = 30 \quad \text{Simplify.}$$

$$x = 30$$

Solution set: $\{30\}$

Example 2: Solve $\frac{y}{8} = 12$ for y.

Solution:

$$\frac{y}{8} = 12$$

$$8 \cdot \frac{y}{8} = 8 \cdot 12 \quad \text{Multiply both sides by 8.}$$

$$y = 96 \quad \text{Simplify.}$$

Solution set: $\{96\}$

Example 3: Solve $-a - 5 = 11$ for a.

Solution:

$$-a - 5 = 11$$

$$-a - 5 + 5 = 11 + 5 \quad \text{Add 5 to both sides.}$$

$$-a = 16 \quad \text{Combine like terms.}$$

$$\frac{-a}{-1} = \frac{16}{-1} \quad \text{Divide both sides by } -1.$$

$$a = -16 \quad \text{Simplify.}$$

Solution set: $\{-16\}$

Example 4: Solve for x: $8x + 3x - 12 + 5 = 4x + 21$

Solution:

$$8x + 3x - 12 + 5 = 4x + 21$$
$$11x - 7 = 4x + 21 \quad \text{Combine like terms.}$$
$$11x - 7 - 4x = 4x + 21 - 4x \quad \text{Subtract } 4x \text{ from both sides.}$$
$$7x - 7 = 21 \quad \text{Combine like terms.}$$
$$7x - 7 + 7 = 21 + 7 \quad \text{Add 7 to both sides.}$$
$$7x = 28 \quad \text{Combine like terms.}$$
$$\frac{7x}{7} = \frac{28}{7} \quad \text{Divide both sides by 7.}$$
$$x = 4 \quad \text{Simplify.}$$

Solution set: $\{4\}$

2.3 EXERCISES

Solve the following equations.

1. $-4x = 32$
2. $-5y = -125$
3. $-x = -20$
4. $2x = 0$
5. $\frac{2}{7}x = 30$
6. $\frac{-x}{6} = 8$
7. $\frac{a}{-2} = 0$
8. $\frac{1}{14}d = \frac{3}{7}$
9. $2c - 9 = 5c + 12$
10. $14t - 1 = 8t + 35$
11. $3y - 7 - y = 11y - 20$
12. $5 + 0.6m = 5$
13. $\frac{x}{3} + 2 = 13$
14. $y - 7y = -18 + y - 17$
15. $6y + \frac{1}{5} = \frac{4}{5}$

2.4 SOLVING LINEAR EQUATIONS

Example 1: Solve $5(6x - 7) + 16 = 2x + 9$.

Solution:

$$5(6x - 7) + 16 = 2x + 9$$
$$30x - 35 + 16 = 2x + 9 \quad \text{Apply the distributive property.}$$
$$30x - 19 = 2x + 9 \quad \text{Combine like terms.}$$
$$30x - 19 - 2x = 2x + 9 - 2x \quad \text{Subtract } 2x \text{ from both sides.}$$

$$28x - 19 = 9 \quad \text{Combine like terms.}$$
$$28x - 19 + 19 = 9 + 19 \quad \text{Add 19 to both sides.}$$
$$28x = 28 \quad \text{Combine like terms.}$$
$$\frac{28x}{28} = \frac{28}{28} \quad \text{Divide both sides by 28.}$$
$$x = 1 \quad \text{Simplify.}$$

Solution set: $\{1\}$

Example 2: Solve for x: $\frac{x}{3} - 2 = \frac{3}{2}x + 5$

Solution:

$$\frac{x}{3} - 2 = \frac{3}{2}x + 5$$
$$6\left(\frac{x}{3} - 2\right) = 6\left(\frac{3}{2}x + 5\right) \quad \text{Multiply both sides by 6.}$$
$$6\left(\frac{x}{3}\right) - 6(2) = 6\left(\frac{3}{2}x\right) + 6(5) \quad \text{Apply the distributive property.}$$
$$2x - 12 = 9x + 30 \quad \text{Simplify.}$$
$$2x - 12 - 2x = 9x + 30 - 2x \quad \text{Subtract } 2x \text{ from both sides.}$$
$$-12 = 7x + 30 \quad \text{Combine like terms.}$$
$$-12 - 30 = 7x + 30 - 30 \quad \text{Subtract 30 from both sides.}$$
$$-42 = 7x \quad \text{Simplify.}$$
$$\frac{-42}{7} = \frac{7x}{7} \quad \text{Divide both sides by 7.}$$
$$-6 = x \quad \text{Simplify.}$$

Solution set: $\{-6\}$

Example 3: Solve $-4(x - 3) + 9 = -2(x + 1) - 2x$

Solution:

$$-4(x - 3) + 9 = -2(x + 1) - 2x$$
$$-4x + 12 + 9 = -2x - 2 - 2x \quad \text{Distribute on both sides.}$$
$$-4x + 21 = -4x - 2 \quad \text{Combine like terms.}$$
$$-4x + 21 + 4x = -4x - 2 + 4x \quad \text{Add } 4x \text{ to both sides.}$$
$$21 = -2 \quad \text{False.}$$

Solution set: $\{ \; \}$

Example 4: Solve $-8(4 - x) = 8x - 32$

Solution:

$$-8(4 - x) = 8x - 32$$
$$-32 + 8x = 8x - 32 \quad \text{Apply the distributive property.}$$
$$-32 + 8x + 32 = 8x - 32 + 32 \quad \text{Add 32 to both sides.}$$
$$8x = 8x \quad \text{Combine like terms.}$$
$$8x - 8x = 8x - 8x \quad \text{Subtract } 8x \text{ from both sides.}$$
$$0 = 0 \quad \text{True.}$$

Solution set: $\{x \mid x \text{ is a real number}\}$

2.4 EXERCISES

Solve each equation.

1. $-3(2x + 1) = 15$
2. $-(4x + 3) = 17$
3. $7(3x - 5) - 2x = 3$
4. $-5(8 + b) + 11 = 29$
5. $10(3y + 2) = 6(y - 1) + 24y$
6. $\frac{4}{5}x - \frac{1}{10} = 2$
7. $\frac{7}{3}x - 4 = \frac{5}{6}$
8. $\frac{8(2 - a)}{3} = 2a$
9. $\frac{-(y + 1)}{5} = 2y + 3$
10. $6x - 3 = 3(5x - 4) + 9(1 - x)$
11. $\frac{x - 2}{3} = \frac{x + 2}{7}$
12. $13 - 6(1 - b) = 3b + 26$

Write each of the following as equations. Then solve.

13. The sum of twice a number and 8 is equal to the sum of the number and 11. Find the number.
14. One-fourth of a number is three-eighths. Find the number.
15. The difference of a number and 10 is twice the number. Find the number.

2.5 AN INTRODUCTION TO PROBLEM SOLVING

Example 1: A 20-inch piece of string is to be cut into two pieces so that the longer piece is 3 times the shorter. Find the length of each piece.

Solution: length of shorter piece: x
length of longer piece: $3x$

length of shorter piece	added to	length of longer	equals	total length of string
x	$+$	$3x$	$=$	20

$4x = 20$ Combine like terms.

$\frac{4x}{4} = \frac{20}{4}$ Divide both sides by 4.

$x = 5$ Simplify.

length of shorter piece: 5 inches
length of longer piece: $3x = 3(5) = 15$ inches

Example 2: Twice the difference of a number and 3 is the same as five times the number increased by 6. Find the number.

Solution: unknown number: x

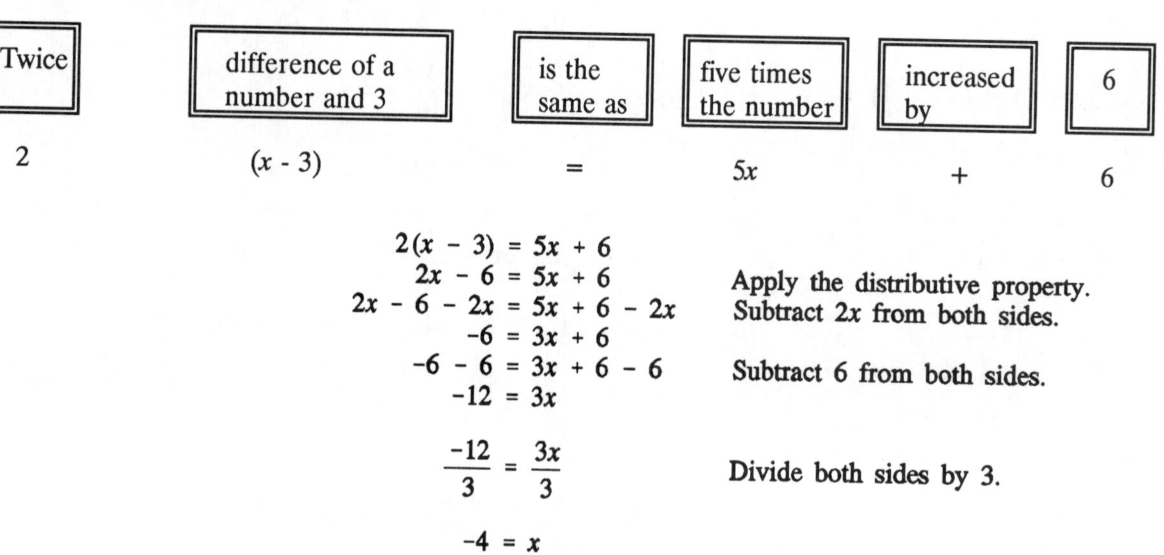

$$2(x - 3) = 5x + 6$$

$$2x - 6 = 5x + 6 \quad \text{Apply the distributive property.}$$

$$2x - 6 - 2x = 5x + 6 - 2x \quad \text{Subtract } 2x \text{ from both sides.}$$

$$-6 = 3x + 6$$

$$-6 - 6 = 3x + 6 - 6 \quad \text{Subtract 6 from both sides.}$$

$$-12 = 3x$$

$$\frac{-12}{3} = \frac{3x}{3} \quad \text{Divide both sides by 3.}$$

$$-4 = x$$

The number is -4.

Example 3: A plumber gave an estimate for the renovation of a bathroom. His hourly pay is \$36 per hour and the plumber's parts will cost \$220. If his total estimate is \$616, how many hours does he expect this job to take?

Solution: number of hours: x

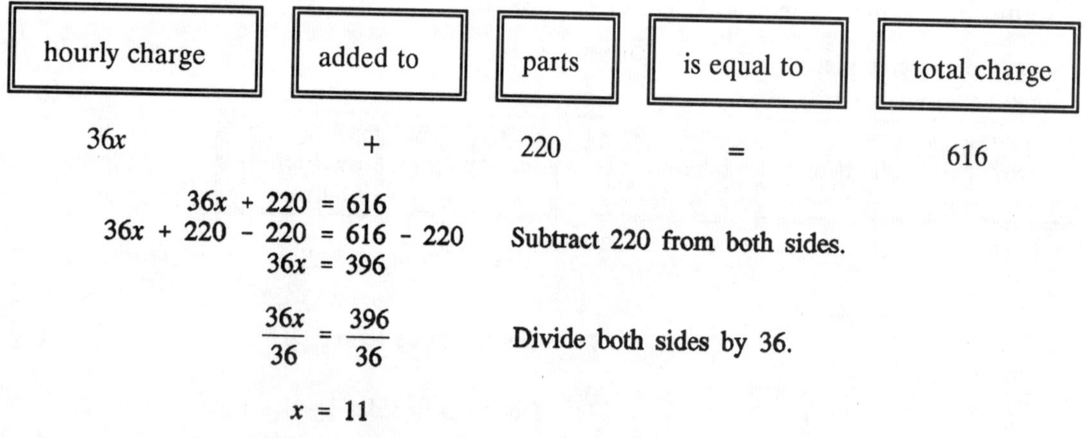

$$36x + 220 = 616$$

$$36x + 220 - 220 = 616 - 220 \quad \text{Subtract 220 from both sides.}$$

$$36x = 396$$

$$\frac{36x}{36} = \frac{396}{36} \quad \text{Divide both sides by 36.}$$

$$x = 11$$

He expects the job to take 11 hours.

2.5 EXERCISES

Solve.

1. John makes three times as much money as Bill. If the total of their salaries is $92,000, find the salary of each.
2. A 17-foot board is to be divided so that the longer piece is 5 feet more than twice the shorter piece. Find the length of both pieces.
3. The product of triple a number and six is the same as the difference of twice the number and $\frac{5}{2}$. Find the number.
4. If the sum of a number and seven is doubled, the result is twenty-six more than triple the number. Find the number.
5. A car rental agency advertised their rate for a mid-size car as $26.95 per day and $0.31 per mile. If you rent this car for 3 days, how many whole miles can be driven on a $150 budget?
6. Two angles are supplementary if their sum is 180°. One angle measures eight times the measure of a smaller angle. Find the measure of each angle.
7. The sum of three consecutive integers is three times the middle integer. Find the integers.
8. Find two consecutive even integers such that four times the larger is 72 more than twice the smaller.
9. Four times the difference of a number and 11 is equal to twice the sum of the number and 12. Find the number.
10. The sum of three consecutive odd integers is 273. What are the numbers?
11. The measures of the two smaller angles in a triangle are the same and the third angle is twice the measure of the smaller angle. Find the measure of each angle.
12. Paula's $36,000 estate is to be divided so that her husband receives three times as much as her daughter. How much money will the husband and daughter each receive?
13. Two angles are complementary if their sum is 90°. One angle measures six less than three times the measure of the other angle. Find the measure of each angle.
14. In a local election Sam Barnes received 260 more votes than William Smith. If a total of 4010 votes were cast, how many did each man receive?
15. Maria invested $8100. Part of the money she invested in a certificate of deposit and the other part was put into a mutual fund. If Maria used three times as much money in the mutual fund as she did in the certificate of deposit, how much did she invest in each type?

2.6 FORMULAS AND PROBLEM SOLVING

Example 1: Sandra plans to enclose her vegetable garden with fencing. The perimeter of her garden is 120 feet. If the width of her garden is 25 feet, find the length.

Solution: length: x

$$P = 2W + 2L$$

$$120 = 2(25) + 2x \qquad \text{Let } P = 120,\ W = 25 \text{ and } L = x.$$

$$120 = 50 + 2x$$

$$120 - 50 = 50 + 2x - 50 \qquad \text{Subtract 50 from both sides.}$$

$$70 = 2x$$

$$\frac{70}{2} = \frac{2x}{2} \qquad \text{Divide both sides by 2.}$$

$$35 = x$$

The length is 35 feet.

Example 2: Solve for b: $A = \frac{1}{2}(B + b)h$

Solution: $A = \frac{1}{2}(B + b)h$

$$2A = 2 \cdot \frac{1}{2}(B + b)h \qquad \text{Multiply both sides by 2.}$$

$$2A = (B + b)h$$

$$2A = Bh + bh \qquad \text{Apply the distributive property.}$$

$$2A - Bh = Bh + bh - Bh \qquad \text{Subtract } Bh \text{ from both sides.}$$

$$2A - Bh = bh$$

$$\frac{2A - Bh}{h} = \frac{bh}{h} \qquad \text{Divide both sides by } h.$$

$$\frac{2A - Bh}{h} = b$$

Example 3: Convert 85° Celsius to degrees Fahrenheit.

Solution: $F = \frac{9}{5}C + 32$

$$F = \frac{9}{5}(85) + 32 \qquad \text{Replace } C \text{ with 85.}$$

$$F = 153 + 32$$

$$F = 185°$$

It is equivalent to 185° F.

2.6 EXERCISES

Substitute the given values into the given formulas and solve for the unknown variable.

1. $A = bh;\ A = 65,\ b = 5$
2. $D = rt;\ D = 220,\ r = 55$
3. $I = PRT;\ I = 1008,\ R = 0.035,\ T = 8$
4. $P = a + b + c;\ P = 62,\ a = 12,\ c = 21$
5. $V = \frac{4}{3}\pi r^3;\ r = 4$ (Use the approximation 3.14 for π.)

Solve each formula for the specified variable.

6. $2x - y = 8$ for y
7. $A = \frac{1}{2}bh$ for h
8. $P = a + b + c$ for b
9. $3x + 5y = 15$ for x
10. $C = 2\pi r$ for r
11. $s = 4lw + 2wh$ for l

Solve.

12. If the length of a rectangularly shaped garden is 18 meters and its width is 7.6 meters, find the amount of fencing required.
13. Convert −15° C to Fahrenheit.
14. Convert 41° F to Celcius.
15. A circle has a 25-foot radius. Find its circumference. (Use the approximation 3.14 for π.)
16. Jack traveled from his home to his brother's home, a distance of 550 miles, at an average speed of 50 miles per hour. How long did the trip take?

2.7 PERCENT AND PROBLEM SOLVING

Example 1: Write 91.6% as a decimal.

Solution: $0_{\wedge}91.6\% = 0.916$

Example 2: Write $\frac{3}{4}$ as a percent.

Solution: $\frac{3}{4} = 0.75 = 75\%$

Example 3: The number 270 is what percent of 1200?

Solution: 1st unknown percent: x

The number 270	is	what percent	of	1200
270	=	x	$\cdot$	1200

$$270 = x \cdot 1200$$

$$270 = 1200x$$

$$\frac{270}{1200} = \frac{1200x}{1200} \quad \text{Divide both sides by 1200.}$$

$$0.225 = x$$

$$22.5\% = x \quad \text{Write as a percent.}$$

Example 4: The cost of a sub at the local deli recently increased from $4.50 to $5.25. Find the percent increase.

Solution: percent increase: x

increase = new price − old price

= 5.25 − 4.50

= 0.75

increase	is	what percent	of	old price
0.75	=	x	$\cdot$	4.50

$$0.75 = x \cdot 4.50$$

$$0.75 = 4.50x$$

$$\frac{0.75}{4.50} = \frac{4.50x}{4.50} \quad \text{Divide both sides by 4.50.}$$

$$0.167 = x \quad \text{Round to 3 decimal places.}$$

$$16.7\% = x \quad \text{Write as a percent.}$$

The percent increase in price is approximately 16.7%.

2.7 EXERCISES

Write each percent as a decimal.

1. 235%
2. 0.32%
3. 1.7%
4. 31.7%

Write each number as a percent.

5. 0.61
6. 7
7. $\frac{7}{8}$
8. $\frac{1}{5}$

Solve the following.

9. What number is 15% of 670?
10. The number 13.12 is what percent of 41?
11. The number 117.6 is 28% of what number?
12. Find 120% of 95.
13. The number 25.2 is 42% of what number.
14. The number 306 is what percent of 450?
15. 483 is what percent of 210?
16. 5 is what percent of 2505?
17. Stop and Eat just increased the price of their $1.75 hamburger by 12%. Find the new price of the hamburger.
18. Betty was shopping and found a sweater on the 30% off sale rack. It originally sold for $32. What sale price did Betty have to pay?
19. Recently Gas Rite decreased their price per gallon of gas from $1.17 to $1.07. Find the percent decrease.
20. At Supergrocery the price of a loaf of bread increased from $1.28 to $1.64. Find the percent increase.

2.8 FURTHER PROBLEM SOLVING

Example 1: The length of a field is 4 yards less than four times its width. Find the dimensions if the perimeter is 202 yards.

Solution: width: x
length: $4x - 4$

$P = 2W + 2L$

$202 = 2(x) + 2(4x - 4)$ Replace P with 202, W with x, and L with $4x - 4$.

$202 = 2x + 8x - 8$ Apply the distributive property.

$202 = 10x - 8$

$202 + 8 = 10x - 8 + 8$ Add 8 to both sides.

$210 = 10x$

$$\frac{210}{10} = \frac{10x}{10}$$ Divide both sides by 10.

$$21 = x$$
$$4x - 4 = 4(21) - 4 = 80$$

The width is 21 yards and the length is 80 yards.

Example 2: The Smith family drove to Orlando at 55 miles per hour and returned on the same route at 45 miles per hour. Find the distance to Orlando if the total driving time was $13\frac{1}{3}$ hours.

Solution: x = time going
Note, 13(1/3) = (40/3)

	rate ·	time =	distance
going	55	x	$55x$
returning	45	$(40/3) - x$	$45[(40/3) - x]$

distance going = distance returning

$$55x = 45\left(\frac{40}{3} - x\right)$$

$$55x = 600 - 45x$$ Apply the distributive property.
$$55x + 45x = 600 - 45x + 45x$$ Add $45x$ to both sides.
$$100x = 600$$

$$\frac{100x}{100} = \frac{600}{100}$$ Divide both sides by 100.

$$x = 6$$
Distance $= 55x = 55(6) = 330$
The distance is 330 miles.

Example 3: Jasvinder invested part of his $30,000 inheritance in a mutual funds account that pays 8% simple interest yearly and the rest in a certificate of deposit that pays 6% simple interest yearly. At the end of one year, Jasvinder's investments earned $2160. Find the amount he invested at each rate.

Solution: amount at 8%: x
amount at 6%: $30000 - x$

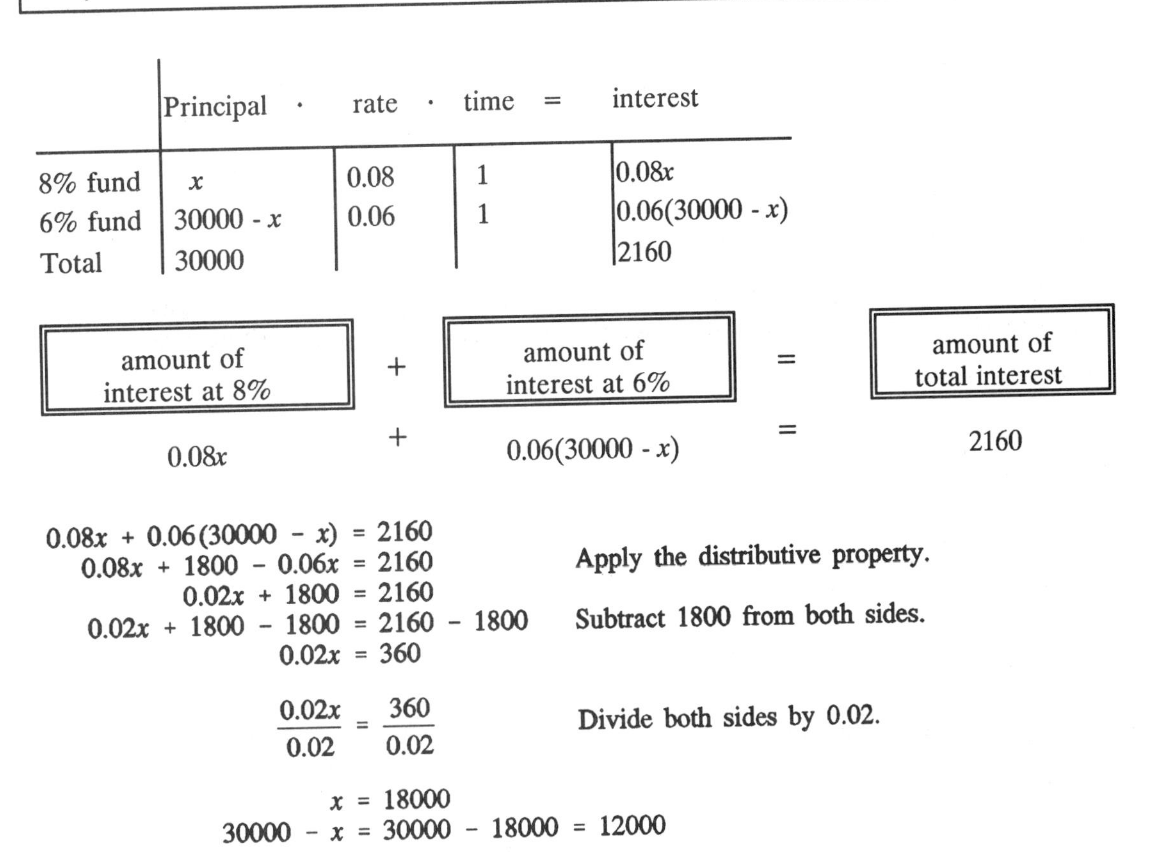

	Principal ·	rate ·	time =	interest
8% fund	x	0.08	1	$0.08x$
6% fund	$30000 - x$	0.06	1	$0.06(30000 - x)$
Total	30000			2160

amount of interest at 8%	+	amount of interest at 6%	=	amount of total interest
$0.08x$	+	$0.06(30000 - x)$	=	2160

$$0.08x + 0.06(30000 - x) = 2160$$

$$0.08x + 1800 - 0.06x = 2160 \qquad \text{Apply the distributive property.}$$

$$0.02x + 1800 = 2160$$

$$0.02x + 1800 - 1800 = 2160 - 1800 \qquad \text{Subtract 1800 from both sides.}$$

$$0.02x = 360$$

$$\frac{0.02x}{0.02} = \frac{360}{0.02} \qquad \text{Divide both sides by 0.02.}$$

$$x = 18000$$

$$30000 - x = 30000 - 18000 = 12000$$

He invested \$18,000 at 8% and \$12,000 at 6%.

2.8 EXERCISES

Solve each word problem.

1. If the length of a rectangular lot is 60 feet less than triple its width, and the perimeter is 712 feet, find the length of the lot.

2. A flower bed is in the shape of a triangle with one side four times the length of the shortest side and the third side is 25 feet more than the length of the shortest side. Find the dimensions if the perimeter is 121 feet.

3. The perimeter of an isosceles triangle is 94 inches. If the shortest side is 5 inches less than the other two sides, find the length of the shortest side. (*Hint*: *An isosceles triangle has two sides the same length.*)

4. A car traveling 65 miles per hour overtakes a truck traveling at 52 miles per hour that had a $1\frac{1}{2}$-hour head start. How far from the starting point are the vehicles?

5. Willy drove to New York at 60 miles per hour and returned on the same route at 50 miles per hour. Find the distance to New York if the total driving time was 8.8 hours.

6. How much pure acid should be mixed with 5 gallons of a 35% acid solution in order to get a 45% acid solution?

7. How many pounds of coffee worth \$8 a pound should be added to 10 pounds of coffee worth \$5 a pound to get a mixture worth \$6 a pound?

8. Kelsey invested some money at 7% annual simple interest and \$400 more than that amount at 8.5% annual simple interest. If her total yearly interest was \$127, how much was invested at each rate?

9. How can \$68,000 be invested, part at 4% annual simple interest and the remaining at 6% annual simple interest so that the interest earned by the two accounts will be equal?

10. How much of an alloy that is 25% copper should be mixed with 300 ounces of alloy that is 55% copper in order to get an alloy that is 35% copper?

11. How much water should be added to 20 gallons of a solution that is 80% antifreeze in order to get a mixture that is 70% antifreeze?

12. Tickets for a local sporting event were \$3.75 for students and \$6.00 for adults. A total of 620 tickets were sold for \$2775. How many adult tickets were sold?

13. Kara can row upstream at 4 miles per hour and downstream at 10 miles per hour. If Kara rows upstream for a while and then rows downstream to her starting point, how far did Kara row if the entire trip took 6 hours?

14. Two hikers are 8 miles apart and walking toward each other. They meet in $1\frac{1}{2}$ hours. Find the rate of each hiker if one hiker walks 1.4 miles per hour faster than the other.

15. Find the break-even quantity for a company that makes x number of VCRs at a cost C given by $C = 3720 + 160x$ and receives revenue R given by $R = 470x$.

2.9 SOLVING LINEAR INEQUALITIES

Example 1: Solve $x + 3 \geq -5$ for x. Graph the solution set.

Solution:

$$x + 3 \geq -5$$
$$x + 3 - 3 \geq -5 - 3 \quad \text{Subtract 3 from both sides.}$$
$$x \geq -8$$

$\{x \mid x \geq -8\}$

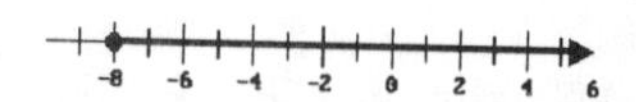

Example 2: Solve $-5x + 8 < -12$ for x, and graph the solution set.

Solution:

$$-5x + 8 < -12$$
$$-5x + 8 - 8 < -12 - 8 \quad \text{Subtract 8 from both sides.}$$

$$-5x < -20$$

$$\frac{-5x}{-5} > \frac{-20}{-5}$$ Divide both sides by -5 and reverse the direction of the inequality sign.

$$x > 4$$

$$\{x \mid x > 4\}$$

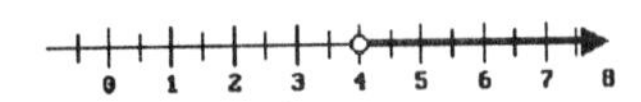

Example 3: Solve $3(x - 4) - 18 \le 6(x + 5) - 12$, and graph the solution set.

Solution:

$3(x - 4) - 18 \le 6(x + 5) - 12$

$3x - 12 - 18 \le 6x + 30 - 12$ Apply the distributive property.

$3x - 30 \le 6x + 18$

$3x - 30 - 3x \le 6x + 18 - 3x$ Subtract $3x$ from both sides.

$-30 \le 3x + 18$

$-30 - 18 \le 3x + 18 - 18$ Subtract 18 from both sides.

$-48 \le 3x$

$\frac{-48}{3} \le \frac{3x}{3}$ Divide both sides by 3.

$-16 \le x$

$\{x \mid x \ge -16\}$

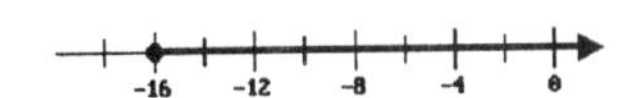

2.9 EXERCISES

Solve each inequality and graph the solution set.

1. $3x > -12$
2. $x - 1 \le 5$
3. $-4x \ge -16$
4. $-2x < 8$
5. $4x - 7 \le 3x - 5$
6. $9 - 3x > 11 - 4x$
7. $-10x + 3 \ge 4(12 - x)$
8. $-(2 - 3x) < 5(x - 4)$

9. $7(x + 1) - 6 > -4(x - 3) + 22$

10. $-3(x + 4) - 2x \leq -(5x + 2) + 6x$

11. $-(x - 8) < 8$

12. $-9 < 3(x - 1) < 12$

13. $-6 \leq 4(x + 3) < 10$

Solve the following.

14. Nine more than triple a number is greater than negative twelve. Find all numbers that make this statement true.

15. The perimeter of a rectangle is to be no greater than 150 centimeters and the width must be 30 centimeters. Find the maximum length of the rectangle.

16. Four times a number decreased by three is between negative eleven and seventeen. Find all such numbers.

CHAPTER 2 PRACTICE TEST

Simplify each of the following expressions.

1. $3y - 7 - y + 5$
2. $3.1x + 4.2 + 5.3x - 6.9$
3. $7(x - 1) - 2(3x - 4)$
4. $-8(y + 3) + 4(2 - 7y)$

Solve each of the following equations.

5. $-\frac{5}{4}x = 5$
6. $5(n - 4) = -(6 - 3n)$
7. $6y - 8 + y = -(4y + y)$
8. $5z + 2 - z = 2 + 3z$
9. $\frac{3(x + 5)}{2} = x - 7$
10. $\frac{9(y - 2)}{4} = 3y + 1$
11. $\frac{1}{3} - x + \frac{5}{3} = x - 7$
12. $\frac{1}{4}(y + 4) = 5y$
13. $-0.2(x - 3) + x = 0.3(5 - x)$
14. $-6(a + 2) - 5a = -4(3a - 2)$

Solve each of the following applications.

15. A number increased by one-third of the number is 32. Find the number.
16. A gallon of water seal covers 250 square feet. How many gallons are needed to paint two coats of water seal on a deck that measures 25 feet by 40 feet?
17. Sedric Augell invested an amount of money in Amoxil stock that earned an annual 8% return, and then he invested triple the original amount in IBM stock that earned an annual 10% return. If his total return from both investments was $2280 find how much he invested in each stock.
18. Two trains leave Los Angeles simultaneously traveling on the same track in opposite directions at speeds of 60 and 72 miles per hour. How long will it take before they are 462 miles apart?
19. Find the value of x if $y = -19$, $m = -3$ and $b = -4$ in the formula $y = mx + b$.

Solve each of the following equations for the indicated variable.

20. $C = \pi d$ for d
21. $2x - 5y = 12$ for y

Solve and graph each of the following inequalities.

22. $2x - 3 > 5x - 9$
23. $x + 4 > 3x - 6$
24. $-5 < 2x - 1 < 6$
25. $0 < 6x - 9 < 15$
26. $\frac{3(4x + 1)}{2} > 5$
27. $-(3 - 5x) \geq 11 + 7x$

3.1 THE RECTANGULAR COORDINATE SYSTEM

Example 1: On a single coordinate system, plot the ordered pairs. State in which quadrant, if any, each point lies.

$A(-2, 3)$; $B(0, 5)$; $C(-1, -4)$;

$D(-4, 0)$; $E(1, 3)$; $F(2, -1)$

Solution:

$A(-2, 3)$:	left 2, up 3; quadrant II
$B(0, 5)$:	up 5; y-axis
$C(-1, -4)$:	left 1, down 4; quadrant III
$D(-4, 0)$:	left 4; x-axis
$E(1, 3)$:	right 1, up 3; quadrant I
$F(2, -1)$:	right 2, down 1; quadrant IV

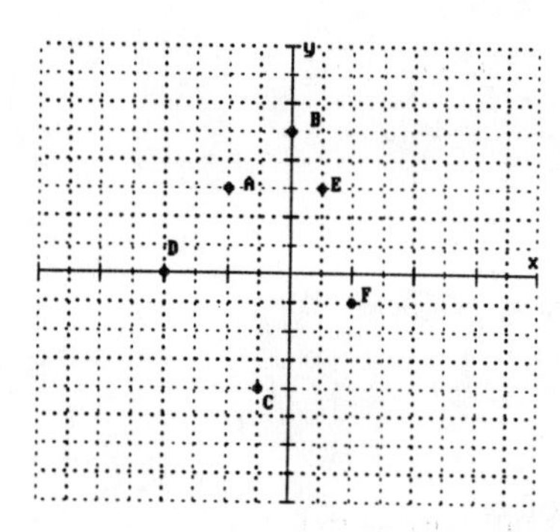

Example 2: Determine whether $(2, -5)$ is a solution to $3x - y = 1$.

Solution:

$$3x - y = 1$$
$$3(2) - (-5) = 1 \quad \text{Replace } x \text{ with 2 and } y \text{ with } -5.$$
$$6 + 5 = 1$$
$$11 = 1$$

False

It is not a solution.

Example 3: Complete the table for the equation $y = -2x + 1$.

x	y
0	
	-3
1	

Solution: $y = -2x + 1$

$y = -2(0) + 1$ Replace x with 0 and solve for y.

$y = 1$ The ordered pair is $(0, 1)$.

$y = -2x + 1$
$-3 = -2x + 1$ Replace y with -3 and solve for x.
$-4 = -2x$
$2 = x$ The ordered pair is $(2, -3)$

$y = -2x + 1$
$y = -2(1) + 1$ Replace x with 1 and solve for y.
$y = -1$ The ordered pair is $(1, -1)$.

x	y
0	1
2	-3
1	-1

3.1 EXERCISES

Plot the ordered pairs. State in which quadrant, if any, that each point lies.

1. $(5, 1)$
2. $(-4, -7)$
3. $(0, -6)$
4. $\left(-\frac{1}{2}, 3\right)$
5. $(6, -5)$
6. $\left(2\frac{3}{4}, 0\right)$

Determine whether each ordered pair is a solution of the given linear equation.

7. $2x - y = 4$; $(0, 4)$, $(0, -4)$, $(1, -2)$
8. $y = 7x$; $(-7, 1)$, $(1, -7)$, $(0, 0)$
9. $x = 3y - 4$; $(2, 2)$, $(3, 5)$, $(-7, -1)$
10. $y = -4$; $(4, -4)$, $(-4, 4)$, $(3, -4)$
11. $x = \frac{3}{4}y$; $(0, 0)$, $(3, 4)$, $\left(1, \frac{3}{4}\right)$
12. $x - 10 = 0$; $(0, 10)$, $(10, -10)$, $(10, -100)$

Complete the table for each given linear equation; then plot each solution. Use a single coordinate system for each equation.

13. $-x + 2y = 5$

x	y
0	
	0
1	

14. $y = -x$

x	y
0	
	-1
2	

15. $x = -6$

x	y
	-2
	0.5
	-4.75

3.2 GRAPHING LINEAR EQUATIONS

Example 1: Graph the linear equation $2x - 5y = 10$.

Solution: Find three ordered pair solutions of $2x - 5y = 10$.

Let $x = 0$: $2(0) - 5y = 10$
$-5y = 10$
$y = -2$
$(0, -2)$

Let $y = 0$: $2x - 5(0) = 10$
$2x = 10$
$x = 5$
$(5, 0)$

Let $y = -4$: $2x - 5(-4) = 10$
$2x + 20 = 10$
$2x = -10$
$x = -5$
$(-5, -4)$

Plot these three points and the line containing them.

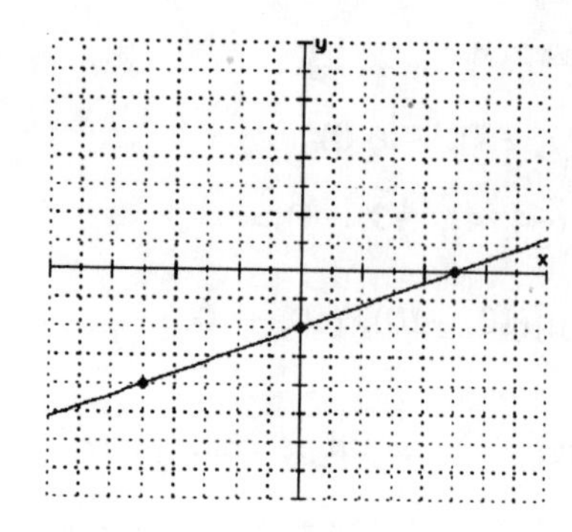

Example 2: Graph $y = -3x - 1$

Solution: Find three ordered pair solutions of $y = -3x - 1$.

Let $x = 0$: $y = -3(0) - 2$
$y = -2$
$(0, -2)$

Let $y = 0$: $0 = -3x - 1$
$1 = -3x$

$$-\frac{1}{3} = x$$

$$\left(-\frac{1}{3}, 0\right)$$

Let $x = -1$: $y = -3(-1) - 1$

$y = 2$

$(-1, 2)$

Plot these three points and the line containing them.

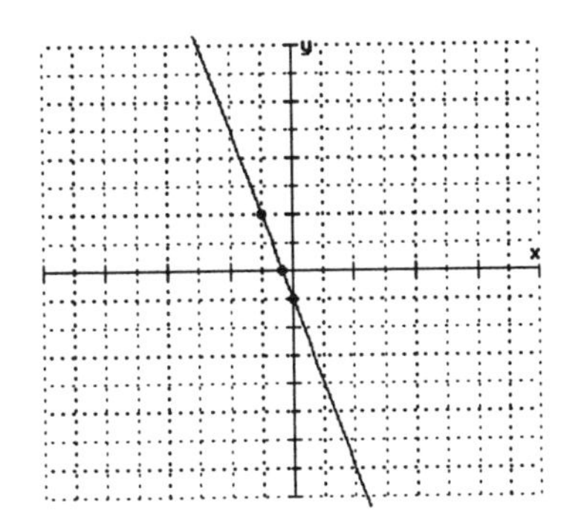

Example 3: Graph the linear equation $y = -\frac{2}{5}x$.

Solution: Find three ordered pair solutions of $y = -\frac{2}{5}x$.

Let $x = 0$: $y = -\frac{2}{5}(0)$

$y = 0$

$(0, 0)$

Let $x = 5$: $y = -\frac{2}{5}(5)$

$y = -2$

$(5, -2)$

Let $x = -5$: $y = -\frac{2}{5}(-5)$

$y = 2$

$(-5, 2)$

Plot these three points and the line containing them.

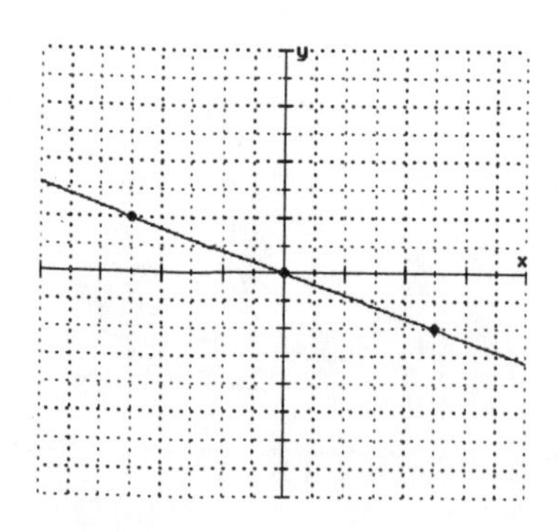

3.2 EXERCISES

Determine whether each equation is a linear equation in two variables.

1. $-7x = 3y + 5$
2. $y = x^2 - 1$
3. $0.6x - 0.8y = 12$

Write each statement as an equation in two variables. Then graph the equation.

4. The y-value is 6 less than the x-value.
5. Three times the x-value added to four times the y-value is 12.
6. The x-value is one-half the y-value.

Graph each linear equation.

7. $x - y = 3$
8. $x + 2y = 6$
9. $y = -4x + 5$
10. $y = -3x - 1$
11. $x = -\frac{3}{4}y$
12. $y = -\frac{1}{2}x + 2$
13. $2y - 6 = 3x$
14. $7x + 1 = 2y$
15. $y = 0.2x$

3.3 INTERCEPTS

Example 1: Identify the x- and y-intercepts and the intercept points.

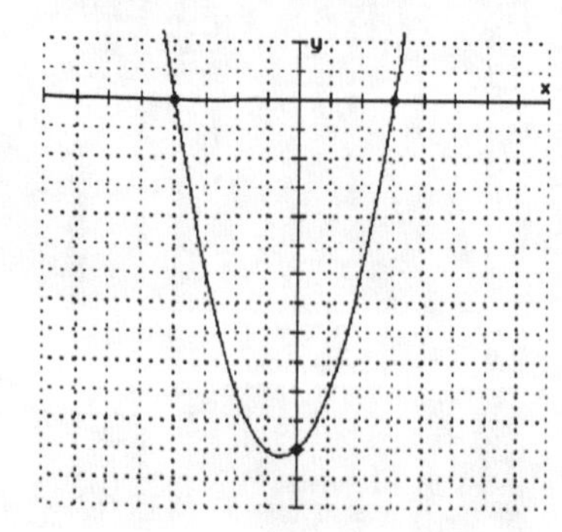

Solution: The graph crosses the x-axis at 3 and -4, so the x-intercepts are 3 and 4. The graph crosses the y-axis at -12, so the y-intercept is -12. The intercept points are (3, 0), (-4, 0), and (0, -12).

Example 2: Graph the line with x-intercept -1 and y-intercept 2.

Solution: The intercept points are $(-1, 0)$ and $(0, 2)$. Plot these points and draw a line through them.

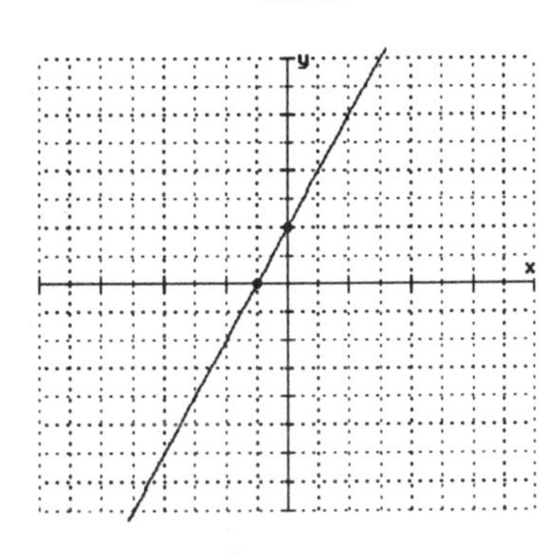

Example 3: Graph $2x - 7y = 14$ by finding and plotting intercept points.

Solution: Let $y = 0$ to find the x-intercept.

$$\begin{aligned} 2x - 7(0) &= 14 \\ 2x &= 14 \\ x &= 7 \end{aligned}$$

(7, 0)

Let $x = 0$ to find the y-intercept.

$$\begin{aligned} 2(0) - 7y &= 14 \\ -7y &= 14 \\ y &= -2 \end{aligned}$$

(0, -2)

Find a third ordered pair to check our work.

Let $y = -4$: $$\begin{aligned} 2x - 7(-4) &= 14 \\ 2x + 28 &= 14 \\ 2x &= -14 \\ x &= -7 \end{aligned}$$

(-7, -4)

Plot these points and draw a line through them.

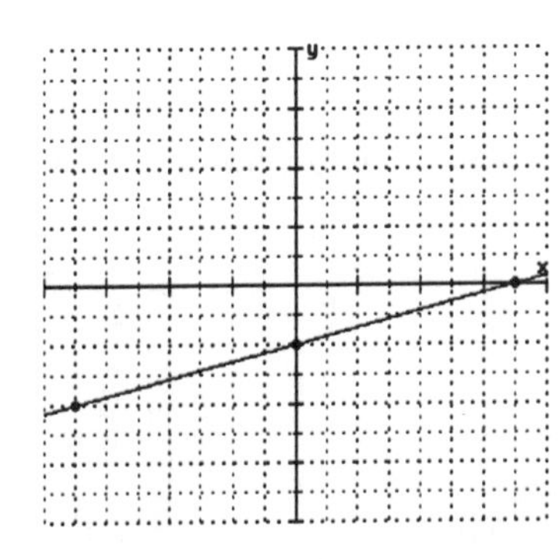

3.3 EXERCISES

Identify the intercepts and intercept points.

1.

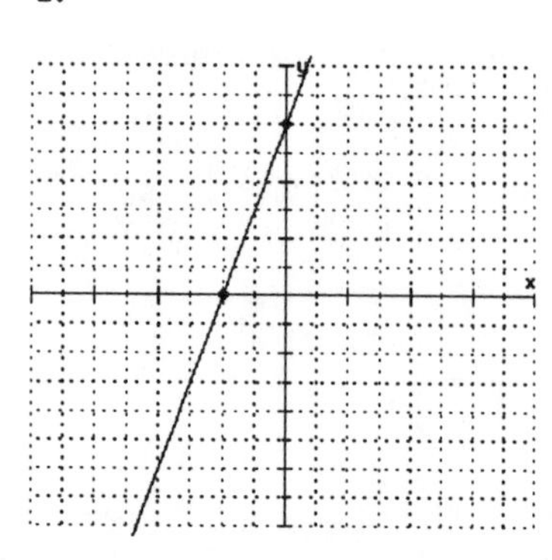

2.

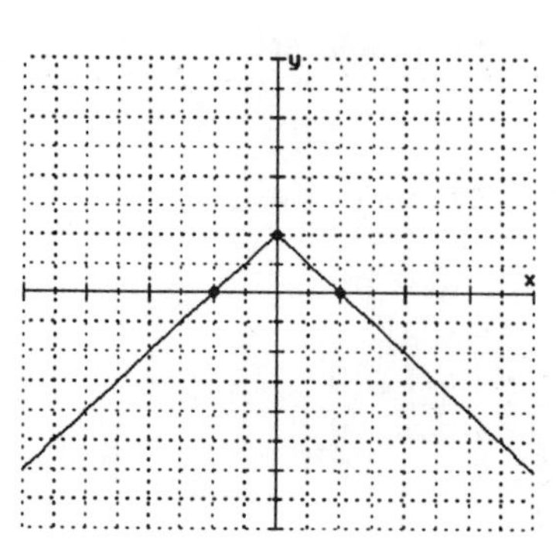

3.

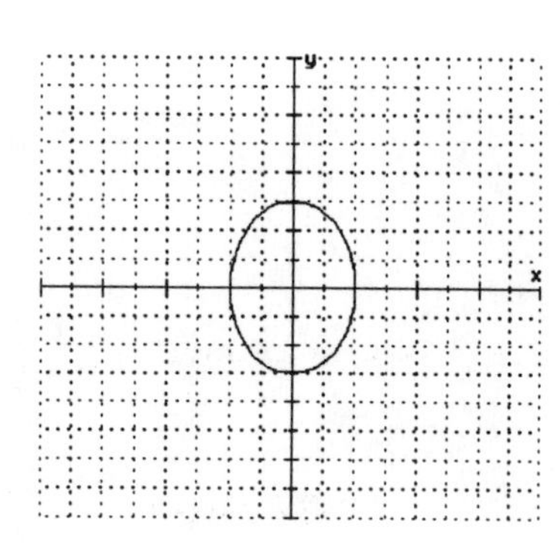

Graph each line with given x- and y-intercepts.

4. x-intercept: -5
   y-intercept: 2

5. x-intercept: -3
   y-intercept: 4

6. x-intercept: $-\frac{3}{2}$
   y-intercept: $\frac{9}{2}$

7. x-intercept: -8
   y-intercept: -7

Graph each linear equation by finding x- and y-intercepts.

8. $x + y = 5$

9. $3x = -y$

10. $2x - y = 7$

11. $3x - 6y = 9$

Graph each linear equation.

12. $x = -7$
13. $y = 5$
14. $x - 2 = 0$
15. $y = -\frac{3}{4}x + \frac{1}{2}$

3.4 SLOPE

Example 1: Find the slope of the line through (-3, -2) and (4, 1). Graph the line.

Solution:

$$m = \frac{y_2 - y_1}{x_2 - x_1} = \frac{-2 - 1}{-3 - 4} = \frac{-3}{-7} = \frac{3}{7}$$

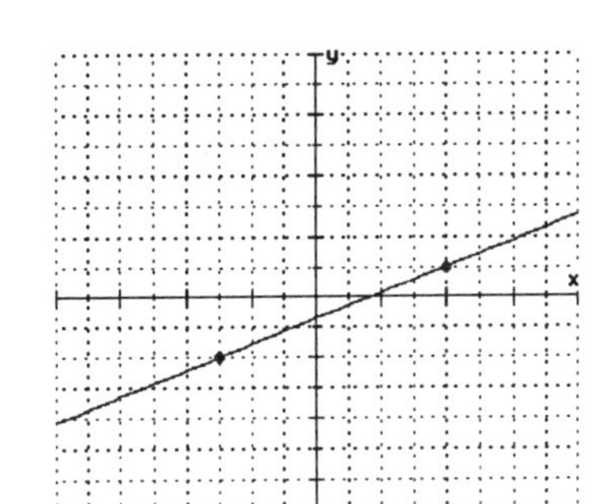

Example 2: Graph the line through (-4, -1) with slope -3.

Solution: $m = -3 = \frac{-3}{1}$

Locate the point (-4, -1) and then move 3 units down and right 1 unit to locate a second point. Draw the line through these two points.

We could also write $m = -3 = \frac{3}{-1}$. Then starting at (-4, -1) move 3 units up and 1 unit left. This will generate the same line.

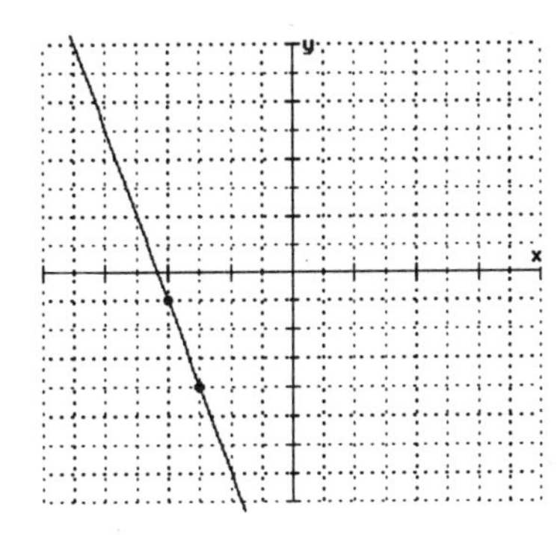

Example 3: Find the slope of the line whose equation is $-6x + 5y = 15$.

Solution: First find two points on the line.

Let $x = 5$:
$$\begin{aligned} -6(5) + 5y &= 15 \\ -30 + 5y &= 15 \\ 5y &= 45 \\ y &= 9 \\ &(5, 9) \end{aligned}$$

Let $x = 0$:
$$\begin{aligned} -6(0) + 5y &= 15 \\ 5y &= 15 \\ y &= 3 \\ &(0, 3) \end{aligned}$$

$$\begin{aligned} m &= \frac{y_2 - y_1}{x_2 - x_1} \\ &= \frac{9 - 3}{5 - 0} \\ &= \frac{6}{5} \end{aligned}$$

Example 4: Find the slope of a line perpendicular to the line passing through the points (0, 3) and (-2, -3).

Solution: First find the slope of the line through (0, 3) and (-2, -3).

$$m = \frac{y_2 - y_1}{x_2 - x_1} = \frac{-3 - 3}{-2 - 0} = \frac{-6}{-2} = 3$$

The slope of a line perpendicular to this line will have slope $-\frac{1}{m} = -\frac{1}{3}$.

3.4 EXERCISES

Find the slope of the line that goes through the given points.

1. (0, -6) and (8, 1)
2. (-4, -9) and (3, 9)
3. (1, 1) and (-6, -6)
4. (12, 0) and (0, -6)
5. (7, 8) and (7, -1)
6. (-2, 3) and (-6, 3)

Graph each line passing through the given point with given slope.

7. Through (0, -5) with slope $\frac{3}{4}$.
8. Through (1, 2) with slope -2.
9. Through (4, -3) with slope $-\frac{2}{5}$.
10. Through (-2, 0) with slope 1.

Find the slope of each line.

11. $2x - 5y = 10$
12. $-x - y = 7$
13. $x - 9 = 0$
14. $y = -12$

Determine whether the lines through each pair of points are parallel, perpendicular, or neither.

15. (0, −3) and (3, 1)
    (7, 2) and (3, −1)
16. (−1, 2) and (4, 8)
    (11, 20) and (17, 15)

3.5 GRAPHING LINEAR INEQUALITIES

Example 1: Determine whether (−3, 2) is a solution of $-x + 2y \geq 7$.

Solution:

$$-x + 2y \geq 7$$
$$-(-3) + 2(2) \geq 7 \quad \text{Replace } x \text{ with } -3 \text{ and } y \text{ with } 2.$$
$$7 \geq 7 \quad \text{True.}$$

The ordered pair (−3, 2) is a solution.

Example 2: Graph $x - 2y < 4$

Solution: First, graph the boundary line $x - 2y = 4$.

x	y
0	-2
4	0
2	-1

Graph this boundary as a dashed line because the inequality sign is <.

Test point: (0, 0)

$$x - 2y < 4$$
$$0 - 2(0) < 4 \quad \text{Replace } x \text{ with } 0 \text{ and } y \text{ with } 0.$$
$$0 < 4 \quad \text{True.}$$

Since the result is a true statement, shade the half-plane containing the test point (0, 0).

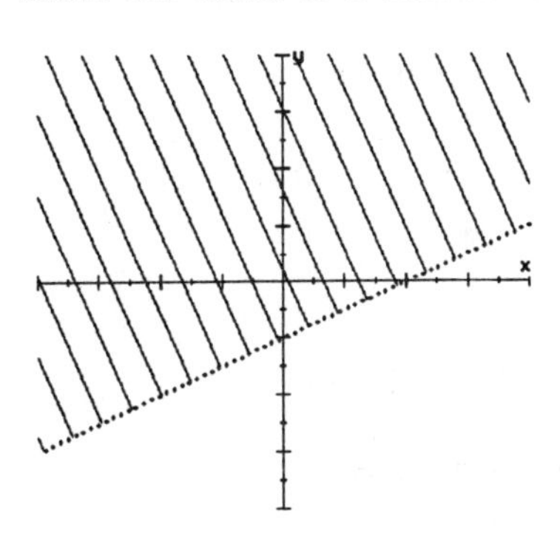

Example 3: Graph $x \leq 5$.

Solution: Graph the solid boundary line $x = 5$. Recall this is a vertical line with x-intercept 5.

Test point: (0, 0)
$x \leq 5$
$0 \leq 5$ Replace x with 0.

Since this result is a true statement, shade the half-plane containing the test point (0, 0).

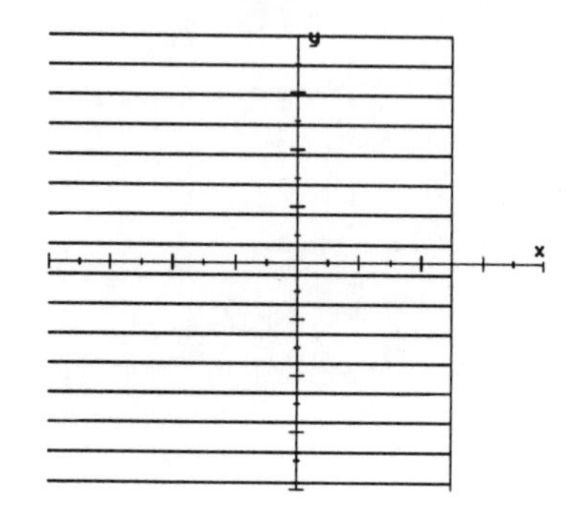

3.5 EXERCISES

Determine which ordered pairs given are solutions of the linear inequality in two variables.

1. $x + y < 2$; (1, -2), (3, -6), (-2, 7)
2. $x \geq -y$; (0, 0), (1, -1), (5, 3)
3. $y > \frac{1}{2}x$; (0, 0), (4, 1), (6, 8)
4. $-x - 3y \leq 12$; (4, 4), (-2, -5), (0, 4)
5. $2x + 7y < 9$; (0, 1), (1, 1), (3, 5)

Graph each inequality.

6. $x + y \geq -1$
7. $2x - y < -5$
8. $7x + 2y > -14$
9. $x \geq 6$
10. $y < -4$
11. $2x + 5y \geq 0$
12. $-\frac{1}{5}x + \frac{1}{2}y \leq 1$
13. $\frac{3}{4}x + \frac{1}{8}y > \frac{1}{6}$
14. $x - 1 \geq 0$
15. $-y < -7$

CHAPTER 3 PRACTICE TEST

Determine whether the ordered pairs are solutions of the equations.

1. $2x - y = 5$; $(1, -3)$
2. $3x + 2y = 12$; $(0, -6)$

Complete the ordered pair solution for the following equations.

3. $17y - 12x = 5$; $(\ , 1)$
4. $y = 6$; $(3, \ )$

Find the slopes of the following lines.

5.

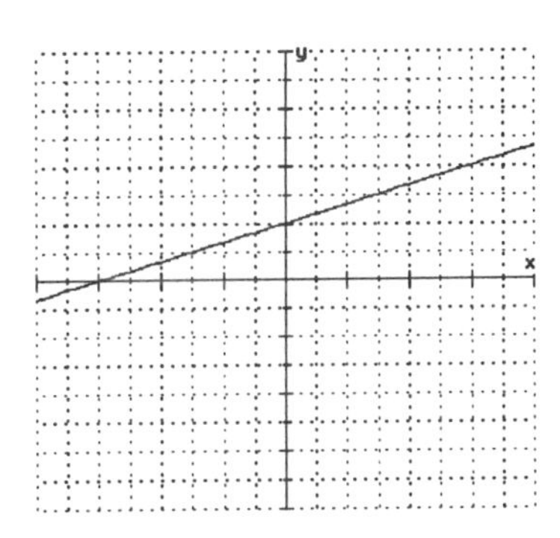

6.

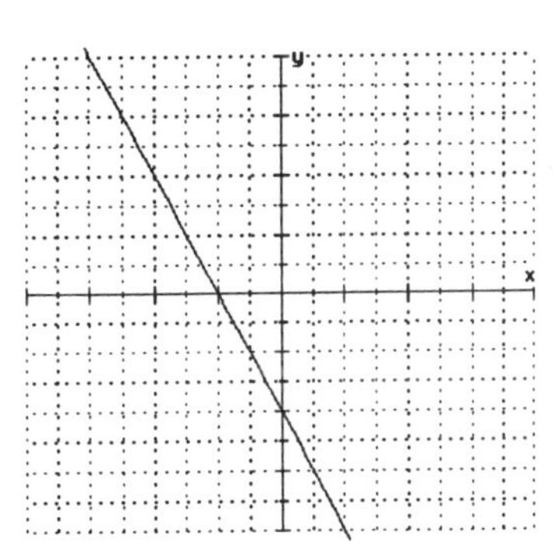

7. Through $(-6, 3)$ and $(-2, 15)$
8. Through $(0, -4)$ and $(-3, 5)$
9. $-7x + y = 4$
10. $x = -4$

Determine whether the lines through the pairs of points are parallel, perpendicular, or neither.

11. $(-2, 4)$ and $(2, -4)$
$(1, 3)$ and $(-1, 2)$
12. $(5, 7)$ and $(-3, 6)$
$(-4, 8)$ and $(4, 9)$

Graph the following.

13. $x + 3y = 6$
14. $-x + 2y = 7$
15. $x - y \le -3$
16. $y \ge -2x$
17. $6x - 5y = 12$
18. $3x - 4y < -8$
19. $5x + y > -2$
20. $y = -4$
21. $x - 4 = 0$
22. $4x - 5y = 20$

Write each statement as an equation in two variables, then graph the equation.

23. The y-value is 2 more than triple the x-value.
24. The x-value added to twice the y-value is less than -6.
25. The perimeter of the parallelogram is 48 meters. Write a linear equation in two variables for the perimeter. Use this equation to find x when y is 4.

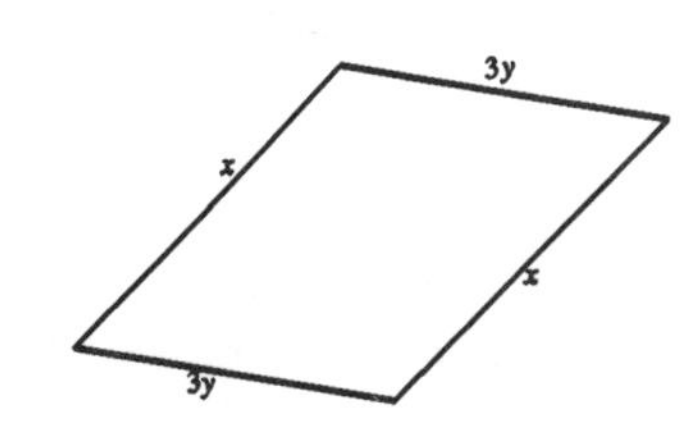

4.1 EXPONENTS

Example 1: Evaluate $(-5)^3$.

Solution:
$$(-5)^3 = (-5)(-5)(-5) = (25)(-5) = -125$$

Example 2: Simplify $x^2 \cdot x^5 \cdot x^{10}$.

Solution:
$$x^2 \cdot x^5 \cdot x^{10} = x^{2+5+10} = x^{17}$$

Example 3: Simplify $\frac{(x^6y^3)^2}{x^4y^3}$.

Solution:
$$\frac{(x^6y^3)^2}{x^4y^3} = \frac{(x^6)^2(y^3)^2}{x^4y^3}$$

$$= \frac{x^{6\cdot 2}y^{3\cdot 2}}{x^4y^3}$$ Use the power of a power rule for exponents.

$$= \frac{x^{12}y^6}{x^4y^3}$$

$$= x^{12-4}y^{6-3}$$ Use the quotient rule.

$$= x^8y^3$$

4.1 EXERCISES

Evaluate each expression.

1. -3^4
2. $\left(\frac{2}{5}\right)^2$
3. $4 \cdot 2^3$

Evaluate each expression given the replacement values for the variables.

4. $2x^3 \cdot y$; $x = -1$ and $y = 7$
5. $\frac{-2a}{b^4}$; $a = 12$ and $b = -2$

Simplify each expression.

6. $(5t)^3$
7. $(2a^2b^5)^4$
8. $\left(\frac{y^3}{-2z^2}\right)^5$
9. $\frac{8x^3y^9}{32xy^6}$
10. $6^0 + 3y^0$
11. $(12x^2y)^2$
12. $\frac{(2xy^2z^6)^3}{4xyz^4}$
13. $\frac{a^{19} \cdot a^6}{a^3}$
14. $\frac{(5cd^2)^3}{-25c^2d}$
15. $\frac{(x^9)^8}{(4x)^3}$
16. The formula $V = x^3$ can be used to find the volume V of a cube with side length x. Find the volume of a cube with side length 8 centimeters.

4.2 ADDING AND SUBTRACTING POLYNOMIALS

Example 1: Find the degree of the following polynomial and tell whether the polynomial is a monomial, binomial, trinomial, or none of these.

$-8t^3 + 3t - 6t^4$

Solution: This polynomial has 3 terms so it is a trinomial. The greatest degree of any of its terms is 4, hence the polynomial has degree 4.

Example 2: Find the value of the polynomial $2x^2 + 5x - 4$ when $x = -3$.

Solution:

$$\begin{aligned} 2x^2 + 5x - 4 &= 2(-3)^2 + 5(-3) - 4 \quad \text{Replace } x \text{ with } -3. \\ &= 2(9) - 15 - 4 \\ &= 18 - 15 - 4 \\ &= -1 \end{aligned}$$

Example 3: Add $(7x^2 - 10x + 4)$ and $(-x^2 - 4x + 7)$.

Solution:

$$\begin{aligned} (7x^2 - 10x + 4) + (-x^2 - 4x + 7) &= 7x^2 - 10x + 4 - x^2 - 4x + 7 \\ &= (7x^2 - x^2) + (-10x - 4x) + (4 + 7) \\ &= 6x^2 - 14x + 11 \end{aligned}$$

Example 4: Subtract: $(4x^3 - 8x^2 + 9) - (6x^2 - 5x - 1)$.

Solution:

$$\begin{aligned} (4x^3 - 8x^2 + 9) - (6x^2 - 5x - 1) &= (4x^3 - 8x^2 + 9) + (-6x^2 + 5x + 1) \\ &= 4x^3 - 8x^2 + 9 - 6x^2 + 5x + 1 \\ &= 4x^3 + (-8x^2 - 6x^2) + 5x + (9 + 1) \\ &= 4x^3 - 14x^2 + 5x + 10 \end{aligned}$$

4.2 EXERCISES

Find the degree of each of the following polynomials and determine whether it is a monomial, binomial, trinomial, or none of these.

1. $2 - x^2$
2. $8a^3 - 5a + 6a^5$
3. $5x^3yz^5$

Find the value of each polynomial when (a) $x = 0$ *and* (b) $x = -2$.

4. $5x - 12$
5. $-x^2 + x + 5$

Perform the indicated operations.

6. $(-8x + 11) + (5x - 4)$
7. $(x^2 - 3x + 4) - (6x^2 + 2x - 1)$
8. Subtract $(15x^2 + 4)$ from $(31x^2 - 9)$.
9. $(17x - 8) - (-3x^2 - 5x + 11)$
10. Subtract $8x$ from $6x + 1$.
11. Subtract $(15y + 8x^2)$ from the sum of $(9y - 2x)$ and $(12 + 4x^2)$.
12. $(14 + 6a) - (-a - 9)$
13. $(-13x^2 + 5x + 1) + (12x^2 + 7x - 1)$

14. $(-9x^5 + 6x^2) + (-9x^5 + 5x^2 + 10)$

15. Subtract $(3y^2 - 5y - 2)$ from the sum of $(9y^2 + 6)$ and $(3y - 7)$.

4.3 MULTIPLYING POLYNOMIALS

Example 1: Use the distributive property to find the product: $-2x^3(4x^2 + 5x - 8)$.

Solution: $-2x^3(4x^2 + 5x - 8) = -2x^3(4x^2) + (-2x^3)(5x) + (-2x^3)(-8)$ Use the distributive property.
$= -8x^5 - 10x^4 + 16x^3$ Multiply.

Example 2: Find the product: $(2x + 3)(5x - 4)$.

Solution: $(2x + 3)(5x - 4) = 2x(5x) + 2x(-4) + 3(5x) + 3(-4)$
$= 10x^2 - 8x + 15x - 12$ Multiply.
$= 10x^2 + 7x - 12$ Combine like terms.

Example 3: Multiply $(y - 3)$ by $(2y^2 + 3y - 1)$.

Solution: $(y - 3)(2y^2 + 3y - 1) = y(2y^2) + y(3y) + y(-1) + (-3)(2y^2) + (-3)(3y) + (-3)(-1)$
$= 2y^3 + 3y^2 - y - 6y^2 - 9y + 3$ Multiply.
$= 2y^3 - 3y^2 - 10y + 3$ Combine like terms.

4.3 EXERCISES

Find the following products.

1. $7a(3a - 5)$
2. $-2x^2(9x - 11)$
3. $(a + 8)(a - 3)$
4. $(4x + y)(x - 5y)$
5. $(3x + 2y)(2x + 3y)$
6. $(5a - 7)^2$
7. $(x^2 - 3)^2$
8. $(x + 3)^3$
9. $(3y - 1)^3$
10. $(x + 2)(x^2 + 7x - 3)$
11. $(b + 5)(b^2 + 8b + 5)$
12. $(9 + d)(1 - 4d - 2d^2)$
13. $(x + 6)(5x^2 - x - 1)$
14. $(4x + 3)(4x - 3)$
15. $(y - 3)(y - 2)$
16. $(x^2 + x + 2)(3x^2 - x + 4)$

4.4 SPECIAL PRODUCTS

Example 1: Find $(x - 4)(x + 3)$ by the FOIL method.

Solution: $(x - 4)(x + 3) = (x)(x) + (x)(3) + (-4)(x) + (-4)(3)$
F 0 I L
$= x^2 + 3x - 4x - 12$
$= x^2 - x - 12$ Collect like terms.

Example 2: Multiply $(4a - 3)^2$.

Solution: $(4a - 3)^2 = (4a)^2 + 2(4a)(-3) + 3^2$
$= 16a^2 - 24a + 9$

Example 3: Multiply $(3x - 7)(3x + 7)$.

Solution: $(3x - 7)(3x + 7) = (3x)^2 - 7^2$
$= 9x^2 - 49$

4.4 EXERCISES

Find each product.

1. $(x - 11)^2$
2. $(3x + 10)^2$
3. $(y + 8)^2$
4. $(2a - 1)^2$
5. $(y + 5)(y - 8)$
6. $(b - 2)(b - 9)$
7. $(x + 9)(x - 9)$
8. $(7y - 6)(7y + 6)$
9. $\left(x + \frac{1}{4}\right)\left(x - \frac{1}{4}\right)$
10. $\left(\frac{2}{5}a^2 + b\right)\left(\frac{2}{5}a^2 - b\right)$
11. $(5b + 2)^2$
12. $(2a + 3)(a - 5)$
13. $(x + 10)(x + 10)$
14. $(5a + 1)(a - 5)$
15. $(4x + 5y)^2$

4.5 NEGATIVE EXPONENTS AND SCIENTIFIC NOTATION

Example 1: Simplify the expression. Write the result with positive exponents. $\left(\frac{3}{5}\right)^{-3}$

Solution: $\left(\frac{3}{5}\right)^{-3} = \frac{1}{\left(\frac{3}{5}\right)^3} = \frac{1}{\frac{3^3}{5^3}} = \frac{5^3}{3^3} = \frac{125}{27}$

Example 2: Simplify the expression. Write the answer with positive exponents. $\frac{x^{-6}}{x^8}$

Solution: $\frac{x^{-6}}{x^8} = x^{-6-8} = x^{-14} = \frac{1}{x^{14}}$

Example 3: Simplify the expression. Write the answer with positive exponents. $(3x^4)(4x)^{-2}$

Solution: $(3x^4)(4x)^{-2} = 3x^4 \cdot 4^{-2}x^{-2} = \frac{3x^{4+(-2)}}{4^2} = \frac{3x^2}{16}$

Example 4: Write the following numbers in scientific notation.

a. 53,000,000,000 b. 0.00000000124

Solution: a. $5_{\wedge}3,000,000,000.$ Move the decimal point until the number is between 1 and 10.

$= 5.3 \times 10^{10}$ The decimal point is moved 10 places to the left, so the count is positive 10.

b. $0.000000001_{\wedge}24$ Move the decimal point until the number is between 1 and 10.

$= 1.24 \times 10^{-9}$ The decimal point is moved 9 places to the right, so the count is -9.

Example 5: Write the following numbers in standard notation, without exponents.

a. 2.65×10^6 b. 9.8×10^{-4}

Solution: a. 2.65×10^6
$= 2.650000_\wedge$ Move the decimal point to the right 6 places.
$= 2,650,000$

b. 9.8×10^{-4}
$= 0_\wedge 0009.8$ Move the decimal point to the left 4 places.

$= 0.00098$

4.5 EXERCISES

Simplify each expression. Write results with positive exponents.

1. 7^{-3}
2. $\dfrac{1}{a^{-4}}$
3. $5^{-1} + 6^{-1}$
4. $\dfrac{a^{-3}}{a^{-8}}$
5. $(x^{-5})^6$
6. $(2x^{-4})^{-3}$
7. $-8^0 - 4x^0$
8. $(2x^{-3}y^5)^{-4}$
9. $\dfrac{9c^3d^{-5}}{9^{-1}c^{-6}d^8}$
10. $\dfrac{(x^2y^4)^6}{(xy^3)^{-5}}$

Write each number in scientific notation.

11. 8,100,000
12. 146,000,000,000
13. 0.007
14. 0.000000912

Write each number in standard notation, without exponents.

15. 3.4×10^{-5}
16. 2.71×10^9
17. 4.0×10^6
18. 5.97×10^{-8}

4.6 DIVISION OF POLYNOMIALS

Example 1: Simplify by performing the indicated division. Write the answer with positive exponents.

$$\frac{45x^3y^8}{9xy^{10}}$$

Solution:

$$\frac{45x^3y^8}{9xy^{10}} = \frac{45}{9}x^{3-1}y^{8-10} = 5x^2y^{-2} = \frac{5x^2}{y^2}$$

Example 2: Simplify $\dfrac{20x^3y^2 - 25x^2y + 5x}{10xy^2}$

Solution:

$$= \frac{20x^3y^2 - 25x^2y + 5x}{10xy^2} = \frac{20x^3y^2}{10xy^2} - \frac{25x^2y}{10xy^2} + \frac{5x}{10xy^2}$$

Divide each term by $10xy^2$.

$$= 2x^2 - \frac{5x}{2y} + \frac{1}{2y^2}$$

Example 3: Divide $x^2 + 6x - 8$ by $x - 2$.

Solution:

$$\begin{array}{rll} & x + 8 & \\ x - 2 \overline{) } & x^2 + 6x - 8 & \text{How many times does } x \text{ divide } x^2\text{?} \\ & \underline{x^2 - 2x} & \text{Multiply } x(x - 2). \\ & 8x - 8 & \text{Subtract and bring down the next term.} \\ & \underline{8x - 16} & \text{Repeat the process.} \\ & 8 & \text{The remainder is 8.} \end{array}$$

The answer is $x + 8 + \dfrac{8}{x - 2}$.

Example 4: Divide $x^3 - 27$ by $x + 3$.

Solution:

$$\begin{array}{rll} & x^2 - 3x + 9 & \\ x + 3 \overline{) } & x^3 + 0x^2 + 0x - 27 & \text{Use 0 coefficients for missing terms.} \\ & \underline{x^3 + 3x^2} & \\ & -3x^2 + 0x & \\ & \underline{-3x^2 - 9x} & \\ & 9x - 27 & \\ & \underline{9x + 27} & \\ & -54 & \end{array}$$

$$x^2 - 3x + 9 - \frac{54}{x + 3}$$

4.6 EXERCISES

Simplify each expression.

1. $\dfrac{-35x^3yz^2}{-7xy^5z^2}$

2. $\dfrac{21a^4b}{-3a^6b^5}$

Perform each division.

3. $\dfrac{38a^4 - 57a^2}{19a^3}$

4. $\dfrac{-10b^7 + 25b^3}{-5b^9}$

5. $\dfrac{x^2 + 5x - 24}{x + 8}$

6. $\dfrac{x^2 - 12x + 35}{x - 7}$

7. $\dfrac{4x^2 - 4x - 3}{2x + 1}$

8. $\dfrac{8x^2 + 22x + 15}{4x + 5}$

9. $\dfrac{x^2 + 11x + 21}{x + 5}$

10. $\dfrac{x^3 - 2x^2 + x + 6}{x - 2}$

11. $\dfrac{x^3 - 125}{x - 5}$

12. $\dfrac{-x^3 - 4x^2 + 6x - 1}{x - 1}$

13. $\dfrac{8a^3 - 12a^2 + 6a - 5}{2a - 1}$

14. $\dfrac{9 - 4x^2}{x + 6}$

15. $\dfrac{x^4 + x}{x^2 + 1}$

CHAPTER 4 PRACTICE TEST

Evaluate each expression.

1. 2^6
2. $(-4)^4$
3. -4^4
4. 3^{-4}

Simplify each exponential expression.

5. $\left(\frac{6x^5y^4}{36x^8y^2}\right)^3$
6. $\frac{8(xy)^5}{(xy)^3}$
7. $2(x^3y^4)^{-5}$

Simplify each expression. Write the result using only positive exponents.

8. $\left(\frac{x^3y^4}{x^4y^{-5}}\right)^{-2}$
9. $\frac{7^3x^{-8}y^{-2}}{7^5x^{-3}y^9}$

Express each number in scientific notation.

10. 8,230,000,000
11. 0.00000714

Write each number in standard form.

12. 2.7×10^{-5}
13. 8.3×10^6

14. Simplify. Write the answer in standard form.

    $(2.1 \times 10^7)(4 \times 10^{-9})$

15. Find the degree of the following polynomial.

    $3x^3y^4 - 9xy^2z + 8x^4yz^3$

16. Simplify by combining like terms.

    $12x^2yz - 7xy^3 - 8x^2yz + 2xy^3$

Perform the indicated operations.

17. $(4x^3 - 6x^2 + 3x - 9) + (5x^3 - 6x - 3)$

18. $$\begin{array}{r} 6x^3 + 2x^2 + x - 3 \\ \underline{-(9x^3 - 3x^2 - 2x + 12)} \end{array}$$

19. Subtract $(5x - 1)$ from the sum of $(6x^2 - 6x + 1)$ and $(x^3 + 5)$.

20. Multiply: $(2x + 3)(x^2 - 4x + 5)$.

21. Multiply $x^3 + x^2 - x - 2$ by $3x^2 - 2x + 6$ using the vertical format.

22. Use the FOIL method to multiply $(x - 6)(2x + 7)$.

Use special products to multiply each of the following.

23. $(2x - 5)(2x + 5)$
24. $(3x - 8)^2$
25. $(7x + 2)^2$
26. $(x^2 - 3b)(x^2 + 3b)$

27. The height of the Bank of China in Hong Kong is 1001 feet. Neglecting air resistance the height of an object dropped from this building at time t seconds is given by the polynomial $-16t^2 + 1001$. Find the height of the object at the given times.

t	2 seconds	4 seconds	6 seconds
$-16t^2 + 1001$			

Divide.

28. $\dfrac{12x^2y^3}{3x^4y^5z^2}$

29. $\dfrac{5x^2 + 10xy - 8x}{15xy}$

30. $(x^2 + 9x + 18) \div (x + 3)$

31. $\dfrac{8x^3 - 27}{2x + 3}$

5.1 THE GREATEST COMMON FACTOR AND FACTORING BY GROUPING

Example 1: Find the GCF of the following list of terms. $a^2,\ a^6,\ a^5$

Solution: The GCF is a^2 since 2 is the smallest exponent to which a is raised.

Example 2: Factor the following polynomial by factoring out the GCF. $8b^4 - 12b^3 + 20b^9$

Solution: $8b^4 - 12b^3 + 20b^9 = 4b^3(2b) + 4b^3(-3) + 4b^3(5b^6)$ The GCF is $4b^3$.
$= 4b^3(2b - 3 + 5b^6)$

Example 3: Factor $4(y - 2) + x(y - 2)$.

Solution: The binomial $(y - 2)$ is the GCF. Use the distributive property to factor out $(y - 2)$.

$4(y - 2) + x(y - 2) = (y - 2)(4 + x)$

Example 4: Factor $xy + 5x - 3y - 15$ by grouping.

Solution: $xy + 5x - 3y - 15$
$= (xy + 5x) + (-3y - 15)$ Group the first two terms and the last two terms.
$= x(y + 5) - 3(y + 5)$ The GCF of the first group is x; the GCF of the second group is -3.
$= (y + 5)(x - 3)$ The GCF is $(y + 5)$.

5.1 EXERCISES

Find the GCF for each list.

1. $16,\ 40,\ 72$
2. $6a^6b^4,\ 3a^3b^9,\ 15a^4b^2$

Factor the following polynomials.

3. $5x - 35$
4. $6x - 15y + 21$
5. $a(b + 4) - 6(b + 4)$
6. $12(6 - y) + x(6 - y)$
7. $c(d^2 - 3) - 9(d^2 - 3)$
8. $-4x^3y^7 - 12x^4y^6 + 6x^5y^2$
9. $xy + 2y + 11x + 22$
10. $3ab + 6a - b - 2$
11. $18 - 45x - 14y + 35xy$
12. $4x^3 - 3x^2 + 4x - 3$
13. $8ac - 8ad + bc - bd$
14. $8xy + 28y - 32x - 112$
15. $42 - 21b - 14a + 7ab$

5.2 FACTORING TRINOMIALS OF THE FORM $x^2 + bx + c$

Example 1: Factor $x^2 + 9x + 20$.

Solution: Positive factors of 20	Sum of Factors
1, 20	21
2, 10	12
4, 5	**9**

$$x^2 + 9x + 20 = (x + 4)(x + 5)$$

Example 2: Factor $x^2 + 6x - 16$.

Solution: Factors of −16	Sum of Factors
1, −16	−15
−1, 16	15
2, −8	−6
−2, 8	**6**
−4, 4	0

$$x^2 + 6x - 16 = (x - 2)(x + 8)$$

Example 3: Factor $4x^2 - 40x + 84$.

Solution: $4x^2 - 40x + 84 = 4(x^2 - 10x + 21)$ The GCF is 4; −7 and −3 have a product of 21 and a sum of −10.

$$= 4(x - 7)(x - 3)$$

5.2 EXERCISES

Factor each trinomial completely.

1. $x^2 - 11x + 28$
2. $x^2 - 6x - 27$
3. $x^2 - 9x + 8$
4. $x^2 - x + 6$
5. $x^2 + x - 30$
6. $x^2 + 2xy - 3y^2$
7. $x^3 - 13x^2 + 42x$
8. $3x^2 + 15xy + 18y^2$
9. $x^2 + x - 110$
10. $x^2 + 14x + 13$
11. $x^2 + 11xy - 12y^2$
12. $x^2 - 7x - 10$
13. $2x^2 - 26x + 80$
14. $4x^2y - 8xy + 16y$
15. $5a^2b - 5ab^2 - 30b^3$

5.3 FACTORING TRINOMIALS OF THE FORM $ax^2 + bx + c$

Example 1: Factor $2x^2 + 5x + 3$.

Solution: Factors of $2x^2$: $2x \cdot x$

$2x^2 + 5x + 3 = (2x \quad)(x \quad)$

Factors of 3: $3 = 1 \cdot 3$

$2x^2 + 5x + 3 = (2x + 1)(x + 3)$
$1x + 6x = 7x$ incorrect middle term.

$$2x^2 + 5x + 3 = (2x + 3)(x + 1)$$
$$2x + 3x = 5x \quad \text{correct middle term.}$$

Hence the factored form of $2x^2 + 5x + 3$ is $(2x + 3)(x + 1)$.

$$2x^2 + 5x + 3 = (2x + 3)(x + 1)$$

Example 2: Factor $9x^2 + 24x + 16$.

Solution:
1. Is the first term a square? Yes, $9x^2 = (3x)^2$.
2. Is the last term a square? Yes, $16 = (4)^2$.
3. Is the middle term twice the product of $3x$ and 4? Yes, $24x = 2(3x)(4)$.

Thus, $9x^2 + 24x + 16 = (3x + 4)^2$.

Example 3: Factor $4x^2 - 5x - 6$ by grouping.

Solution: Find two numbers whose product is $a \cdot c$ or $4 \cdot (-6) = -24$, and whose sum is b or -5.

Factors of −24	Sum of Factors
$1 \cdot -24$	−23
$-1 \cdot 24$	23
$2 \cdot -12$	−10
$-2 \cdot 12$	10
$3 \cdot -8$	**−5**
$-3 \cdot 8$	5
$4 \cdot -6$	−2
$-4 \cdot 6$	2

Write $-5x$ as $3x - 8x$.

$$\begin{aligned} 4x^2 - 5x - 6 &= 4x^2 + 3x - 8x - 6 \\ &= x(4x + 3) - 2(4x + 3) \\ &= (4x + 3)(x - 2) \end{aligned}$$

EXERCISE SET 5.3

Factor completely.

1. $5x^2 + 7x + 2$
2. $2x^2 + 9x + 9$
3. $2x^2 - 5x - 12$
4. $3a^2 - 5a - 2$
5. $9y^2 + y - 8$
6. $4b^2 + b + 5$
7. $16a^2 + 8a + 1$
8. $49a^2 - 28a + 4$
9. $10x^2 + 58x - 12$
10. $10x^2 - 55x + 45$
11. $9y^2 + 6y - 4$
12. $9x^2 + 36x + 54$
13. $8d^2 - 26d + 15$
14. $6x^3 + 25x^2 + 4x$
15. $9x^2y^2 - 30xy + 25$
16. $100 - 180a + 81a^2$
17. $49x^2 + 84xy + 36y^2$

5.4 FACTORING BINOMIALS

Example 1: Factor $9x^2 - 4$.

Solution: $9x^2 - 4 = (3x)^2 - (2)^2$ Each term is a perfect square.
$= (3x + 2)(3x - 2)$

Example 2: Factor $49y^2 - z^2$.

Solution: $49y^2 - z^2 = (7y)^2 - (z)^2$ Each term is a perfect square.
$= (7y + z)(7y - z)$

Example 3: Factor $x^3 + 1$.

Solution: $x^3 + 1 = (x)^3 + (1)^3$ Write in the form $a^3 + b^3$.

$= (x + 1)\left[(x)^2 - (x)(1) + (1)^2\right]$

$= (x + 1)(x^2 - x + 1)$

Example 4: Factor $2x^3 - 16y^3$.

Solution: $2x^3 - 16y^3 = 2(x^3 - 8y^3)$ Factor out the GCF, 2.

$= 2\left[(x)^3 - (2y)^3\right]$ Write in the form $x^3 - y^3$.

$= 2\{(x - 2y)\left[(x)^2 + (x)(2y) + (2y)^2\right]\}$

$= 2(x - 2y)(x^2 + 2xy + 4y^2)$

5.4 EXERCISES

Factor the binomials completely.

1. $9x^2 - 25$
2. $100x^2 + 9$
3. $144a^2 - 49b^2$
4. $y^3 - 64$
5. $b^3 + 125$
6. $12x^3 + 27$
7. $2 - 8a^2$
8. $49x^2 - 121y^2$
9. $24x^3 + 1029$
10. $27c^3 - 125d^3$
11. $27x^2y^3 + 4x^2y$
12. $4x^2y^2 + 9z^2$
13. $486 - 18a^3$
14. $16x^4y + 64xy^3$
15. $7 + 56z^3$

5.5 CHOOSING A FACTORING STRATEGY

Example 1: Factor $8a^2 - 50b^2$.

Solution: The terms of this binomial contain a greatest common factor of 2.

$8a^2 - 50b^2 = 2(4a^2 - 25b^2)$ Factor out the GCF.
$= 2\left[(2a)^2 - (5b)^2\right]$ The binomial is a difference of squares.
$= 2(2a + 5b)(2a - 5b)$ Factor the difference of squares.

Example 2: Factor $10x^2 - 3x - 4$.

Solution: The terms of this trinomial have no common factor other than 1. It is not a perfect square trinomial.
Look for factors of $(10)(-4) = -40$ whose sum is -3.

Factors of −40	Sum of Factors
1, −40	−39
−1, 40	39
2, −20	−18
−2, 20	18
4, −10	−6
−4, 10	6
5, −8	−3
−5, 8	3

Rewrite $-3x$ as $5x - 8x$.

$$\begin{aligned} 10x^2 - 3x - 4 &= 10x^2 + 5x - 8x - 4 \\ &= 5x(2x + 1) - 4(2x + 1) \\ &= (2x + 1)(5x - 4) \end{aligned}$$

Example 3: Factor $3x^3 + 5x^2 - 12x - 20$.

Solution: The terms of this polynomial have no common factor other than 1. There are four terms so try factoring by grouping.

$3x^3 + 5x^2 - 12x - 20 = x^2(3x + 5) - 4(3x + 5)$ Factor out the GFC for each pair of terms.

$= (3x + 5)(x^2 - 4)$ Factor out $(3x + 5)$.

$= (3x + 5)[(x)^2 - (2)^2]$

$= (3x + 5)(x + 2)(x - 2)$ Factor $x^2 - 4$ as a difference of squares.

5.5 EXERCISES

Factor the following completely.

1. $x^2 - 4xy + 4y^2$
2. $a^2 - 7a + 12$
3. $b^2 + 3b + 2$
4. $x^2 + 6x + 9$
5. $2x^2 + 2x - 112$
6. $x^3 - 9x^2 + 20x$
7. $4x^2 + 49y^2$
8. $10 - 7x + x^2$
9. $4a^2 - 49b^2$
10. $d^3 - 64c^3$
11. $6x^3 - 150x$
12. $3ab - 5a + 6b - 10$
13. $x^2 + 7x + 8$
14. $25x^2 + 40xy + 16y^2$
15. $x - y + a(x - y)$
16. $x^3 - 16x + 2x^2 - 32$

5.6 SOLVING QUADRATIC EQUATIONS BY FACTORING

Example 1: Solve $(x - 7)(2x + 3) = 0$.

Solution: Use the zero factor theorem; set each factor equal to 0 and solve the resulting linear equations.

$$(x - 7)(2x + 3) = 0$$
$$x - 7 = 0 \quad \text{or} \quad 2x + 3 = 0$$
$$x = 7 \quad \text{or} \quad 2x = -3$$
$$x = -\frac{3}{2}$$

The solution set is $\left\{7, -\frac{3}{2}\right\}$.

Example 2: Solve $x^2 - 7x = -6$.

Solution:

$x^2 - 7x = -6$	
$x^2 - 7x + 6 = 0$	Write the equation in standard form by adding 6 to both sides.
$(x - 6)(x - 1) = 0$	Factor.
$x - 6 = 0$ or $x - 1 = 0$	Use the zero factor theorem.
$x = 6$ or $x = 1$	Solve each linear equation.

The solution set is $\{6, 1\}$.

Example 3: Solve $6x^3 - 150x = 0$.

Solution:

$6x^3 - 150x = 0$	
$6x(x^2 - 25) = 0$	Factor out the GCF of $6x$.
$6x(x + 5)(x - 5) = 0$	Factor $x^2 - 25$, a difference of squares.
$6x = 0$ or $x + 5 = 0$ or $x - 5 = 0$	Set each factor equal to 0.
$x = 0$ or $x = -5$ or $x = 5$	Solve.

The solution set is $\{0, -5, 5\}$.

5.6 EXERCISES

Solve each equation.

1. $(x - 6)(x + 7) = 0$
2. $3x(2x - 9) = 0$
3. $(8y + 1)(7y + 2) = 0$
4. $x^2 + 8x + 12 = 0$
5. $b^2 + 4b - 45 = 0$
6. $2x^2 + 5x - 7 = 0$
7. $z^2 - 3z = -2$
8. $x^2 + 21 = -10x$
9. $25(x + 1) - 6x^2$
10. $a(11 - a) = 0$
11. $25b^2 - 1 = 0$
12. $x^3 + 8x^2 + 7x = 0$
13. $9t + 8 = 12$
14. $12x^2 + 25x + 12 = 0$
15. $x^2 - 10x + 10 = -15$

5.7 QUADRATIC EQUATIONS AND PROBLEM SOLVING

Example 1: A rock is dropped from a cliff 144 feet above the ground. Neglecting air resistance, the height h in feet of the rock above the ground after t seconds is given by the quadratic equation

$$h = -16t^2 + 144$$

Find how long it takes for the rock to hit the ground.

Solution: We must find for what value of t will $h = 0$.
Replace h with 0.

$h = -16t^2 + 144$
$0 = -16t^2 + 144$
$0 = -16(t^2 - 9)$ Factor out the GCF of -16.
$0 = -16(t + 3)(t - 3)$ Factor $t^2 - 9$, difference of squares.
$t + 3 = 0$ or $t - 3 = 0$ Use the zero factor theorem.
$t = -3$ or $t = 3$ Solve.

Since t represents time, t cannot be negative. Hence $t = 3$ is the solution.

It will take 3 seconds for the rock to hit the ground.

Example 2: The height of a triangular sign is 1 meter less than twice the length of the base. If the sign has an area of 14 square meters, find the length of its base and the height.

Solution: Let x = length of the base. Then
$2x - 1$ = the height.

$$\text{Area} = \frac{1}{2} \cdot \text{base} \cdot \text{height}$$

$14 = \frac{1}{2}(x)(2x - 1)$ Replace Area with 14, base with x and height with $2x - 1$.

$2(14) = 2\left(\frac{1}{2}\right)(x)(2x - 1)$ Multiply both sides by 2 to clear the fractions.

$28 = x(2x - 1)$
$28 = 2x^2 - x$
$0 = 2x^2 - x - 28$ Write in standard form.
$0 = (2x + 7)(x - 4)$ Factor.

$2x + 7 = 0$ or $x - 4 = 0$ Set each factor equal to 0.
$2x = -7$ or $x = 4$
$x = -\frac{7}{2}$

Since x represents the length of the base, discard the solution $-\frac{7}{2}$. The base of a triangle cannot be negative.

$$x = 4$$
$$2x - 1 = 2(4) - 1 = 7$$

The base is 4 meters and the height is 7 meters.

Example 3: Find the lengths of the sides of a right triangle if the lengths can be expressed by three consecutive integers.

Solution: Let the lengths of the sides be represented by

x = one leg
$x + 1$ = other leg
$x + 2$ = hypotenuse (longest side).

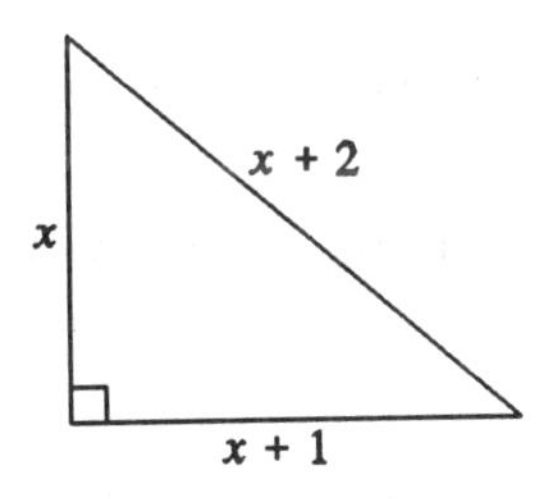

By the Pythagorean theorem, we have that

$$(\text{hypotenuse})^2 = (\text{leg})^2 + (\text{leg})^2$$

$$(x + 2)^2 = (x)^2 + (x + 1)^2$$
$$x^2 + 4x + 4 = x^2 + x^2 + 2x + 1 \quad \text{Multiply.}$$
$$x^2 + 4x + 4 = 2x^2 + 2x + 1$$
$$0 = x^2 - 2x - 3 \quad \text{Write in standard form.}$$
$$0 = (x - 3)(x + 1) \quad \text{Factor.}$$
$$x - 3 = 0 \quad \text{or} \quad x + 1 = 0 \quad \text{Set each factor equal to 0.}$$
$$x = 3 \quad \text{or} \quad x = -1$$

Discard $x = -1$ since length cannot be negative. If $x = 3$, then $x + 1 = 3 + 1 = 4$, and $x + 2 = 3 + 2 = 5$.

The sides of the right triangle have lengths 3 units, 4 units, and 5 units.

5.7 EXERCISES

Represent the given conditions using a single variable x.

1. Two numbers whose sum is 18.
2. A man's age now and his age 11 years from now.
3. The three sides of a triangle if the first side is 5 inches less than triple the second side. The third side is 9 inches longer than the second side.

4. Three consecutive even integers.

Solve the following.

5. The length of a rectangle is 2 centimeters less than twice the width. Its area is 84 square centimeters. Find the dimensions of the rectangle.

6. A hang-glider pilot accidentally drops her compass from the top of a 576-foot cliff. The height h of the compass after t seconds is given by the quadratic equation

$$h = -16t^2 + 576.$$

When will the compass hit the ground?

7. The base of a triangle is 5 meters less than four times the length of the altitude. If the triangle has an area of 57 square meters, find the length of its base and altitude.

8. If the sides of a square are decreased by 4 inches the area becomes 25 square inches. Find the length of the sides of the original square.

9. Find the lengths of the sides of a right triangle if the hypotenuse is 14 centimeters longer than the short leg and 7 centimeters longer than the long leg.

10. The sum of a number and its square is 272. Find the number.

11. The sum of two numbers is 17, and the sum of their squares is 157. Find the numbers.

12. A rectangle has a perimeter of 52 feet and an area of 153 square feet. Find the dimensions of the rectangle.

13. The sum of the squares of two consecutive odd integers is 18 less than 20 times the larger integer. Find the integers.

14. Find two consecutive even numbers whose product is 1088.

15. The altitude of a triangle is 7 inches less than the base. If the area is 99 square inches, find the base.

CHAPTER 5 PRACTICE TEST

Factor each polynomial completely. If a polynomial cannot be factored, write "prime".

1. $8x^3 + 29x^2 + 20x$
2. $x^2 + x - 9$
3. $y^2 - 13y - 48$
4. $2a^2 + 2ab - 5a - 5b$
5. $3x^2 - 7x + 2$
6. $x^2 + 19x + 80$
7. $x^2 + 11xy + 24y^2$
8. $40x^3 + 10x^2 - 16x$
9. $175 - 7x^2$
10. $27x^3 - 64$
11. $4t^2 + 12t - 7$
12. $xy^2 - 11y^2 - 9x + 99$
13. $16x - x^5$

Solve each equation.

14. $x^2 - 2x = 35$
15. $x^2 - 121 = 0$
16. $2x(5x - 1)(4x + 3) = 0$
17. $8x^2 = -48x$

Solve each problem.

18. Find the dimensions of a rectangular garden whose length is 4 feet longer than its width and whose area is 525 square feet.

19. The sum of two numbers is 20, and the sum of their squares is 202. Find the numbers.

20. The length of the base of a triangle is 5 feet longer than its altitude. If the area of the triangle is 33 square feet, find the length of the base.

6.1 SIMPLIFYING RATIONAL EXPRESSIONS

Example 1: Find the value of $\frac{x-5}{3x+4}$ for the given replacement values.

a. $x = -1$ b. $x = 3$

Solution:

a. $\frac{x-5}{3x+4} = \frac{-1-5}{3(-1)+4}$ Replace x with -1.

$= \frac{-6}{-3+4}$

$= \frac{-6}{1}$

$= -6$

b. $\frac{x-5}{3x+4} = \frac{3-5}{3(3)+4}$ Replace x with 3.

$= \frac{-2}{9+4}$

$= -\frac{2}{13}$

Example 2: Are there any values for x for which the given expression is undefined? $\frac{x^2+1}{x^2-5x+6}$

Solution:

$x^2 - 5x + 6 = 0$ Set the denominator equal to 0.

$(x-2)(x-3) = 0$ Factor.

$x - 2 = 0$ or $x - 3 = 0$ Set each factor equal to 0.

$x = 2$ or $x = 3$ Solve.

The expression $\frac{x^2+1}{x^2-5x+6}$ is undefined when $x = 2$ or when $x = 3$.

Example 3: Simplify $\frac{3x+6}{x^2+6x+8}$.

Solution:

$\frac{3x+6}{x^2+6x+8} = \frac{3(x+2)}{(x+2)(x+4)}$ Factor the numerator and the denominator.

$= \frac{3}{x+4}$ Apply the Fundamental Principle.

Example 4: Simplify $\frac{9-x^2}{2x^2-5x-3}$.

Solution: $\dfrac{9 - x^2}{2x^2 - 5x - 3} = \dfrac{(3 - x)(3 + x)}{(2x + 1)(x - 3)}$ Factor.

$= \dfrac{(-1)(x - 3)(3 + x)}{(2x + 1)(x - 3)}$ Write $3 - x$ as $-1(x - 3)$.

$= \dfrac{(-1)(3 + x)}{2x + 1}$ Apply the Fundamental Priciple.

or $\dfrac{-3 - x}{2x + 1}$

6.1 EXERCISES

Find the value of the following expressions when $x = 3$, $y = -1$, *and* $z = -4$.

1. $\dfrac{x + 6}{x - 9}$
2. $\dfrac{2y + 3}{-8 - y}$
3. $\dfrac{2z}{z^2 + 3z + 7}$
4. $\dfrac{-y^3}{y^2 + 1}$

Find any real numbers for which each rational expression is undefined.

5. $\dfrac{x + 4}{x - 8}$
6. $\dfrac{2x - 3}{7}$
7. $\dfrac{y^3 + 8y}{16y - 32}$
8. $\dfrac{10y^5 + y^4}{y^2 + 100}$

Simplify each expression.

9. $\dfrac{18x^3y^9}{-9x^2y^{12}}$
10. $\dfrac{(x + 2)(x - 5)}{4(x - 5)}$
11. $\dfrac{-7(a - 8)}{(a - 8)^2}$
12. $\dfrac{7c - 6d}{6d - 7c}$
13. $\dfrac{x + 10}{x^2 + 19x + 90}$
14. $\dfrac{x^2 - 9}{x^2 - 6x + 9}$
15. $\dfrac{-a^2 - ab}{5a + 5b}$
16. $\dfrac{6d^2 + 24d}{d + 4}$
17. $\dfrac{x^2 + 5x - 24}{x^2 + 15x + 56}$
18. $\dfrac{14 - 5y - y^2}{y^2 - 9y + 14}$
19. $\dfrac{7x^2 - 20x - 3}{6x^2 - 17x - 3}$
20. $\dfrac{4 - x}{x^3 - 64}$

6.2 MULTIPLYING AND DIVIDING RATIONAL EXPRESSIONS

Example 1: Multiply $\dfrac{-8x^3}{7y} \cdot \dfrac{9y^2}{16x^3}$

Solution: $\dfrac{-8x^3}{7y} \cdot \dfrac{9y^2}{16x^3} = \dfrac{-8x^3 \cdot 9y^2}{7y \cdot 16x^3}$ Multiply.

$$= \frac{-8 \cdot 9 \cdot x^3 \cdot y \cdot y}{7 \cdot 2 \cdot 8 \cdot x^3 \cdot y}$$ Factor.

$$= -\frac{9y}{14}$$ Apply the Fundamental Principle.

Example 2: Multiply $\frac{x^2 + 2x}{5x} \cdot \frac{15}{7x + 14}$

Solution: $\frac{x^2 + 2x}{5x} \cdot \frac{15}{7x + 14} = \frac{x(x + 2)}{5x} \cdot \frac{5 \cdot 3}{7(x + 2)}$ Factor numerators and denominators.

$$= \frac{x(x + 2) \cdot 5 \cdot 3}{5x \cdot 7(x + 2)}$$ Multiply.

$$= \frac{3}{7}$$ Simplify.

Example 3: Divide: $\frac{12a^4b^{10}}{65} \div \frac{4b^9}{13a^2}$

Solution: $\frac{12a^4b^{10}}{65} \div \frac{4b^9}{13a^2} = \frac{12a^4b^{10}}{65} \cdot \frac{13a^2}{4b^9}$ Multiply by the reciprocal of $\frac{4b^9}{13a^2}$.

$$= \frac{12 \cdot 13a^6b^{10}}{5 \cdot 13 \cdot 4b^9}$$ Factor and multiply.

$$= \frac{3a^6b}{5}$$ Simplify.

Example 4: Divide $\frac{(x - 3)(x - 5)}{9}$ by $\frac{4x - 12}{3}$.

Solution: $\frac{(x - 3)(x - 5)}{9} \div \frac{4x - 12}{3}$

$$= \frac{(x - 3)(x - 5)}{9} \cdot \frac{3}{4x - 12}$$ Multiply by the reciprocal of $\frac{4x - 12}{3}$.

$$= \frac{(x - 3)(x - 5) \cdot 3}{3 \cdot 3 \cdot 4(x - 3)}$$ Factor and multiply.

$$= \frac{x - 5}{12}$$ Simplify.

6.2 EXERCISES

Perform the indicated operations.

1. $\frac{10x^2}{3y} \cdot \frac{9y}{5x^2}$

2. $-\frac{7a^3b^2}{49a^3b^4} \cdot b^5$

3. $\frac{4x^2 - 9y^2}{2x + 3y} \cdot \frac{x}{x^2 - xy}$

4. $\frac{8x^{10}}{3x^4} \div \frac{24x^8}{15x}$

5. $\frac{(x + 1)^2}{6} \div \frac{3x + 3}{18}$

6. $\frac{c^2 - 100d^2}{c + 10d} \div \frac{2c - 20d}{4c - 8}$

7. $\frac{y^2 + 11y}{7} \cdot \frac{49}{6y + 66}$

8. $\frac{x^2 - x - 2}{x^2 + 4x + 3} \cdot \frac{x^2 + 10x + 21}{x^2 + 9x + 14}$

9. $\frac{x^2 - 36}{5y} \div \frac{6 - x}{10xy}$

10. $\frac{b^2 + 8b + 16}{b^2 - b - 20} \cdot \frac{b^2 - 13b + 40}{b^2 + 10b + 24}$

11. $\frac{d^2 - 9d + 14}{d^2 + 2d + 1} \div \frac{d^2 - d - 42}{d^2 + 7d + 6}$

12. $\frac{6x^2 + 5x - 4}{2x^2 + 9x - 5} \div \frac{12x^2 + 7x - 12}{7x^2 + 36x + 5}$

13. $\frac{a^2 - 9b^2}{a^2 - 6ab + 9b^2} \cdot \frac{3b - a}{a + 3b}$

14. $\frac{x^2 - xz + xy - yz}{x + z} \div \frac{x + y}{x - z}$

15. $\frac{x^3 + 27}{x^2 - 3x + 9} \cdot \frac{6}{x^2 - 9}$

6.3 ADDING AND SUBTRACTING RATIONAL EXPRESSIONS WITH COMMON DENOMINATORS AND LEAST COMMON DENOMINATOR

Example 1: Add $\frac{7x}{3y} + \frac{2x}{3y}$.

Solution: $\frac{7x}{3y} + \frac{2x}{3y} = \frac{7x + 2x}{3y}$ Add the numerators.

$= \frac{9x}{3y}$ Simplify the numerator by combining like terms.

$= \frac{3x}{y}$ Apply the Fundamental Principle.

Example 2: Subtract: $\frac{7x^2 + 2x}{x - 3} - \frac{21x + 6}{x - 3}$.

Solution: $\frac{7x^2 + 2x}{x - 3} - \frac{21x + 6}{x - 3}$

$= \frac{7x^2 + 2x - (21x + 6)}{x - 3}$ Subtract the numerators. Notice the parentheses.

$= \dfrac{7x^2 + 2x - 21x - 6}{x - 3}$ Use the distributive property.

$= \dfrac{7x^2 - 19x - 6}{x - 3}$ Combine like terms.

$= \dfrac{(7x + 2)(x - 3)}{x - 3}$ Factor.

$= 7x + 2$ Simplify.

Example 3: Find the LCD of $\dfrac{8a^2}{5a + 5}$ and $\dfrac{4}{a^2 + 6a + 9}$.

Solution: $5a + 5 = 5(a + 3)$ Factor each denominator.
$a^2 + 6a + 9 = (a + 3)^2$

The greatest number of times that the factor 5 appears is 1. The greatest number of times that the factor $a + 3$ appears in any one denominator is 2.

LCD $= 5(a + 3)^2$

Example 4: Write $\dfrac{6x}{13y}$ as an equivalent fraction with denominator $65x^2y^3$.

Solution: "What do we multiply $13y$ by to get $65x^2y^3$?"

The answer is $5x^2y^2$.

$$\frac{6x}{13y} = \frac{6x(5x^2y^2)}{13y(5x^2y^2)} = \frac{30x^3y^2}{65x^2y^3}$$

6.3 EXERCISES

Add or subtract as indicated.

1. $\dfrac{x}{21} + \dfrac{8}{21}$
2. $\dfrac{10}{2 + y} + \dfrac{y + 7}{2 + y}$
3. $\dfrac{9a}{a - 3} - \dfrac{27}{a - 3}$
4. $\dfrac{12}{x + 7} - \dfrac{5 - x}{x + 7}$
5. $\dfrac{4}{yz} + \dfrac{10}{yz}$
6. $\dfrac{8}{a - b} - \dfrac{7}{a - b}$
7. $\dfrac{x^2 + 3x}{x + 6} + \dfrac{6x + 18}{x + 6}$
8. $\dfrac{2x^2 + xy}{x - y} - \dfrac{2xy + y^2}{x - y}$

Find the LCD for the following lists of rational expressions.

9. $\dfrac{8}{9}, \dfrac{7}{15}$
10. $\dfrac{12}{5x^3}, \dfrac{2}{15x^2}$
11. $\dfrac{18x + 7}{2x - 10}, \dfrac{9}{x^2 - x - 20}$
12. $\dfrac{4x + 1}{8x(x - 7)}, \dfrac{5x - 4}{12(x - 7)^3}$

Rewrite each rational expression as an equivalent rational expression whose denominator is the given polynomial.

13. $\frac{4}{3x}$; $12x^3$

14. $\frac{7}{x - 2}$; $3(x - 2)$

15. $\frac{5a + 2}{4a + 20}$; $4a(a + 5)$

16. $\frac{b}{b^3 + 4b^2 - 5b}$; $b(b + 3)(b + 5)(b - 1)$

6.4 ADDING AND SUBTRACTING RATIONAL EXPRESSIONS WITH UNLIKE DENOMINATORS

Example 1: Add $\frac{9}{8x} + \frac{15}{12x^2}$.

Solution: Since $8x = 2 \cdot 2 \cdot 2 \cdot x = 2^3x$
and $12x^2 = 2 \cdot 2 \cdot 3 \cdot x^2 = 2^2 \cdot 3 \cdot x^2$,
the LCD $= 2^3 \cdot 3 \cdot x^2 = 24x^2$.

Write each fraction as an equivalent fraction with a denominator of $24x^2$.

$$\frac{9}{8x} + \frac{15}{12x^2} = \frac{9(3x)}{8x(3x)} + \frac{15(2)}{(12x^2)(2)}$$

$$= \frac{27x}{24x^2} + \frac{30}{24x^2}$$

$$= \frac{27x + 30}{24x^2}$$

$$= \frac{3(9x + 10)}{3 \cdot 8x^2}$$ Factor the numerator.

$$= \frac{9x + 10}{8x^2}$$ Simplify.

Example 2: Subtract: $\frac{4x}{x^2 - 49} - \frac{5}{x + 7}$.

Solution: Since $x^2 - 49 = (x + 7)(x - 7)$,
LCD $= (x + 7)(x - 7)$.

$$\frac{4x}{x^2 - 49} - \frac{5}{x + 7} = \frac{4x}{(x + 7)(x - 7)} - \frac{5(x - 7)}{(x + 7)(x - 7)}$$ Write as equivalent expression with the LCD as denominators.

$$= \frac{4x - 5(x - 7)}{(x + 7)(x - 7)}$$ Subtract numerators and use the common denominator.

$$= \frac{4x - 5x + 35}{(x + 7)(x - 7)}$$ Apply the distributive property in the numerator.

$$= \frac{-x + 35}{(x + 7)(x - 7)}$$ Combine like terms in the numerator.

Example 3: Add $5 + \frac{2x}{x - 3}$.

Solution: $5 = \frac{5}{1}$

The LCD of $\frac{5}{1}$ and $\frac{2x}{x - 3}$ is $x - 3$.

$$5 + \frac{2x}{x - 3} = \frac{5}{1} + \frac{2x}{x - 3}$$ Write 5 as $\frac{5}{1}$.

$$= \frac{5(x - 3)}{1(x - 3)} + \frac{2x}{x - 3}$$ Multiply both the numerator and the denominator of $\frac{5}{1}$ by $x - 3$.

$$= \frac{5(x - 3) + 2x}{x - 3}$$ Add numerators, and use the common denominator.

$$= \frac{5x - 15 + 2x}{x - 3}$$ Apply the distributive property in the numerator.

$$= \frac{7x - 15}{x - 3}$$ Combine like terms in the numerator.

6.4 EXERCISES

Perform the indicated operations.

1. $\frac{7}{3y} + \frac{9}{4y}$
2. $\frac{16a}{b} + \frac{3b}{4}$
3. $\frac{8}{x + 2} + \frac{5}{3x + 6}$
4. $\frac{12}{x - 1} - \frac{17}{5x - 5}$
5. $\frac{4}{x - 2} - \frac{4}{(x - 2)^2}$
6. $\frac{7}{b} - 4$
7. $\frac{7x + 6}{(x + 2)(x + 3)} - \frac{6}{x + 3}$
8. $\frac{9}{5 - x} + \frac{x}{3x - 15}$
9. $\frac{-4}{b - 6} + \frac{12}{12 - 2b}$
10. $\frac{-12}{x^2 + 5x + 4} + \frac{4}{x + 1}$
11. $\frac{8}{y^2 - 16} + \frac{2}{3(y + 4)}$
12. $\frac{162}{a^2 - 81} - \frac{9}{a - 9}$
13. $\frac{5}{x^2 + 9x + 14} + \frac{x}{x^2 + 10x + 21}$
14. $\frac{2y}{y^2 + 11y + 18} - \frac{y + 1}{y^2 + 8y - 9}$

15. The length of a rectangle is $\frac{2}{x + 4}$ feet while its width is $\frac{3}{x}$ feet. Find its perimeter and then find its area.

6.5 SIMPLIFYING COMPLEX FRACTIONS

Example 1: Simplify $\dfrac{\frac{4x}{9}}{\frac{2x^2}{3}}$

Solution: $\dfrac{\frac{4x}{9}}{\frac{2x^2}{3}} = \dfrac{4x}{9} \cdot \dfrac{3}{2x^2}$ Multiply by the reciprocal of the denominator.

$= \dfrac{4x \cdot 3}{9 \cdot 2x^2}$ Multiply.

$= \dfrac{2}{3x}$ Simplify.

Example 2: Simplify $\dfrac{\frac{x - 2}{3y}}{\frac{x}{y} + 4}$.

Solution: The LCD for $\frac{x - 2}{3y}$ and $\frac{x}{y}$ is $3y$.

$\dfrac{\frac{x - 2}{3y}}{\frac{x}{y} + 4} = \dfrac{3y\left(\frac{x - 2}{3y}\right)}{3y\left(\frac{x}{y} + 4\right)}$ Multiply by the LCD.

$= \dfrac{x - 2}{3y\left(\frac{x}{y}\right) + 3y(4)}$ Simplify the numerator and apply the distributive property in the denominator.

$= \dfrac{x - 2}{3x + 12y}$ Simplify.

Example 3: Simplify $\dfrac{\frac{3}{x} + \frac{x}{3}}{\frac{3}{x} - \frac{x}{3}}$.

Solution: The LCD for $\frac{3}{x}$ and $\frac{x}{3}$ is $3x$.

$$\frac{\frac{3}{x} + \frac{x}{3}}{\frac{3}{x} - \frac{x}{3}} = \frac{3x\left(\frac{3}{x} + \frac{x}{3}\right)}{3x\left(\frac{3}{x} - \frac{x}{3}\right)}$$ Multiply the numerator and the denominator by the LCD.

$$= \frac{3x\left(\frac{3}{x}\right) + 3x\left(\frac{x}{3}\right)}{3x\left(\frac{3}{x}\right) - 3x\left(\frac{x}{3}\right)}$$ Apply the distributive property.

$$= \frac{9 + x^2}{9 - x^2}$$ Simplify.

6.5 EXERCISES

Simplify each complex fraction.

1. $\dfrac{\frac{2}{7}}{-\frac{8}{21}}$
2. $\dfrac{\frac{-4y}{13}}{\frac{10y}{26}}$
3. $\dfrac{\frac{10x + 1}{9x^2}}{\frac{5x + 1}{18x}}$
4. $\dfrac{\frac{(a + 3)(a - 3)}{12}}{\frac{(a + 3)(a - 4)}{18}}$
5. $\dfrac{\frac{1}{3} + \frac{2}{5}}{\frac{3}{4} - \frac{1}{2}}$
6. $\dfrac{\frac{b}{3} + 3}{\frac{b}{3} - 3}$
7. $\dfrac{\frac{1}{6} - \frac{2}{x}}{\frac{5}{12} + \frac{1}{x^2}}$
8. $\dfrac{1 + \frac{1}{x - 5}}{x + \frac{1}{x - 5}}$
9. $\dfrac{\frac{-8}{15x^2}}{\frac{24}{5x^2}}$
10. $\dfrac{\frac{7y - 21}{14}}{\frac{2y - 6}{6}}$
11. $\dfrac{4}{3 + \frac{1}{5}}$
12. $\dfrac{\frac{6}{7a} + 4}{\frac{6}{7a} - 4}$
13. $\dfrac{\frac{ax - ab}{x^2 - b^2}}{\frac{x - b}{x + b}}$
14. $\dfrac{\frac{5}{x - 3} + 7}{\frac{10}{x - 3} - 7}$
15. $\dfrac{\frac{-4 + y}{6}}{\frac{4 - y}{18}}$

6.6 SOLVING EQUATIONS CONTAINING RATIONAL EXPRESSIONS

Example 1: Solve $\frac{x - 2}{2} - \frac{5x - 6}{16} = \frac{1}{8}$.

Solution: $$\frac{x - 2}{2} - \frac{5x - 6}{16} = \frac{1}{8}$$

$$16\left(\frac{x - 2}{2} - \frac{5x - 6}{16}\right) = 16\left(\frac{1}{8}\right)$$ Multiply both sides of the equation by the LCD of 16.

$$16\left(\frac{x-2}{2}\right) - 16\left(\frac{5x-6}{16}\right) = 16\left(\frac{1}{8}\right)$$ Apply the distributive property.

$8(x-2) - (5x-6) = 2$ Simplify.
$8x - 16 - 5x + 6 = 2$ Use the distributive property.
$3x - 10 = 2$ Combine like terms.
$3x = 12$
$x = 4$ Solve for x.

The solution set is $\{4\}$.

Example 2: Solve $2 - \frac{4}{x} = x - 3$.

Solution: In this equation, 0 cannot be a solution because if $x = 0$, the rational expression $\frac{4}{x}$ is undefined.

$$2 - \frac{4}{x} = x - 3$$

$x\left(2 - \frac{4}{x}\right) = x(x-3)$ Multiply both sides of the equation by x.

$x(2) - x\left(\frac{4}{x}\right) = x(x) - (x)(3)$ Apply the distributive property.

$2x - 4 = x^2 - 3x$ Simplify.
$0 = x^2 - 5x + 4$ Write the quadratic equation in standard form.
$0 = (x-4)(x-1)$ Factor.
$x - 4 = 0$ or $x - 1 = 0$ Set each factor equal to 0.
$x = 4$ or $x = 1$

Notice that neither 4 or 1 makes the denominator in the original equation equal to 0.

The solution set is $\{4, 1\}$.

Example 3: Solve for x: $\frac{3x}{a} + 4 = \frac{2x}{b} + 5a$

Solution: $$\frac{3x}{a} + 4 = \frac{2x}{b} + 5a$$

$ab\left(\frac{3x}{a} + 4\right) = ab\left(\frac{2x}{b} + 5a\right)$ Multiply both sides of the equation by the LCD of ab.

$ab\left(\frac{3x}{a}\right) + ab(4) = ab\left(\frac{2x}{b}\right) + ab(5a)$ Apply the distributive property.

$3xb + 4ab = 2xa + 5a^2b$ Simplify.

$$3xb - 2xa = 5a^2b - 4ab$$
$$x(3b - 2a) = 5a^2b - 4ab$$

Isolate all terms containing x on one side of the equation. Factor out x from each term on the left side.

$$\frac{x(3b - 2a)}{3b - 2a} = \frac{5a^2b - 4ab}{3b - 2a}$$

Divide both sides by $3b - 2a$.

$$x = \frac{5a^2b - 4ab}{3b - 2a}$$

Simplify.

6.6 EXERCISES

Solve each equation.

1. $\frac{x}{4} - 2 = 7$
2. $\frac{x}{3} + \frac{4x}{6} = \frac{x}{18}$
3. $\frac{5b}{2} = \frac{b - 3}{4}$
4. $\frac{y}{7} = \frac{y + 1}{6}$
5. $\frac{x - 4}{2} + \frac{x - 8}{3} = \frac{1}{6}$
6. $\frac{10}{3a - 1} = -5$
7. $\frac{z}{z + 6} + \frac{5}{z + 6} = 2$
8. $\frac{3a}{a - 5} + 4 = \frac{2a}{a - 5}$
9. $1 + \frac{2}{x - 1} = \frac{x}{x + 1}$
10. $\frac{9}{y^2 + 7y + 10} + \frac{3}{y + 5} = \frac{5}{y + 2}$
11. $\frac{1 - a}{1 + a} + \frac{4}{a^2 - 1} = \frac{a + 7}{a - 1}$
12. $\frac{8}{x^2 + 4x} + \frac{2}{x + 4} = 2$

Solve each equation for the indicated variable.

13. $P = 2w + 2l$; for l
14. $\frac{7}{x} = \frac{4y}{x - 5}$; for x
15. $\frac{1}{5} - \frac{2}{x} = \frac{4}{y}$; for x

6.7 RATIO AND PROPORTION

Example 1: Write a ratio for the phrase:

The ratio of 22 inches to 3 feet.

Solution: Use the same unit of measurement in the numerator and the denominator.

3 feet = 3(12) inches = 36 inches

$$\frac{22}{36} = \frac{11}{18}$$

Example 2: Solve for x: $\frac{54}{x} = \frac{6}{5}$

Solution: $\dfrac{54}{x} = \dfrac{6}{5}$

$54 \cdot 5 = x \cdot 6$ Cross multiply.

$270 = 6x$

$\dfrac{270}{6} = \dfrac{6x}{6}$ Divide both sides by 6.

$45 = x$

The solution set is $\{45\}$.

Example 3: Four boxes of cereal cost \$12.48. How much should 7 boxes cost?

Solution: $\dfrac{\text{4 boxes}}{\text{7 boxes}} = \dfrac{\text{price of 4 boxes}}{\text{price of 7 boxes}}$

$\dfrac{4}{7} = \dfrac{12.48}{x}$ x = price of 7 boxes

$4 \cdot x = 7 \cdot 12.48$

$4x = 87.36$

$\dfrac{4x}{4} = \dfrac{87.36}{4}$

$x = 21.84$

7 boxes should cost \$21.84.

6.7 EXERCISES

Write each ratio in fractional notation in lowest terms.

1. 4 megabytes to 16 megabytes
2. 7 quarts to 5 gallons
3. 3 dimes to 6 dollars
4. 200 minutes to 2 hours

Solve each proportion.

5. $\dfrac{5}{7} = \dfrac{x}{35}$
6. $\dfrac{y}{4} = \dfrac{36}{48}$
7. $\dfrac{3x}{8} = \dfrac{9}{4}$
8. $\dfrac{y}{y-11} = \dfrac{2}{3}$
9. $\dfrac{2x}{x+1} = \dfrac{20}{12}$
10. $\dfrac{x+1}{x-2} = \dfrac{7}{5}$

Solve.

11. On a map, 1 inch corresponds to 150 miles. Find the distance represented by a line that is $4\frac{3}{5}$ inches long on the map.

12. Terry's car gets 28 miles per gallon. Find how far she can drive if the tank contains 10.5 gallons of gas.

13. A human factors expert recommends that there be at least 9 square feet of floor space in a college classroom for every student in the class. Find the minimum floor space that 33 students need.

14. If Andrea can travel 416 miles in 8 hours, find how far she can travel if she maintains the same speed for 3 hours.

15. Sam earned $208 for working 32 hours. How much would he earn at the same rate of pay for working 17 hours?

6.8 RATIONAL EQUATIONS AND PROBLEM SOLVING

Example 1: The quotient of a number and 3 minus $\frac{1}{6}$ is the quotient of the number and 4. Find the number.

Solution: x = the unknown number

The quotient of $\frac{x}{3}$ minus $\frac{1}{6}$ is the quotient of x and 4.

$$\frac{x}{3} \quad - \quad \frac{1}{6} \quad = \quad \frac{x}{4}$$

$$\frac{x}{3} - \frac{1}{6} = \frac{x}{4}$$

$$12\left(\frac{x}{3} - \frac{1}{6}\right) = 12\left(\frac{x}{4}\right)$$ Multiply both sides of the equation by the LCD 12.

$$12\left(\frac{x}{3}\right) - 12\left(\frac{1}{6}\right) = 12\left(\frac{x}{4}\right)$$ Apply the distributive property.

$$4x - 2 = 3x$$ Simplify.

$$-2 = -x$$

$$2 = x$$

The unknown number is 2.

Example 2: Janice and Paul clean houses. Janice can complete a job in 6 hours, while it takes Paul 9 hours to complete the same job. How long would it take Janice and Paul working together to complete this job?

Solution: Summarize the information in a chart. x = time to do whole job working together

	Hours to Complete Total Job	Part of Job Completed in 1 hr
Janice	6	1/6
Paul	9	1/9
Together	x	$1/x$

Part of job Janice completed in 1 hour	added to	part of job Paul completed in 1 hour	is equal to	part of job they completed together in 1 hour
$\frac{1}{6}$	+	$\frac{1}{9}$	=	$\frac{1}{x}$

$$\frac{1}{6} + \frac{1}{9} = \frac{1}{x}$$

$$18x\left(\frac{1}{6} + \frac{1}{9}\right) = 18x\left(\frac{1}{x}\right)$$ Multiply both sides of the equation by the LCD $18x$.

$$18x\left(\frac{1}{6}\right) + 18x\left(\frac{1}{9}\right) = 18x\left(\frac{1}{x}\right)$$ Apply the distributive property.

$$3x + 2x = 18$$ Simplify.

$$5x = 18$$

$$x = \frac{18}{5}$$

Together it will take them $\frac{18}{5}$ or $3\frac{3}{5}$ hours.

Example 3: A car travels 275 miles in the same time that a truck travels 225 miles. If the car's speed is 10 miles per hour faster than the truck's, find the car's speed and the truck's speed.

Solution: x = speed of the truck
$x + 10$ = speed of the car

Recall, time $= \dfrac{\text{distance}}{\text{rate}}$

	distance =	rate ·	time
truck	225	x	$225/x$
car	275	$x + 10$	$275/(x + 10)$

car's time equals truck's time

$$\frac{225}{x} = \frac{275}{x + 10}$$

$$\frac{225}{x} = \frac{275}{x + 10}$$

$225(x + 10) = x(275)$ Cross multiply.
$225x + 2250 = 275x$ Use the distributive property.

$2250 = 50x$ Subtract $225x$ from both sides.
$45 = x$ Divide both sides by 50.

Since $x = 45$, $x + 10 = 45 + 10 = 55$.

The truck's speed is 45 mph and the car's speed is 55 mph.

6.8 EXERCISES

Solve the following.

1. Five times the reciprocal of a number equals 10 times the reciprocal of 8. Find the number.

2. Six divided by the sum of x and 3 equals the quotient of 4 and the difference of x and 3. Find x.

3. A number added to the product of 20 and the reciprocal of the number equals 12. Find the number.

4. An experienced bricklayer constructs a wall in 5 hours. The apprentice completes the job in 8 hours. Find how long it takes if they work together.

5. A fisherman rows 6 miles downstream in the same amount of time he rows 4 miles upstream. If the current of the river is 3 miles per hour, find the speed of the boat in still water.

6. A tank is being filled by two pipelines. Using both pipelines together it takes 20 hours to fill it, while using pipeline A alone it takes 30 hours. How long would it take pipeline B alone to fill the tank?

7. A seamstress wishes to make a doll's triangular diaper that will have the same shape as a full-size diaper. Use the following diagram to find the missing dimensions. (Round answers to the nearest tenth.)

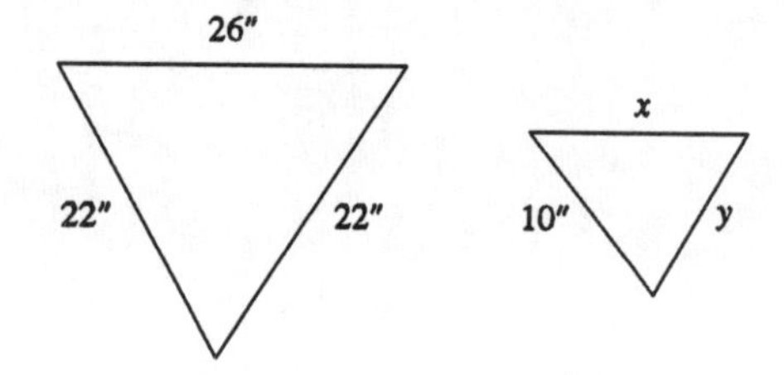

8. Sally travels 1440 miles in a jet and then an additional 360 miles by car. If the car ride takes 2 hours longer than the jet ride takes and if the rate of the jet is 6 times the rate of the car, find the time Sally travels by jet and find the time Sally travels by car.

9. A cyclist rides 15 miles per hour on level ground on a still day. He finds that he rides 42 miles with the wind behind him in the same amount of time that he rides 18 miles into the wind. Find the rate of the wind.

10. The ratio of a number minus 1 to the number plus 2 is $\frac{2}{3}$. Find the number.

11. The difference of 2 and the reciprocal of a number is 3 times the reciprocal of the square of the number. Find the number.

12. Juan wishes to produce a triangular sign that will be the same shape as his model sign. Use the following diagram to find the missing dimensions.

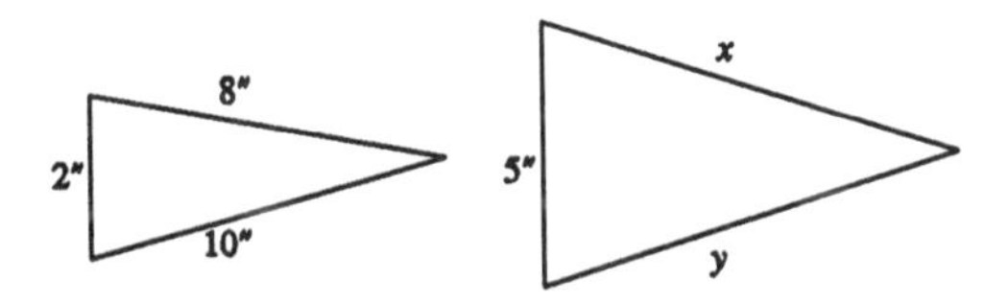

13. Mrs. Rath balances the company books in 6 hours. It takes her assistant twice as long to do the same job. If they work together, find how long it takes them to balance the books.

14. It took Mr. Scott 8 hours to drive 455 miles to St. Augustine. Before noon he was averaging 60 miles per hour while after noon the traffic was heavier and he only averaged 55 miles per hour. How much time did he drive after noon?

15. Kara takes 7 hours longer to do an audit than Wilma, and together they can do the work in 9 hours less than Wilma alone. How long does it take Kara and Wilma working together to do the audit?

CHAPTER 6 PRACTICE TEST

1. Find any real numbers for which the following expression is undefined.

$$\frac{x - 4}{x^2 - 3x + 2}$$

2. For a certain computer desk, the manufacturing cost C per desk (in dollars) is

$C = \dfrac{100x + 4000}{x}$ where x is the number of desks manufactured.

a.) Find the average cost per desk when manufacturing 300 computer desks.

b.) Find the average cost per desk when manufacturing 2000 computer desks.

Simplify each rational expression.

3. $\dfrac{2x - 6}{7x - 21}$

4. $\dfrac{x + 8}{x^2 - 64}$

5. $\dfrac{x + 5}{x^2 + 10x + 25}$

6. $\dfrac{x - 2}{x^3 - 8}$

7. $\dfrac{3m^3 + 15m^2 + 18m}{m^2 + 7m + 10}$

8. $\dfrac{ay + 4a - 2y - 8}{ay + 4a + 6y + 24}$

9. $\dfrac{y + x}{x^2 - y^2}$

Perform the indicated operation and simplify if possible.

10. $\dfrac{x^2 + x - 30}{x^2 + 7x + 10} \div \dfrac{x^2 + 3x - 4}{x^2 + x - 2}$

11. $\dfrac{2}{x - 3} \cdot (8x - 24)$

12. $\dfrac{y^2 + 5y + 6}{2y - 4} \cdot \dfrac{y - 2}{5y + 15}$

13. $\dfrac{4}{7x - 3} - \dfrac{5}{7x - 3}$

14. $\dfrac{6a}{a^2 - 5a + 6} - \dfrac{1}{a - 2}$

15. $\dfrac{8}{x^2 - 9} + \dfrac{5}{x + 3}$

16. $\dfrac{x^2 - 16}{x^2 - 4x} \div \dfrac{xy + 7x + 4y + 28}{3x + 21}$

17. $\dfrac{x + 3}{x^2 + 11x + 24} + \dfrac{6}{x^2 - x - 12}$

18. $\dfrac{2y}{y^2 + 3y - 4} - \dfrac{9}{y^2 + 4y - 5}$

Solve each equation.

19. $\frac{3}{y} - \frac{4}{7} = \frac{-1}{6}$

20. $\frac{4}{y + 3} = \frac{5}{y + 4}$

21. $\frac{a}{a - 6} = \frac{2}{a - 6} - \frac{1}{3}$

22. $\frac{9}{x^2 - 16} = \frac{5}{x + 4} + \frac{2}{x - 4}$

Simplify each complex fraction.

23. $\dfrac{\frac{8x^3}{y^2z}}{\frac{16x^2}{z^3}}$

24. $\dfrac{\frac{a}{b} + \frac{b}{a}}{\frac{a}{b} + \frac{a}{b}}$

25. $\dfrac{4 - \frac{2}{y^2}}{\frac{3}{y} + \frac{5}{y^2}}$

26. In a sample of 110 florescent bulbs, 6 were found to be defective. At this rate, how many defective bulbs should be found in 1210 bulbs?

27. One number plus three times its reciprocal is equal to four. Find the number.

28. A pleasure boat traveling down the Red River takes the same time to go 16 miles upstream as it takes to go 19 miles downstream. If the current of the river is 3 miles per hour, find the speed of the boat in still water.

29. An inlet pipe can fill a tank in 16 hours. A second pipe can fill the tank in 20 hours. If both pipes are used, find how long it takes to fill the tank.

30. Decide which is the better buy in crackers:

8 ounces for $1.80
12 ounces for $2.43
15 ouces for $2.97

7.1 THE SLOPE-INTERCEPT FORM

Example 1: Find the slope and the y-intercept of the line whose equation is $4x + y = 8$.

Solution:

$4x + y = 8$

$y = -4x + 8$ Subtract $4x$ from both sides.

The coefficient of x, -4, is the slope and the constant term, 8, is the y-intercept.

Example 2: Determine whether the graphs of $y = -\frac{3}{4}x + 3$ and $-4x + 3y = 24$ are parallel lines, perpendicular lines, or neither.

Solution: The graph of $-\frac{3}{4}x + 3$ is a line with slope $-\frac{3}{4}$ and with y-intercept 3.

$-4x + 3y = 24$

$3y = 4x + 24$ Add $4x$ to both sides.

$y = \frac{4}{3}x + 8$ Divide both sides by 3.

The graph of this equation is a line with slope $\frac{4}{3}$ and y-intercept 8.

Since the product of their slopes is -1, $-\frac{3}{4} \cdot \frac{4}{3} = -1$, the lines are perpendicular.

Example 3: Find an equation of the line with y-intercept 11 and the slope of $\frac{1}{2}$.

Solution: $y = mx + b$

$y = \frac{1}{2}x + 11$ Replace m with $\frac{1}{2}$ and b with 11.

Example 4: Use the slope-intercept form to graph the equation $3x + y = 2$.

Solution:

$3x + y = 32$

$y = -3x + 2$

Locate the intercept point (0, 2). To find another point use the slope -3, which can be written as $\frac{-3}{1}$. Start at the point (0, 2) and move 3 units down (since the numerator is -3), then move 1 unit to the right (since the denominator is 1). We arrive at the point (1, -1). The line through (0, 2) and (1, -1) is the graph of $3x + y = 2$.

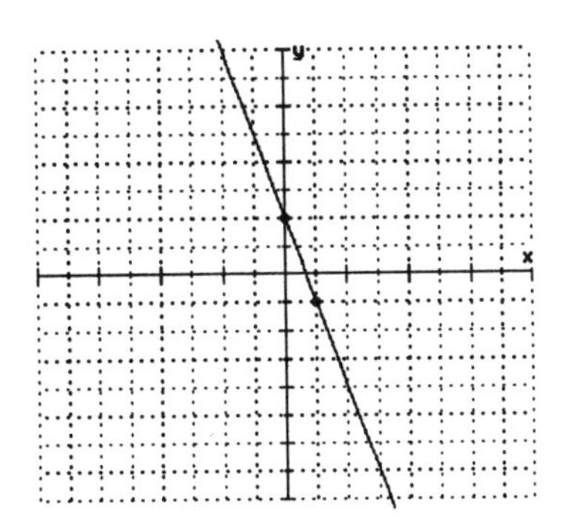

7.1 EXERCISES

Determine the slope and the y-intercept of the graph of each equation.

1. $7x + y = 10$
2. $-12x - y = 7$
3. $2x - 5y = -10$
4. $6x + 7y = 42$
5. $x - 3 = 0$
6. $y = 15$

Determine whether the lines are parallel lines, perpendicular lines, or neither.

7. $2x + y = 8$; $3x + 6y = 12$
8. $5x + 2y = 20$; $2y - 16 = -5x$
9. $13x - 26y = 52$; $8x + 4y = 7$
10. $y = 10x + 9$; $2y + 20x = -3$

Use the slope-intercept form of the linear equation to write an equation of each line with given slope and y-intercept.

11. Slope -2; y-intercept 5
12. Slope $-\frac{3}{4}$; y-intercept 2
13. Slope $\frac{5}{11}$; y-intercept $-\frac{1}{6}$
14. Slope 8; y-intercept 0

Use the slope-intercept form to graph each equation.

15. $y = \frac{3}{4}x - 1$
16. $y = 4x$
17. $2x + 3y = 12$
18. $4x - 3y = 6$

7.2 THE POINT-SLOPE FORM

Example 1: Find an equation of the line passing through (-2, -3) with slope 7. Write the equation in standard form: $Ax + By = C$.

Solution:

$y - y_1 = m(x - x_1)$

$y - (-3) = 7[x - (-2)]$ Let $m = 7$ and $(x_1, y_1) = (-2, -3)$.

$y + 3 = 7(x + 2)$ Simplify.

$y + 3 = 7x + 14$ Use the distributive property.

$3 = 7x - y + 14$ Subtract y from both sides.

$-11 = 7x - y$ Subtract 14 from both sides.

In standard form, the equation is $7x - y = -11$.

Example 2: Find an equation of the line through (1, 6) and (2, -3). Write the equation in standard form.

Solution: $m = \dfrac{y_2 - y_1}{x_2 - x_1} = \dfrac{6 - (-3)}{1 - 2} = \dfrac{9}{-1} = -9$

$y - y_1 = m(x - x_1)$

$y - 6 = -9(x - 1)$ Let $m = -9$ and $(x_1, y_1) = (1, 6)$.

$y - 6 = -9x + 9$ Use the distributive property.

$9x + y - 6 = 9$ Add $9x$ to both sides.

$9x + y = 15$ Add 6 to both sides.

In standard form, the equation is $9x + y = 15$.

Example 3: The Hits Galore Company has learned that by pricing a newly released CD at $14, sales will reach 2500 CDs per day. Raising the price to $18 will cause the sales to fall to 2000 CDs per day.

a. Assume that the relationship between sales price and number of CDs sold is linear and write an equation describing this relationship. Write the equation in slope-intercept form.

b. Predict the daily sales of CDs if the price is $15.

Solution: a. (sales price, number sold): (14, 2500) and (18, 2000)

$m = \dfrac{2500 - 2000}{14 - 18} = \dfrac{500}{-4} = -125$

$y - y_1 = m(x - x_1)$ Use point-slope form.

$y - 2000 = -125(x - 18)$ Let $x_1 = 18$ and $y_1 = 2000$.

$y - 2000 = -125x + 2250$ Use the distributive property.

$y = -125x + 4250$ Write in slope-intercept form.

b. $y = -125x + 4250$

$y = -125(15) + 4250$ Let $x = 15$.

$y = 2375$ Simplify.

If the price is $15, sales will reach 2375 CDs per day.

7.2 EXERCISES

Use the point-slope form of the linear equation to find the equation of each line with the given slope and passing through the given point. Then write the equation in standard form.

1. Slope -6; through (3, 5).
2. Slope $\frac{2}{3}$; through (-4, -1).
3. Slope $-\frac{5}{2}$; through (-2, -7).

Find the equation of the line through the given points. Write the equation in standard form.

4. Through (2, 1) and (3, 6).

5. Through (−7, −8) and (−5, 4).

6. Through (0, 0) and $\left(\frac{2}{3}, \frac{3}{4}\right)$.

Find the equation of each line.

7. Vertical line through (5, 15).

8. Horizontal line through (−8, 16).

Find the equation of each line. Write each equation in standard form.

9. Parallel to $y = 10$, through (7, 13).

10. Perpendicular to $y = -8$, through (4, 6).

11. With slope 0, through (8.1, −7.6).

12. With undefined slope, through $\left(\frac{1}{7}, \frac{1}{6}\right)$.

13. With slope $-\frac{3}{7}$ and y-intercept −2.

14. Through (6, 1), parallel to the x-axis.

15. The value of a computer bought in 1992 depreciates as time passes. Three years after the computer was bought, it was worth $2200 and 6 years after it was bought, it was worth $1675.

    a. If the relationship between number of years past 1992 and value of computer is linear, write an equation describing this relationship. (Use ordered pairs of the form (years past 1992, value of computer).)

    b. Use this equation to estimate the value of the computer in the year 2002.

7.3 GRAPHING NONLINEAR EQUATIONS

Example 1: Graph $y = -2x^2$.

Solution:

x	$y = -2x^2$
-3	$-2(-3)^2 = -18$
-2	$-2(-2)^2 = -8$
-1	$-2(-1)^2 = -2$
0	$-2(0)^2 = 0$
1	$-2(1)^2 = -2$
2	$-2(2)^2 = -8$
3	$-2(3)^2 = -18$

Pick some x-values and solve for y.

Plot these ordered pairs and connect with a smooth curve.

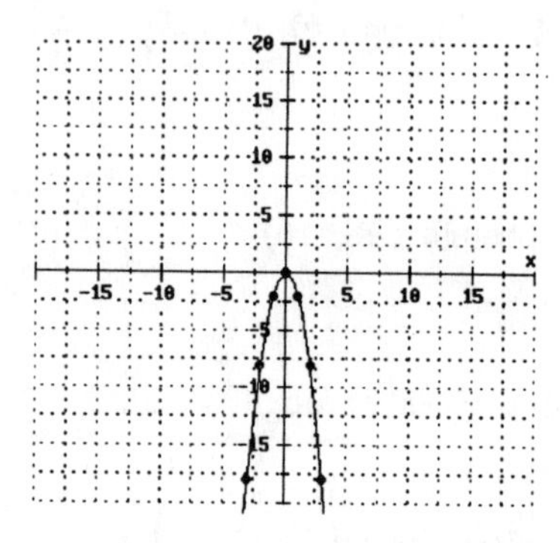

Example 2: Graph the equation $y = |x| + 2$.

Solution:

x	$y = \|x\| + 2$
-3	$\|-3\| + 2 = 3 + 2 = 5$
-2	$\|-2\| + 2 = 2 + 2 = 4$
-1	$\|-1\| + 2 = 1 + 2 = 3$
0	$\|0\| + 2 = 0 + 2 = 2$
1	$\|1\| + 2 = 1 + 2 = 3$
2	$\|2\| + 2 = 2 + 2 = 4$
3	$\|3\| + 2 = 3 + 2 = 5$

Plot these ordered pairs and connect with a smooth curve.

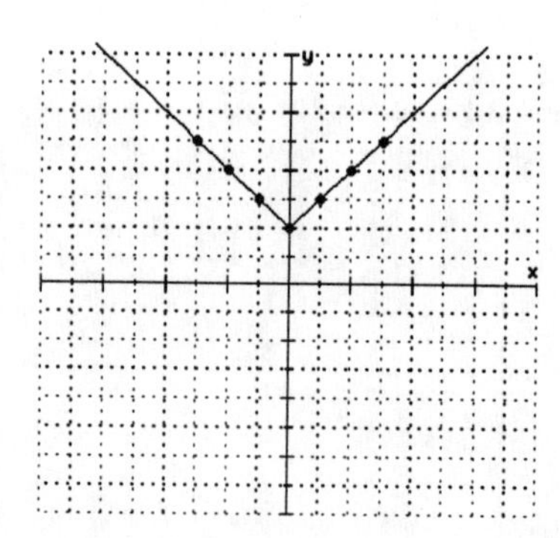

Example 3: Find the missing coordinate so that each ordered pair is a solution of the equation graphed below.

a. $(-1,\ \ )$ b. $(\ \ , 3)$

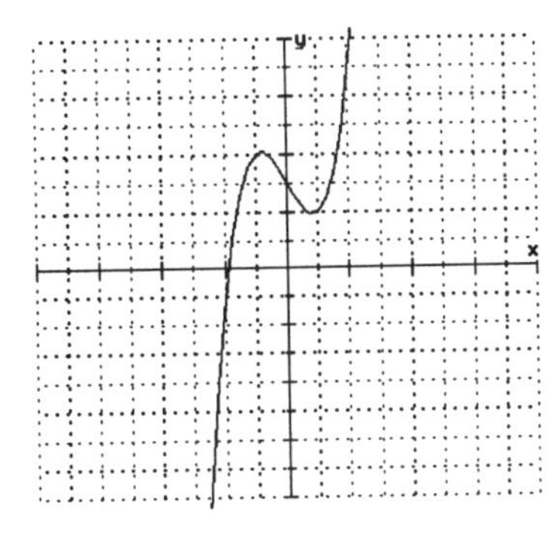

Solution: a. To find the y-coordinate that corresponds to an x-value of -1, find -1 on the x-axis and move vertically until the graph is reached. From the point on the graph move horizontally until the y-axis is reached and read the y-value. The y-value is 4, so the ordered pair is $(-1, 4)$.

b. The completed ordered pair is $(0, 3)$.

7.3 EXERCISES

Graph each nonlinear equation.

1. $y = x^2 - 3$
2. $y = 3x^2 + 1$
3. $y = |x| + 4$
4. $y = |x - 3|$
5. $y = -(x - 1)^2$
6. $y = 2 - |x|$

Use the given graph to complete each ordered pair.

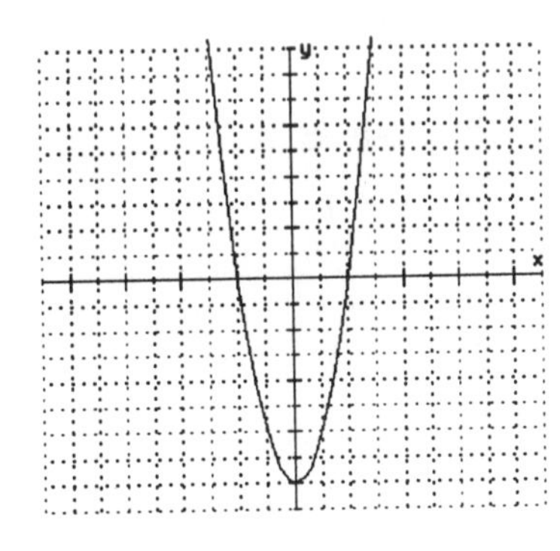

7. $(0,\ \ )$
8. $(\ \ , 0)$
9. $(1,\ \ )$

Determine whether each equation is linear or not. Then graph each equation.

10. $x + 2y = 6$
11. $y = |x| - 7$
12. $y = (x + 1)^2$
13. $y = -5x + 4$
14. $y = x^2 + 6x$
15. $y = -x^2 - 2x$

7.4 AN INTRODUCTION TO FUNCTIONS

Example 1: Find the domain and range of the relation{(1, 3), (-2, 5), (4, 7)}.

Solution: The domain is the set of all x-values or {-2, 1, 4}, and the range is the set of all y-values, or {3, 5, 7}.

Example 2: Which of the following relations are also functions?

a. {(-2, 5), (6, 4), (-2, 3), (0, 7)}

b. {(1, 8), (-2, 9), (3, 8), (5, 0)}

Solution:

a. The x-value -2 is assigned to two y-values, 5 and 3, so this set of ordered pairs is not a function.

b. Although the ordered pairs (1, 8) and (3, 8) have the same y-value, each x-value is assigned to only one y-value, so this set of ordered pairs is a function.

Example 3: Which of the following graphs are graphs of functions?

a.

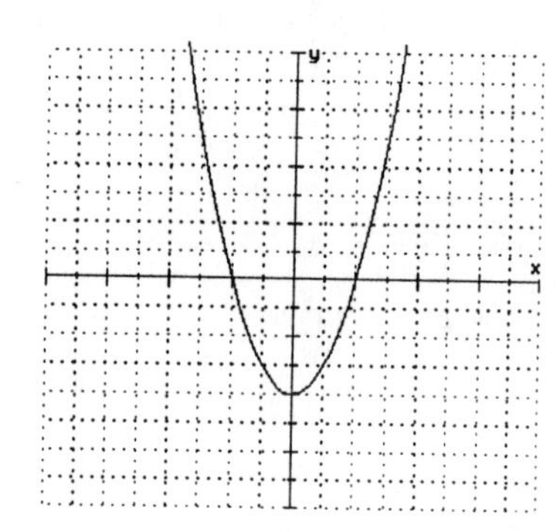

b.

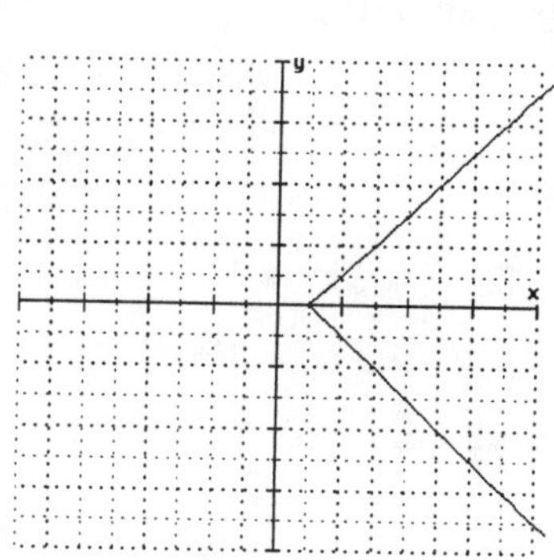

c.

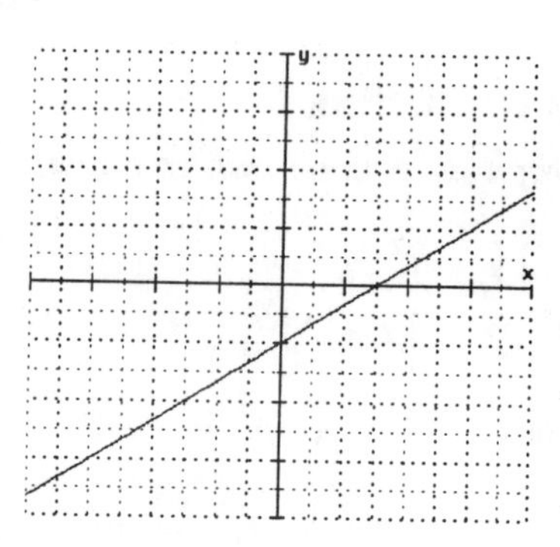

Solution: a. This graph is the graph of a function since no vertical line will intersect this graph more than once.

b. This graph is not the graph of a function. Note that vertical lines can be drawn that intersect the graph in two points.

c. This graph is the graph of a function.

Example 4: Given $f(x) = -3x^2 + 1$, find the following.

a. $f(-1)$ b. $f(2)$ c. $f(0)$

Solution: a. $f(x) = -3x^2 + 1$

$$\begin{aligned} f(-1) &= -3(-1)^2 + 1 \\ &= -3(1) + 1 \\ &= -3 + 1 \\ &= -2 \end{aligned}$$

b. $f(x) = -3x^2 + 1$

$$\begin{aligned} f(2) &= -3(2)^2 + 1 \\ &= -3(4) + 1 \\ &= -12 + 1 \\ &= -11 \end{aligned}$$

c. $f(x) = -3x^2 + 1$

$$\begin{aligned} f(0) &= -3(0)^2 + 1 \\ &= -3(0) + 1 \\ &= 0 + 1 \\ &= 1 \end{aligned}$$

7.4 EXERCISES

Find the domain and the range of each relation.

1. $\{(-3, 7), (1, 6), (0, 2), (4, -8)\}$
2. $\{(0, 1), (2, 1), (3, -1)\}$

Determine which relations are functions.

3. $\{(0, 6), (1, 5), (3, 6), (6, 9)\}$
4. $\{(-9, 10), (8, 7), (-9, 0)\}$

Decide whether the equation describes a function.

5. $y = 3x - 5$
6. $x = -2y^2$
7. $x = 4$
8. $y < 3x - 2$

Use the vertical line test to determine whether each graph is the graph of a function.

9.

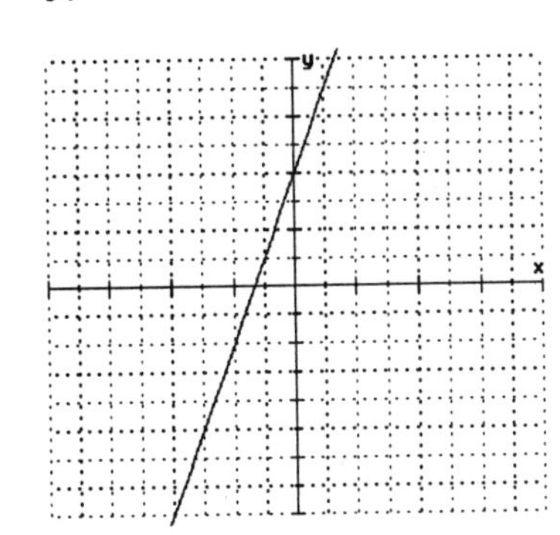

10.

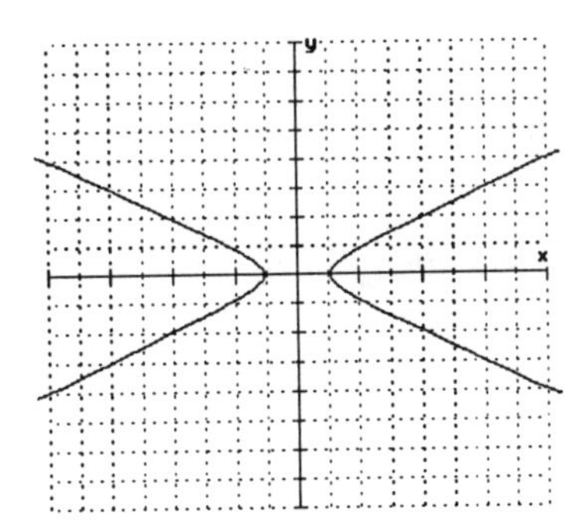

11.

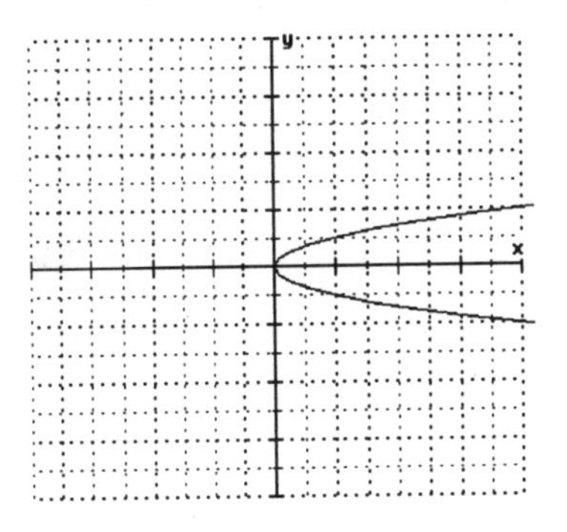

Given the following functions, find $g(-1)$, $g(0)$ and $g(4)$.

12. $g(x) = 4 - 3x$
13. $g(x) = 2|x| + 1$
14. $g(x) = -x^3 + 5$
15. $g(x) = -10$

CHAPTER 7 PRACTICE TEST

1. Determine the slope and y-intercept of the graph of $6x - 5y = 8$.
2. Determine whether the graphs of $y = 3x - 7$ and $-9x = 3y$ are parallel lines, perpendicular lines, or neither.

Find equations of the following lines. Write the equation in standard form.

3. With slope $-\frac{1}{3}$, through $(6, -6)$.
4. Through the origin and $(-1, 3)$.
5. Through $(8, -5)$ and $(2, 4)$.
6. Through $(-12, 5)$ and parallel to $x = 4$.
7. With slope $\frac{1}{9}$ and y-intercept 6.

Graph each equation.

8. $x - 3y = 6$
9. $y = -5x$
10. $y = |x - 1|$
11. $y = x^2 + 2$

Which of the following are functions?

12. $8x - 3y = 7$
13. $y = \dfrac{1}{x - 2}$

14.

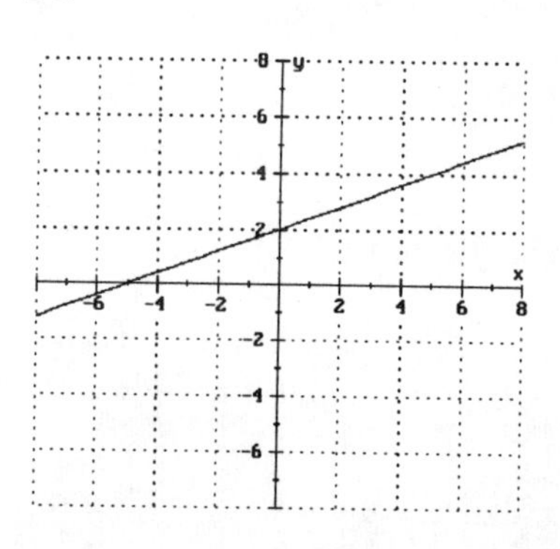

15.

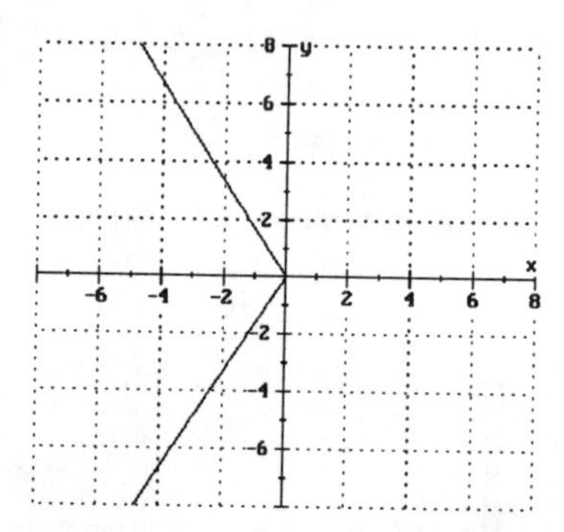

Given the following functions, find the indicated function values.

16. $f(x) = 3x + 5$ (a) $f(-1)$ (b) $f(0.3)$ (c) $f(0)$

17. $h(x) = -x^3 + 2x$ (a) $h(0)$ (b) $h(-1)$ (c) $h(3)$

18. $g(x) = -7$ (a) $g(0)$ (b) $g(-12)$ (c) $g(a)$

19. Find the domain of $y = \dfrac{3}{x - 4}$.

20. Find the domain and the range of the following function.

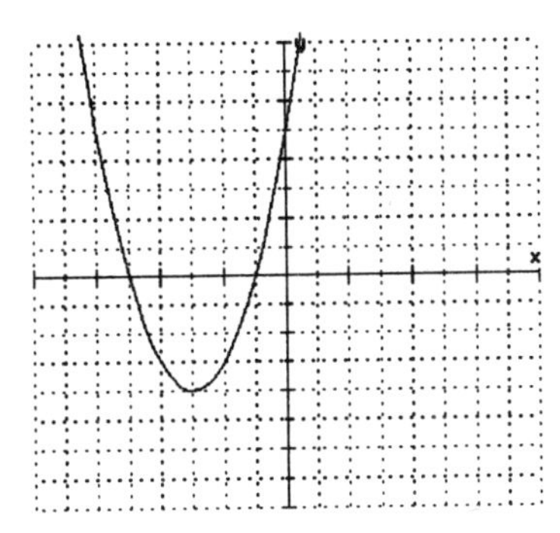

8.1 SOLVING SYSTEMS OF LINEAR EQUATIONS BY GRAPHING

Example 1: Determine whether (−1, 4) is a solution of the given system.

$$\begin{cases} 4x + 5y = 16 \\ x = -4y \end{cases}$$

Solution: Replace x with −1 and y with 4 in both equations.

$$\begin{aligned} 4x + 5y &= 16 \\ 4(-1) + 5(4) &= 16 \\ -4 + 20 &= 16 \\ 16 &= 16 \end{aligned}$$
True

$$\begin{aligned} x &= -4y \\ -1 &= -4(4) \\ -1 &= -16 \end{aligned}$$
False

While (−1, 4) is a solution of the first equation, it is not a solution of the second equation, so it is not a solution of the system.

Example 2: Solve the system of equations by graphing.

$$\begin{cases} -2x + y = -7 \\ x - y = 5 \end{cases}$$

Solution: $-2x + y = -7$

x	y
0	-7
1	-5
2	-3

$x - y = 5$

x	y
0	-5
5	0
1	-4

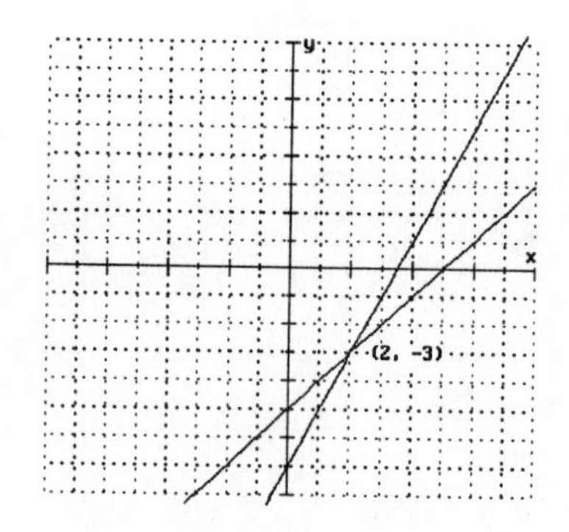

The solution is (2, −3).

Example 3: Without graphing, determine the number of solutions of the system.

$$\begin{cases} \frac{1}{3}x + y = 6 \\ 2x = 1 - 6y \end{cases}$$

Solution: First write each equation in slope-intercept form.

$$\frac{1}{3}x + y = 6 \qquad\qquad 2x = 1 - 6y$$
$$2x - 1 = -6y$$
$$y = -\frac{1}{3}x + 6 \qquad\qquad -\frac{1}{3}x + \frac{1}{6} = y$$

The slope of each line is $-\frac{1}{3}$, but they have different y-intercepts. This tells us that the lines representing these equations are parallel. Hence the system has no solution and is inconsistent.

8.1 EXERCISES

Determine whether any of the following ordered pairs satisfy the system of linear equations.

1. $\begin{cases} x + y = 6 \\ 5x + 3y = 26 \end{cases}$
   a. (4, 2)
   b. (−3, 9)

2. $\begin{cases} x - 9y = -1 \\ 3x + 2y = 5 \end{cases}$
   a. (9, 1)
   b. (1, 0)

3. $\begin{cases} 3y = 6x \\ 3x - y = 1 \end{cases}$
   a. (1, 2)
   b. (2, 4)

4. $\begin{cases} 5y = 10x \\ y - 2x = 0 \end{cases}$
   a. (−2, −4)
   b. (3, 6)

Solve each system of equations by graphing the equations on the same set of axis. Tell whether the system is consistent or inconsistent and whether the equations are dependent or independent.

5. $\begin{cases} y = x - 1 \\ y = -3x + 19 \end{cases}$

6. $\begin{cases} y = 3x + 1 \\ x + y = -3 \end{cases}$

7. $\begin{cases} x - 2y = 4 \\ y = -1 \end{cases}$

8. $\begin{cases} x - y = 4 \\ y - x = 5 \end{cases}$

9. $\begin{cases} x = y \\ x = 3 \end{cases}$

10. $\begin{cases} y = \frac{1}{2}x - 3 \\ x = 2y + 6 \end{cases}$

11. $\begin{cases} \frac{2}{3}x + y = 5 \\ 3x - 4y = -20 \end{cases}$

12. $\begin{cases} 4x - 3y = 6 \\ -8x + 6y = 12 \end{cases}$

Without graphing, decide:

a. *Are the graphs of the equations identical lines, parallel lines, or lines intersecting at a single point?*

b. *How many solutions does the system have?*

13. $\begin{cases} x + y = 10 \\ x - y = 6 \end{cases}$

14. $\begin{cases} 4x - y = 2 \\ 2x - \frac{1}{2}y = 4 \end{cases}$

15. $\begin{cases} 3x + 2y = 1 \\ 4y = 2 - 6x \end{cases}$

16. $\begin{cases} x = 3 \\ y = -8 \end{cases}$

8.2 SOLVING SYSTEMS OF LINEAR EQUATIONS BY SUBSTITUTION

Example 1: Solve the system:

$$\begin{cases} x - 3y = -3 \\ x = y + 1 \end{cases}$$

Solution:

$x - 3y = -3$ First equation.
$(y + 1) - 3y = -3$ Substitute $y + 1$ for x since $x = y + 1$.
$y + 1 - 3y = -3$
$-2y + 1 = -3$
$-2y = -4$
$y = 2$

The y-value of the ordered pair solution of the system is 2. To find the corresponding x-value, replace y with 2 in either equation and solve for x.

$x = y + 1$
$x = 2 + 1$
$x = 3$

The solution is (3, 2).

Example 2: Solve the system $\begin{cases} 3x + y = 5 \\ -6x + 5y = 18. \end{cases}$

Solution:

$3x + y = 5$ First equation.
$y = -3x + 5$ Solve for y.

$-6x + 5y = 18$ Second equation.
$-6x + 5(-3x + 5) = 18$ Let $y = -3x + 5$.
$-6x - 15x + 25 = 18$ Distributive property.
$-21x + 25 = 18$ Simplify.
$-21x = -7$ Subtract 25 from both sides.
$x = \frac{1}{3}$ Divide both sides by -21.

To find y, let $x = \frac{1}{3}$ in the equation $y = -3x + 5$.

$y = -3x + 5$

$y = -3\left(\frac{1}{3}\right) + 5$

$y = -1 + 5$
$y = 4$

The solution is $\left(\frac{1}{3}, 4\right)$.

Example 3: Use substitution to solve the system.

$$\begin{cases} 14x - 7y = -1 \\ -28x + 14y = -2 \end{cases}$$

Solution:

$$14x - 7y = -1 \qquad \text{First equation.}$$
$$-7y = -14x - 1$$
$$y = 2x + \frac{1}{7} \qquad \text{Solve for } y.$$

$$-28x + 14y = -2 \qquad \text{Second equation.}$$

$$-28x + 14\left(2x + \frac{1}{7}\right) = -2 \qquad \text{Let } y = 2x + \frac{1}{7}.$$

$$-28x + 28x + 2 = -2 \qquad \text{Distributive property.}$$
$$2 = -2 \qquad \text{Simplify.}$$

The false statement $2 = -2$ indicates that this system has no solution and is inconsistent.

8.2 EXERCISES

Solve each system by the substitution method.

1. $\begin{cases} x + y = -5 \\ y = -2x + 2 \end{cases}$

2. $\begin{cases} x - y = -3 \\ y = 2x \end{cases}$

3. $\begin{cases} x + 3y = 6 \\ -2x = 6y + 12 \end{cases}$

4. $\begin{cases} -x + y = 4 \\ 5x - y = -4 \end{cases}$

5. $\begin{cases} 3x - 2y = 0 \\ x - y = 1 \end{cases}$

6. $\begin{cases} \frac{1}{2}x + 2y = -1 \\ x + 4y = -2 \end{cases}$

7. $\begin{cases} 8x + 7y = -15 \\ x - 3y = 2 \end{cases}$

8. $\begin{cases} 2x + 3y = 0 \\ 4x = 6y - 4 \end{cases}$

9. $\begin{cases} 3x + 9y = 18 \\ 4x + 12y = 24 \end{cases}$

10. $\begin{cases} -5x + y = 3 \\ 7x - 2y = -3 \end{cases}$

11. $\begin{cases} 5x = 10y - 15 \\ 2x - 4y = -6 \end{cases}$

12. $\begin{cases} y = 3x \\ x + y = 1 \end{cases}$

13. $\begin{cases} 2x - y = -4 \\ x + y = 28 \end{cases}$

14. $\begin{cases} 6x + 3y = -1 \\ x = -y - 1 \end{cases}$

15. $\begin{cases} 6x - y = 0 \\ -9x + 4y = 0 \end{cases}$

8.3 SOLVING SYSTEMS OF LINEAR EQUATIONS BY ADDITION

Example 1: Solve the system $\begin{cases} x + y = -3 \\ -x + y = 7 \end{cases}$ by the addition method.

Solution:

$x + y = -3$ First equation.
$-x + y = 7$ Second equation.
$2y = 4$ Add the equations.
$y = 2$ Divide both sides by 2.

The y-value of the solution is 2. To find the corresponding x-value, let $y = 2$ in either equation of the system.

$x + y = -3$ First equation.
$x + 2 = -3$ Let $y = 2$.
$x = -5$ Solve for x.

The solution is $(-5, 2)$.

Example 2: Use the addition method to solve

$$\begin{cases} 3x + 2y = 11 \\ x - y = -3 \end{cases}$$

Solution:

$\begin{cases} 3x + 2y = 11 \\ 2(x - y) = 2(-3) \end{cases}$ Multiply both sides of the second equation by 2.

$3x + 2y = 11$
$2x - 2y = -6$ Distributive property.
$5x = 5$ Add.
$x = 1$ Divide both sides by 5.

$x - y = -3$ Second equation.
$1 - y = -3$ Let $x = 1$.
$-y = -4$
$y = 4$

The solution is $(1, 4)$.

Example 3: Solve the system $\begin{cases} 4x + 3y = -18 \\ 5x + 2y = -19 \end{cases}$ by the addition method.

Solution: $\begin{cases} -2(4x + 3y) = -2(-18) \\ 3(5x + 2y) = 3(-19) \end{cases}$ Multiply both sides of the first equation by −2. Multiply both sides of the second equation by 3.

$$\begin{aligned} -8x - 6y &= 36 \\ \underline{15x + 6y} &\underline{= -57} \\ 7x \quad &= -21 \\ x &= -3 \end{aligned}$$

Add the equations.

$$\begin{aligned} 5x + 2y &= -19 \\ 5(-3) + 2y &= -19 \\ -15 + 2y &= -19 \\ 2y &= -4 \\ y &= -2 \end{aligned}$$

Second equation. Let $x = -3$.

The solution is $(-3, -2)$.

8.3 EXERCISES

Solve each system of equations by the addition method.

1. $\begin{cases} 2x - y = -5 \\ 3x + y = 10 \end{cases}$

2. $\begin{cases} x - 4y = 24 \\ -x + 3y = -20 \end{cases}$

3. $\begin{cases} 5x + 2y = -6 \\ 7x - 2y = 6 \end{cases}$

4. $\begin{cases} -7x + 3y = 8 \\ 7x + 4y = -22 \end{cases}$

5. $\begin{cases} x + 3y = 16 \\ 2x - 5y = -23 \end{cases}$

6. $\begin{cases} 3x + 6y = 18 \\ 2x + 4y = 5 \end{cases}$

7. $\begin{cases} 2x + 4y = 2 \\ 3x + 6y = 3 \end{cases}$

8. $\begin{cases} 8x - 32y = 16 \\ -\frac{1}{2}x + 2y = -1 \end{cases}$

9. $\begin{cases} \frac{x}{3} - \frac{y}{6} = 1 \\ -\frac{x}{9} + \frac{y}{4} = 2 \end{cases}$

10. $\begin{cases} 3x - 6y = 0 \\ 12x + 28y = 13 \end{cases}$

11. $\begin{cases} -6x + 5y = -42 \\ 4x + 2y = 28 \end{cases}$

12. $\begin{cases} 2x + 10y = 5 \\ 7x + 35y = 3 \end{cases}$

13. $\begin{cases} 20x + 30y = -9 \\ 5x + 10y = -2 \end{cases}$

14. $\begin{cases} \frac{x}{7} - y = -5 \\ x + \frac{y}{6} = 8 \end{cases}$

15. $\begin{cases} 8x - 3y = -14 \\ -12x + 2y = 11 \end{cases}$

8.4 SYSTEMS OF LINEAR EQUATIONS AND PROBLEM SOLVING

Example 1: Find two numbers whose sum is 46 and whose difference is 18.

Solution: x = first number
y = second number

two numbers whose sum	is	46
$x + y$	$=$	46

two numbers whose difference	is	18
$x - y$	$=$	18

$$\begin{cases} x + y = 46 \\ x - y = 18 \end{cases}$$

$$\begin{aligned} x + y &= 46 \\ \underline{x - y} &\underline{= 18} \\ 2x \quad &= 64 \\ x &= 32 \end{aligned}$$

$$\begin{aligned} x + y &= 46 \\ 32 + y &= 46 \\ y &= 14 \end{aligned}$$

The numbers are 32 and 14.

Example 2: The Family Fun Fair is in town. Admission for 3 adults and 2 children is \$26, while admission for 5 adults and 6 children is \$52. Find the price of an adult's ticket and the price of a child's ticket.

Solution: A = the price of an adult's ticket
C = the price of a child's ticket

admission for 3 adults	and	admission for 2 children	is	\$26
$3A$	$+$	$2C$	$=$	26

admission for 5 adults	and	admission for 6 children	is	\$52
$5A$	$+$	$6C$	$=$	52

$$\begin{cases} 3A + 2C = 26 \\ 5A + 6C = 52 \end{cases}$$

$$\begin{cases} -3(3A + 2C) = -3(26) \\ 5A + 6C = 52 \end{cases}$$

$$\begin{array}{r} -9A - 6C = -78 \\ \underline{5A + 6C = 52} \\ -4A \qquad = -26 \\ A = 6.50 \end{array}$$

$$\begin{aligned} 3A + 2C &= 26 \\ 3(6.50) + 2C &= 26 \\ 19.50 + 2C &= 26 \\ 2C &= 6.50 \\ C &= 3.25 \end{aligned}$$

The price of an adult's ticket is \$6.50 and the price of a child's ticket is \$3.25.

Example 3: Paul and Tom are 12 miles away from each other when they start walking toward one another. After 1.5 hours they meet. If Paul walks one mile per hour slower than Tom, find both walking speeds.

Solution: x = Paul's rate in miles per hour
y = Tom's rate in miles per hour

	rate ·	time =	distance
Paul	x	1.5	$1.5x$
Tom	y	1.5	$1.5y$

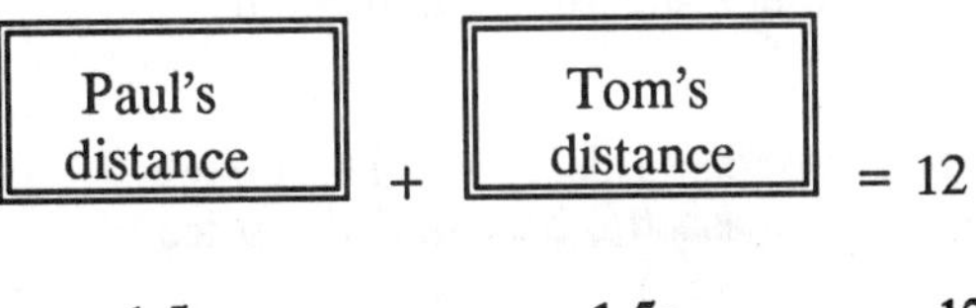

$$1.5x \quad + \quad 1.5y \quad = \quad 12$$

Paul's rate | is | one mile per hour slower than Tom's rate

$$x \quad = \quad y - 1$$

$$\begin{cases} 1.5x + 1.5y = 12 \\ x = y - 1 \end{cases}$$

Substituting,

$$
\begin{aligned}
1.5(y - 1) + 1.5y &= 12 \\
1.5y - 1.5 + 1.5y &= 12 \\
3y - 1.5 &= 12 \\
3y &= 13.5 \\
y &= 4.5
\end{aligned}
$$

$$
\begin{aligned}
x &= y - 1 \\
x &= 4.5 - 1 \\
x &= 3.5
\end{aligned}
$$

Paul's rate is 3.5 mph and Tom's rate is 4.5 mph.

8.4 EXERCISES

Write a system of equation describing the known facts. Do not solve the system.

1. Two numbers add up to 28 and have a difference of 14.
2. Suzanne has a total of $9300, which she invested in two accounts. The larger account is $1200 greater than the smaller account.
3. The total of two numbers is 31. The first number plus 7 more than four times the second equals 71.

Solve.

4. Two numbers total 89 and have a difference of 33. Find the two numbers.
5. One number is 5 more than twice the second number. Their sum is −1. Find the numbers.
6. A first number plus three times a second number is 43. Four times the first number plus the second totals 40. Find the numbers.
7. Last month Betsy purchased 4 cassettes and 3 compact disc for $84. This month she bought 6 cassettes and 2 compact disc for $86. Find the price for each cassette, and find the price of each compact disc.
8. Andrea has 75 coins in a jar, all of which are either dimes or quarters. The total value is $13.80. How many of each type of coin does she have?
9. Judi purchased 66 stamps, a mixture of 32 cent and 19 cent stamps. Find the number of each type of stamp if she spent $18.78.
10. Alan can row 24 miles down the Delaware River in 3 hours, but the return trip took him 12 hours. Find the rate Alan could row in still water, and find the rate of the current.
11. With a strong wind behind it, a United Airlines jet flies 1686 miles in 3 hours and 15 minutes. The return trip takes 4 hours, as the plane flies into the wind. Find the speed of the plane in still air, and find the wind speed to the nearest tenth.

12. Wendy needs to prepare 15 ounces of a 12.2% hydrochloric acid solution. To get this solution, find the amount of 5% and the amount of 14% solution she should mix.

13. Jose blends coffee for Nescafe. He needs to prepare 150 pounds of blended coffee beans selling for $5.68 per pound. He intends to do this by blending together a high-quality bean costing $8.25 per pound and a cheaper bean costing $4.75 per pound. To the nearest pound, find how much high-quality coffee and how much cheaper coffee he should blend.

14. Grace Church held its annual supper and fed a total of 342 people. They charged $6.25 for adults and $3.75 for children. If they took in $1932.50, find how many adults and how many children attended the dinner.

15. Tammi plans to erect 122 feet of fencing around her rectangular pasture. A river bank serves as one side of the rectangle. If each width is 5 feet shorter than half the length opposite the river, find the dimensions.

8.5 SYSTEMS OF LINEAR INEQUALITIES

Example 1: Graph the solution of the system

$$\begin{cases} 2x \geq y \\ 3x + y \leq 5 \end{cases}$$

Solution: Graph each inequality on the same set of axes.

$2x \geq y$ (solid line)

x	y
-1	-2
0	0
1	2

Test point: (1, 0)

$$\begin{aligned} 2x &\geq y \\ 2(1) &\geq 0 \\ 2 &\geq 0 \end{aligned}$$

True

Shade the half-plane containing (1, 0).

$3x + y \leq 5$ (solid line)

x	y
0	5
1	2
2	-1

Test point: (0, 0)

$$\begin{aligned} 3x + y &\leq 5 \\ 3(0) + 0 &\leq 5 \\ 0 &\leq 5 \end{aligned}$$

True

Shade the half-plane containing (0, 0).

The solution set to the system is where the shadings overlap.

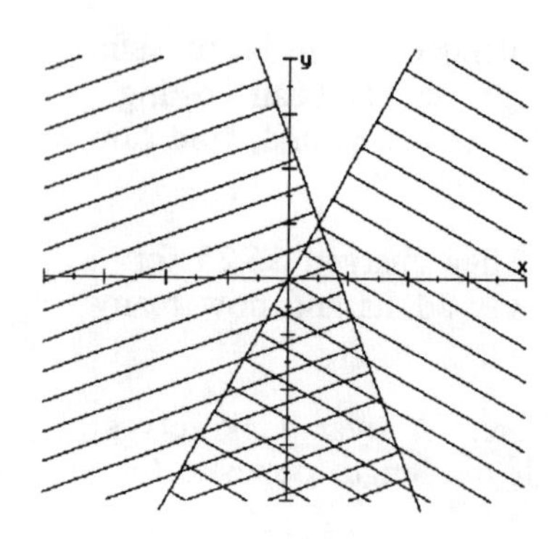

Example 2: Graph the solution of the system

$$\begin{cases} -2x + 5y > 10 \\ y \le 1 \end{cases}$$

Solution: $-2x + 5y > 10$ (dashed line)

$-2x + 5y = 10$

x	y
0	2
-5	0
5	4

Test point: (0, 0)

$-2x + 5y > 10$
$-2(0) + 5(0) > 10$
$0 > 10$
False

Shade the half-plane not containing (0, 0).

$y \le 1$ (solid line)

$y = 1$
horizontal line

Test point: (0, 0)

$y \le 1$
$0 \le 1$
True

Shade the half-plane containing (0, 0).

The solution set to the system is where the shadings overlap.

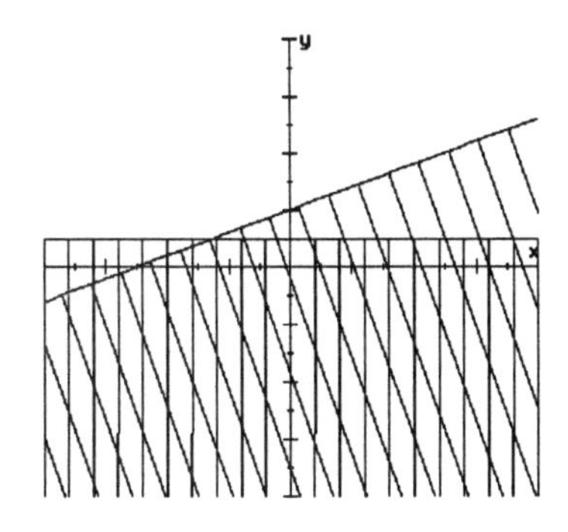

8.5 EXERCISES

Graph the solution of each system of linear inequalities.

1. $\begin{cases} y \geq x + 2 \\ y \geq 4 - x \end{cases}$

2. $\begin{cases} y \geq x - 5 \\ y \geq -3 - x \end{cases}$

3. $\begin{cases} y < 4x - 3 \\ y \leq x + 3 \end{cases}$

4. $\begin{cases} y \leq 3x + 2 \\ y > x + 4 \end{cases}$

5. $\begin{cases} y \leq -5x - 5 \\ y \geq x + 6 \end{cases}$

6. $\begin{cases} y \leq 4x + 8 \\ y \geq -x - 7 \end{cases}$

7. $\begin{cases} y \geq -x + 1 \\ y \leq 4x + 1 \end{cases}$

8. $\begin{cases} y \geq x - 4 \\ y \leq -2x + 2 \end{cases}$

9. $\begin{cases} x \geq 4y \\ x + 2y \leq 8 \end{cases}$

10. $\begin{cases} -3x < y \\ 2x + y < 5 \end{cases}$

11. $\begin{cases} y + 4x \geq 0 \\ 7x - 2y \leq 14 \end{cases}$

12. $\begin{cases} x \leq 3 \\ y \geq -4 \end{cases}$

13. $\begin{cases} x > -5 \\ y > -1 \end{cases}$

14. $\begin{cases} x + y \leq 4 \\ x < 3 \end{cases}$

15. $\begin{cases} y \geq \frac{1}{3}x + 3 \\ y \leq 4 \end{cases}$

CHAPTER 8 PRACTICE TEST

Is the ordered pair a solution of the given linear system?

1. $\begin{cases} 3x + 4y = 2 \\ 5x - y = 9 \end{cases}$; $(2, -1)$

2. $\begin{cases} 7x - 2y = -24 \\ 7x + 3y = 1 \end{cases}$; $(-2, 5)$

3. Use graphing to find the solutions of the system:

   $\begin{cases} y - x = 4 \\ y + 3x = 16 \end{cases}$

4. Use the substitution method to solve the system:

   $\begin{cases} 4x + y = -11 \\ x - 2y = 4 \end{cases}$

5. Use the substitution method to solve the system:

   $\begin{cases} \frac{1}{3}x + 3y = -\frac{53}{3} \\ 6x = -y \end{cases}$

6. Use the addition method to solve the system:

   $\begin{cases} 2x + 7y = 23 \\ -3x + 4y = 9 \end{cases}$

7. Use the addition method to solve the system:

   $\begin{cases} 9x - 7y = 46 \\ 8x - 5y = 36 \end{cases}$

Solve each system using the substitution method or the addition method.

8. $\begin{cases} 4x - y = 22 \\ 5x + 3y = 53 \end{cases}$

9. $\begin{cases} 2(3x + y) = 5x - 3 \\ 2x - y = -1 \end{cases}$

10. $\begin{cases} \dfrac{x - 11}{2} = \dfrac{2 - 7y}{6} \\ \dfrac{45 - 4x}{18} = \dfrac{y}{2} \end{cases}$

11. Lisa has a bundle of money consisting of \$1 bills and \$5 bills. There are 41 bills in the bundle. The total value of the bundle is \$117. Find the number of \$1 bills and the number of \$5 bills.

12. Don has invested \$6000, part at 4% simple annual interest and the rest at 7%. Find how much he invested at each rate if the total interest after 1 year is \$375.

Graph the solutions of the following systems of linear inequalities.

13. $\begin{cases} y - 3x \le 6 \\ y \ge 4 \end{cases}$

14. $\begin{cases} 2y + x \ge -1 \\ x - y \ge 3 \end{cases}$

9.1 INTRODUCTION TO RADICALS

Example 1: Find each square root.

a. $\sqrt{4}$ b. $-\sqrt{100}$ c. $\sqrt{\frac{4}{81}}$

Solution: a. $\sqrt{4} = 2$ because $2^2 = 4$ and 2 is positive.

b. $-\sqrt{100} = -10$. The negative sign in front of the radical indicates the negative square root of 100.

c. $\sqrt{\frac{4}{81}} = \frac{2}{9}$, because $\left(\frac{2}{9}\right)^2 = \frac{4}{81}$ and $\frac{2}{9}$ is positive.

Example 2: Simplify the following expressions.

a. $\sqrt[4]{256}$ b. $\sqrt[3]{-64}$ c. $\sqrt[5]{\frac{1}{32}}$

Solution: a. $\sqrt[4]{256} = 4$, because $4^4 = 256$ and 4 is positive.

b. $\sqrt[3]{-64} = -4$, because $(-4)^3 = -64$.

c. $\sqrt[5]{\frac{1}{32}} = \frac{1}{2}$, because $\left(\frac{1}{2}\right)^5 = \frac{1}{32}$.

Example 3: Simplify $\sqrt{25x^8}$

Solution: $\sqrt{25x^8} = 5x^4$, because $(5x^4)^2 = 25x^8$.

9.1 EXERCISES

Find each root.

1. $\sqrt{121}$
2. $\sqrt{\frac{4}{49}}$
3. $-\sqrt{169}$
4. $\sqrt[3]{27}$
5. $-\sqrt[3]{64}$
6. $\sqrt[5]{\frac{1}{243}}$
7. $\sqrt[4]{0}$
8. $-\sqrt[4]{-16}$
9. $-\sqrt[4]{256}$

Find each root. Assume that each variable represents a nonnegative real number.

10. $\sqrt{a^8}$
11. $\sqrt{25x^4}$
12. $\sqrt[3]{b^{15}}$
13. $-\sqrt[3]{8m^6n^9}$
14. $\sqrt{x^{10}y^2}$
15. $\sqrt[3]{-c^{12}d^9}$
16. $-\sqrt{144x^6}$
17. $\sqrt{-144x^6}$
18. $-\sqrt[3]{-27b^3}$

9.2 SIMPLIFYING RADICALS

Example 1: Simplify $\sqrt{52}$.

Solution: $\sqrt{52} = \sqrt{4 \cdot 13}$ Factor.
$= \sqrt{4} \cdot \sqrt{13}$ Apply the product rule.
$= 2\sqrt{13}$ Write $\sqrt{4}$ as 2.

Example 2: Simplify $\sqrt{\dfrac{100}{121}}$.

Solution: $\sqrt{\dfrac{100}{121}} = \dfrac{\sqrt{100}}{\sqrt{121}}$ Use the quotient rule.

$= \dfrac{10}{11}$

Example 3: Simplify $\sqrt{12x^7}$. Assume that variables represent positive numbers only.

Solution: $\sqrt{12x^7} = \sqrt{4 \cdot 3 \cdot x^6 \cdot x}$

$= \sqrt{4x^6 \cdot 3x}$

$= \sqrt{4x^6} \cdot \sqrt{3x}$

$= 2x^3\sqrt{3x}$

Example 4: Simplify $\sqrt[3]{540}$.

Solution: $\sqrt[3]{540} = \sqrt[3]{27 \cdot 20}$

$= \sqrt[3]{27} \cdot \sqrt[3]{20}$

$= 3\sqrt[3]{20}$

9.2 EXERCISES

Simplify each expression. Assume that all variables represent positive numbers only.

1. $\sqrt{28}$
2. $\sqrt{108}$
3. $\sqrt[3]{32}$
4. $\sqrt[3]{-54}$
5. $\sqrt{40}$
6. $-\sqrt[3]{40}$
7. $\sqrt{\dfrac{8}{18}}$
8. $-\sqrt{\dfrac{180}{125}}$
9. $\sqrt[3]{\dfrac{9}{8}}$
10. $\sqrt{x^{11}}$

11. $\sqrt{a^5b^4}$ 12. $-\sqrt{m^{16}n^7}$ 13. $\sqrt[3]{y^{17}}$ 14. $-\sqrt[3]{b^7}$ 15. $\sqrt[3]{8a^{10}b^{20}}$

16. $\sqrt{\dfrac{4x}{y^6}}$ 17. $\sqrt{\dfrac{12m^6}{n^8}}$ 18. $\sqrt[3]{\dfrac{b^5c^9}{8d^3}}$ 19. $\sqrt[4]{80}$ 20. $\sqrt[4]{162a^6b^8}$

9.3 ADDING AND SUBTRACTING RADICAL EXPRESSIONS

Example 1: Simplify by combining like radical terms.

$3\sqrt{11} - 4\sqrt{11} + 18\sqrt{11}$

Solution: $3\sqrt{11} - 4\sqrt{11} + 18\sqrt{11} = (3 - 4 + 18)\sqrt{11}$

$= 17\sqrt{11}$

Example 2: Simplify by first simplifying each radical.

$\sqrt{28} - 4\sqrt{63} + 3\sqrt{7}$

Solution: $\sqrt{28} - 4\sqrt{63} + 3\sqrt{7}$

$= \sqrt{4 \cdot 7} - 4\sqrt{9 \cdot 7} + 3\sqrt{7}$ Factor radicands.

$= \sqrt{4} \cdot \sqrt{7} - 4 \cdot \sqrt{9} \cdot \sqrt{7} + 3\sqrt{7}$ Apply the product rule.

$= 2\sqrt{7} - 4 \cdot 3\sqrt{7} + 3\sqrt{7}$ Simplify $\sqrt{4}$ and $\sqrt{9}$.

$= 2\sqrt{7} - 12\sqrt{7} + 3\sqrt{7}$ Multiply.

$= -7\sqrt{7}$ Combine like radicals.

Example 3: Simplify $4\sqrt[3]{48x^5} - 9x\sqrt[3]{162x^2}$

Solution: $4\sqrt[3]{48x^5} - 9x\sqrt[3]{162x^2}$

$= 4\sqrt[3]{8x^3 \cdot 6x^2} - 9x\sqrt[3]{27 \cdot 6x^2}$ Factor.

$= 4\sqrt[3]{8x^3} \cdot \sqrt[3]{6x^2} - 9x \cdot \sqrt[3]{27} \cdot \sqrt[3]{6x^2}$ Apply the product rule.

$= 4 \cdot 2x \cdot \sqrt[3]{6x^2} - 9x \cdot 3 \cdot \sqrt[3]{6x^2}$ Simplify.

$= 8x\sqrt[3]{6x^2} - 27x\sqrt[3]{6x^2}$ Multiply.

$= -19x\sqrt[3]{6x^2}$ Subtract like radicals.

9.3 EXERCISES

Simplify the following. Assume that all variables represent nonnegative real numbers.

1. $9 - 5\sqrt{6} + 12 - 2\sqrt{6}$
2. $4\sqrt[3]{5} + 8\sqrt[3]{5} - 6\sqrt{5}$
3. $4\sqrt{12} - 6\sqrt{27}$
4. $\sqrt{200} + 3\sqrt{18}$
5. $2\sqrt[3]{9} - 8\sqrt[3]{243}$
6. $14x + 6\sqrt{x} - 5\sqrt{x^2}$
7. $\sqrt{25x} - 5\sqrt{x^3}$
8. $19 - 3\sqrt{2} - \sqrt{8}$
9. $\sqrt{\dfrac{7}{25}} + \sqrt{\dfrac{7}{64}}$
10. $\sqrt{9x^3} + 2\sqrt{81x^3} - 5\sqrt{x}$
11. $x\sqrt{36x^2} + \sqrt{4x^4}$
12. $-10a\sqrt{98b} - 4\sqrt{2a^2b}$
13. $\sqrt{49y^2} + 2\sqrt[3]{49y^2} + \sqrt{9x^2}$
14. $5\sqrt{6} - \sqrt{16} + 3\sqrt{24} + \sqrt{12}$
15. $\sqrt{\dfrac{10}{45}} - \sqrt{\dfrac{8}{225}}$
16. $\sqrt{15} - \sqrt{60}$
17. $\sqrt[3]{27} - \sqrt[3]{135} + 6$
18. $\sqrt{20x} + \sqrt[3]{189x^4} - 2\sqrt{5x} - \sqrt[3]{7x}$

9.4 MULTIPLYING AND DIVIDING RADICAL EXPRESSIONS

Example 1: Find the product: $\sqrt{5} \cdot \sqrt{35}$.

Solution: $\sqrt{5} \cdot \sqrt{35} = \sqrt{5 \cdot 35} = \sqrt{175}$

Next, simplify $\sqrt{175}$.

$\sqrt{175} = \sqrt{25 \cdot 7} = \sqrt{25} \cdot \sqrt{7} = 5\sqrt{7}$

Example 2: Find the product and simplify.

$(\sqrt{y} + \sqrt{5})(\sqrt{2} - \sqrt{5})$

Solution: Use the FOIL method of multiplication.

$(\sqrt{y} + \sqrt{5})(\sqrt{2} - \sqrt{5})$

$= \sqrt{y} \cdot \sqrt{2} - \sqrt{y} \cdot \sqrt{5} + \sqrt{5} \cdot \sqrt{2} - \sqrt{5} \cdot \sqrt{5}$

$= \sqrt{2y} - \sqrt{5y} + \sqrt{10} - \sqrt{25}$ — Apply the product rule.

$= \sqrt{2y} - \sqrt{5y} + \sqrt{10} - 5$ — Simplify.

Example 3: Find the quotient and simplify.

$$\frac{\sqrt{45x^5}}{\sqrt{5x}}$$

Solution: $\dfrac{\sqrt{45x^5}}{\sqrt{5x}} = \sqrt{\dfrac{45x^5}{5x}} = \sqrt{9x^4} = 3x^2$

Example 4: Rationalize each denominator and simplify.

a. $\dfrac{3}{\sqrt{11}}$ b. $\dfrac{4}{\sqrt[3]{16}}$ c. $\dfrac{12}{\sqrt{6}+2}$

Solution a. $\dfrac{3}{\sqrt{11}} = \dfrac{3 \cdot \sqrt{11}}{\sqrt{11} \cdot \sqrt{11}} = \dfrac{3\sqrt{11}}{11}$

b. $\dfrac{4}{\sqrt[3]{16}} = \dfrac{4}{\sqrt[3]{8 \cdot 2}}$ Factor.

$= \dfrac{4}{\sqrt[3]{8} \cdot \sqrt[3]{2}}$ Apply the product rule.

$= \dfrac{4}{2\sqrt[3]{2}}$ $\sqrt[3]{8} = 2$.

$= \dfrac{2}{\sqrt[3]{2}}$ Reduce.

$= \dfrac{2 \cdot \sqrt[3]{4}}{\sqrt[3]{2} \cdot \sqrt[3]{4}}$

$= \dfrac{2\sqrt[3]{4}}{\sqrt[3]{8}}$ Multiply.

$= \dfrac{2\sqrt[3]{4}}{2}$ Simplify.

$= \sqrt[3]{4}$ Reduce.

c. $\dfrac{12}{\sqrt{6}+2} = \dfrac{12(\sqrt{6}-2)}{(\sqrt{6}+2)(\sqrt{6}-2)}$ The conjugate of $\sqrt{6}+2$ is $\sqrt{6}-2$.

$= \dfrac{12(\sqrt{6}-2)}{(\sqrt{6})^2-(2)^2}$

$= \dfrac{12(\sqrt{6}-2)}{6-4}$

$$= \frac{12(\sqrt{6} - 2)}{2}$$

$$= 6(\sqrt{6} - 2)$$
$$= 6\sqrt{6} - 12$$

9.4 EXERCISES

Find each product and simplify.

1. $2\sqrt{3} \cdot 4\sqrt{12}$
2. $\sqrt[3]{20} \cdot \sqrt[3]{6}$
3. $\sqrt{14}(\sqrt{7} - \sqrt{2})$
4. $\sqrt{20}(2\sqrt{5} + \sqrt{2})$
5. $(2\sqrt{6} - \sqrt{5})(3 + 4\sqrt{5})$
6. $(\sqrt{2} + 3)^2$
7. $(3 + \sqrt{x})(3 - \sqrt{x})$
8. $(\sqrt{5} + 3\sqrt{y})(\sqrt{5} - 3\sqrt{y})$
9. $(\sqrt{a} + 2\sqrt{b})^2$
10. $(9\sqrt{x})^2$

Find each quotient and simplify.

11. $\dfrac{\sqrt{50}}{\sqrt{2}}$
12. $\dfrac{\sqrt{56}}{\sqrt{7}}$
13. $\dfrac{\sqrt{125a^5}}{\sqrt{5a}}$

Rationalize each denominator and simplify.

14. $\sqrt{\dfrac{2}{5}}$
15. $\dfrac{1}{\sqrt{7x}}$
16. $\sqrt{\dfrac{3}{80x}}$
17. $\dfrac{7}{\sqrt[3]{4}}$
18. $\sqrt[3]{\dfrac{6}{25x}}$
19. $\dfrac{2}{\sqrt{3} - 1}$
20. $\dfrac{\sqrt{2} + 3}{\sqrt{2} - 3}$

9.5 SOLVING EQUATIONS CONTAINING RADICALS

Example 1: Solve for x: $\sqrt{x + 5} = 7$.

Solution:

$\sqrt{x + 5} = 7$	Original equation.
$(\sqrt{x + 5})^2 = 7^2$	Square both sides.
$x + 5 = 49$	Simplify.
$x = 44$	Subtract 5 from both sides.

Check: $\sqrt{x + 5} = 7$ Original equation.

$\sqrt{44 + 5} = 7$ Let $x = 44$.

$\sqrt{49} = 7$

$7 = 7$ True.

Since a true statement results, 44 is the solution.

The solution set is $\{44\}$.

Example 2: Solve the equation $\sqrt{y} = \sqrt{4y - 3}$.

Solution:

$\sqrt{y} = \sqrt{4y - 3}$	Original equation.
$(\sqrt{y})^2 = (\sqrt{4y - 3})^2$	Square both sides.
$y = 4y - 3$	Simplify.
$-3y = -3$	Subtract $4y$ from both sides.
$y = 1$	Divide both sides by -3.
Check: $\sqrt{y} = \sqrt{4y - 3}$	Original equation.
$\sqrt{1} = \sqrt{4(1) - 3}$	Let $y = 1$.
$\sqrt{1} = \sqrt{1}$	True.

Solution set: $\{1\}$

Example 3: Solve the equation $\sqrt{x - 2} + 2 = x$.

Solution:

$\sqrt{x - 2} + 2 = x$	
$\sqrt{x - 2} = x - 2$	Subtract 2 from both sides.
$(\sqrt{x - 2})^2 = (x - 2)^2$	Square both sides.
$x - 2 = x^2 - 4x + 4$	Simplify.
$0 = x^2 - 5x + 6$	Write the equation in standard form.
$0 = (x - 2)(x - 3)$	Factor.
$0 = x - 2$ or $0 = x - 3$	Set each factor equal to 0.
$2 = x$ or $3 = x$	Solve for x.

Check both of the proposed solutions:

Let $x = 2$

$\sqrt{x - 2} + 2 = x$
$\sqrt{2 - 2} + 2 = 2$
$\sqrt{0} + 2 = 2$
$2 = 2$
True

Let $x = 3$

$\sqrt{x - 2} + 2 = x$
$\sqrt{3 - 2} + 2 = 3$
$\sqrt{1} + 2 = 3$
$1 + 2 = 3$
$3 = 3$
True

Solution set: $\{2, 3\}$

9.5 EXERCISES

Solve the following for x.

1. $\sqrt{x} = 11$
2. $\sqrt{x - 5} = 6$
3. $\sqrt{x} - 6 = 3$
4. $\sqrt{3x + 4} = 5$
5. $\sqrt{2x + 7} = \sqrt{x - 8}$
6. $\sqrt{7x + 3} = \sqrt{9x - 1}$
7. $\sqrt{5x} - \sqrt{x + 2} = 0$
8. $\sqrt{10 - x} + 8 = 3$
9. $\sqrt{2x^2 + 8x + 15} = x$
10. $\sqrt{2x - 4} = x - 6$
11. $\sqrt{21 - 2x} + 3 = x$
12. $2 = \sqrt{12x^2 - 20x + 7}$
13. $8\sqrt{x} + 7 = 2$
14. A number is 20 more than its principal square root. Find the number.
15. A number is twice the principal square root of triple the number. Find the number.

9.6 RADICAL EQUATIONS AND PROBLEM SOLVING

Example 1: Find the length of the hypotenuse c of a right triangle whose legs are $a = 12$ cm and $b = 16$ cm long.

Solution:

$a^2 + b^2 = c^2$	Pythagorean formula.
$12^2 + 16^2 = c^2$	Substitute the lengths of the legs.
$144 + 256 = c^2$	Simplify.
$400 = c^2$	
$\sqrt{400} = \sqrt{c^2}$	Apply the definition of principal square root.
$20 = c$	Simplify.

The hypotenuse has a length of 20 cm.

Example 2: Find the distance between $(2, -5)$ and $(-4, 1)$.

Solution: Use the distance formula with $(x_1, y_1) = (2, -5)$ and $(x_2, y_2) = (-4, 1)$.

$d = \sqrt{(x_2 - x_1)^2 + (y_2 - y_1)^2}$ The distance formula.

$= \sqrt{(-4 - 2)^2 + [1 - (-5)]^2}$ Substitute known values.

$= \sqrt{(-6)^2 + 6^2}$ Simplify.

$= \sqrt{36 + 36}$

$= \sqrt{72} = 6\sqrt{2}$ Simplify the radical.

The distance is **exactly** $\sqrt{72}$ units or **approximately** 8.5 units.

Example 3: Paul and Lorraine leave their home at the same time. Paul drives southward at a rate of 45 miles per hour, while Lorraine drives east at 58 miles per hour. Find how far apart they are after 2 hours to the nearest mile.

Solution:

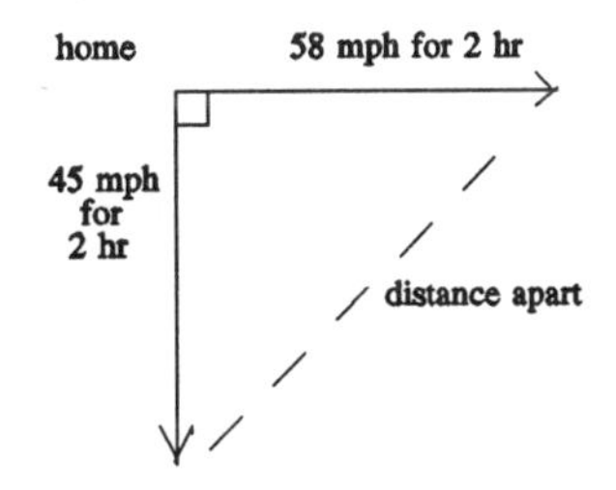

Recall $d = rt$.
Lorraine's distance: $d = (58)(2) = 116$ miles
Paul's distance: $d = (45)(2) = 90$ miles

Paul's and Lorraine's distances are the leg lengths, a and b, in a right triangle. Use the Pythagorean theorem to find c which corresponds to their distance apart.

$a^2 + b^2 = c^2$ Pythagorean formula.

$116^2 + 90^2 = c^2$ Substitute known values.

$13456 + 8100 = c^2$

$21556 = c^2$

$\sqrt{21556} = \sqrt{c^2}$

$147 \approx c$ Rounded to the nearest mile.

They are 147 miles apart.

9.6 EXERCISES

Find the length of the unknown side of each right triangle with sides a, b, and c, where c is the hypotenuse.

1. $a = 15, \ b = 20$
2. $a = 21, \ c = 35$
3. $b = 2, \ c = 6$
4. $a = \sqrt{3}, \ b = 2$

Use the distance formula to find the distance between the points given.

5. (5, 7), (3, -2)
6. (-9, -1), (0, 4)
7. (-6, -4), (-2, -8)
8. (0, -8), (6, 0)
9. $\left(\frac{3}{2}, 1\right)$, (4, -1)
10. $\left(\frac{1}{3}, 3\right), \left(\frac{4}{3}, -7\right)$

Solve the following.

11. For a square-based pyramid, the formula $b = \sqrt{\frac{3V}{H}}$ describes the relationship between the length b of one side of the base, the volume V, and the height H. Find the volume if each edge of the base is 8 feet long and the pyramid is 6 feet high.

12. Sam needs to attach a diagonal brace to a rectangular frame in order to make it structurally sound. If the framework is 5 feet by 12 feet, find how long the brace needs to be to the nearest tenth of a foot.

13. Police use the formula $s = \sqrt{30fd}$ to estimate the speed s of a car in miles per hour. In this formula, d represents the distance the car skidded in feet and f represents the coefficient of friction. The value of f depends in the type of road surface, and for wet concrete is 0.35. Find how fast a car was moving if it skidded 320 feet on wet concrete, to the nearest mile per hour.

14. Dave and Dan leave their office at the same time. Dave drives southward at a rate of 40 miles per hour, while Dan drives west at 56 miles per hour. Find how far apart they are after 45 minutes to the nearest mile.

15. Find the length of x:

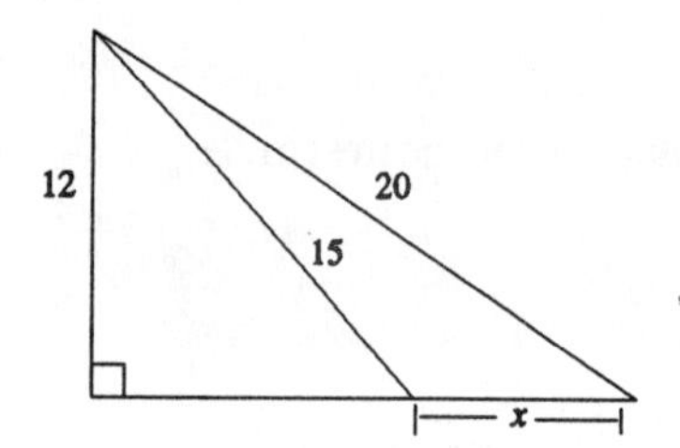

9.7 RATIONAL EXPONENTS

Example 1: Write in radical notation and then simplify.

a. $49^{1/2}$ b. $\left(\frac{1}{27}\right)^{1/3}$ c. $-81^{1/4}$

Solution:

a. $49^{1/2} = \sqrt{49} = 7$

b. $\left(\frac{1}{27}\right)^{1/3} = \sqrt[3]{\frac{1}{27}} = \frac{1}{3}$

c. $-81^{1/4} = -\sqrt[4]{81}$ In $-81^{1/4}$, the base of the exponent is 81.

$= -3$

Example 2: Write each expression with a positive exponent and then simplify.

a. $36^{-3/2}$ b. $81^{-1/4}$ c. $-4^{5/2}$

Solution:

a. $36^{-3/2} = \frac{1}{36^{3/2}} = \frac{1}{(\sqrt{36})^3} = \frac{1}{6^3} = \frac{1}{216}$

b. $81^{-1/4} = \frac{1}{81^{1/4}} = \frac{1}{\sqrt[4]{81}} = \frac{1}{3}$

c. $-4^{5/2} = -(\sqrt{4})^5 = -(2)^5 = -32$

Example 3: Simplify each expression. Write results with positive exponents only. Assume that all variables represent positive numbers.

a. $5^{1/3} \cdot 5^{5/3}$ b. $\frac{x^{3/4}}{x^{-5/8}}$ c. $(y^{3/5})^{15}$

Solution:

a. $5^{1/3} \cdot 5^{5/3} = 5^{1/3+5/3} = 5^{6/3} = 5^2 = 25$

b. $\frac{x^{3/4}}{x^{-5/8}} = x^{3/4-(-5/8)} = x^{3/4+5/8} = x^{11/8}$

c. $(y^{3/5})^{15} = y^{(3/5)(15)} = y^9$

9.7 EXERCISES

Simplify each expression. Write each answer with positive exponents. Assume that all variables represent positive numbers.

1. $8^{4/3}$
2. $100^{3/2}$
3. $64^{2/3}$
4. $-49^{1/2}$
5. $16^{-3/4}$
6. $-32^{-3/5}$
7. $\left(\frac{36}{25}\right)^{-1/2}$
8. $\left(\frac{64}{125}\right)^{-2/3}$
9. $\frac{x^{1/7}}{x^{5/7}}$
10. $(y^{3/4})^{20}$
11. $x^{2/5} \cdot x^{7/5}$
12. $y^{3/4} \cdot y^{1/3}$
13. $\frac{x^{5/6}}{x^{-2/3}}$
14. $\left(\frac{x^{3/4}}{y^{1/6}}\right)^{2}$
15. $\left(\frac{x^{2/5}}{y^{7/5}}\right)^{10}$
16. $\frac{7^{-2/9}}{7^{5/9}}$
17. $(x^{3/8})^{1/4}$
18. $3^{2/5} \cdot 3^{8/5}$

CHAPTER 9 PRACTICE TEST

Simplify the following. Indicate if the expression is not a real number.

1. $\sqrt{49}$
2. $\sqrt[3]{-125}$
3. $81^{1/4}$
4. $\left(\frac{16}{9}\right)^{3/2}$
5. $\sqrt[4]{-256}$
6. $64^{-3/2}$

Simplify each radical expression. Assume that variables represent positive numbers only.

7. $\sqrt{112}$
8. $\sqrt{12x^3y^9}$
9. $\sqrt[3]{27x^{12}y^7}$
10. $\sqrt{\frac{8}{50}}$
11. $\sqrt{3x^2} + \sqrt[3]{56} - 2x\sqrt{12}$

Rationalize the denominator.

12. $\sqrt{\frac{10}{7}}$
13. $\sqrt[3]{\frac{11}{9x}}$
14. $\frac{4}{\sqrt{6} - 2}$

Solve each of the following radical equations.

15. $\sqrt{x} + 7 = 19$
16. $\sqrt{2x - 4} = \sqrt{x + 1}$
17. Find the length of the unknown leg of a right triangle if the other leg is 9 inches long and the hypotenuse is 15 inches long.
18. Find the distance between (2, 6) and (−8, 3).

Simplify each expression using positive exponents only.

19. $25^{-3/2} \cdot 25^{-1/2}$
20. $(x^{3/7})^4$

10.1 SOLVING QUADRATIC EQUATIONS BY THE SQUARE ROOT METHOD

Example 1: Use the square root property to solve $x^2 - 121 = 0$.

Solution:

$$x^2 - 121 = 0$$

$$x^2 = 121 \qquad \text{Add 121 to both sides.}$$

$$x = \pm\sqrt{121} \qquad \text{Apply the square root property.}$$

$$x = \pm 11$$

The solution set is $\{\pm 11\}$.

Example 2: Solve $(x + 5)^2 = 49$.

Solution:

$$(x + 5)^2 = 49$$

$$x + 5 = \pm\sqrt{49} \qquad \text{Use the square root property.}$$

$$x + 5 = \pm 7 \qquad \text{Simplify.}$$

$$x = -5 \pm 7$$

$$x = -5 + 7 \quad \text{or} \quad x = -5 - 7$$

$$x = 2 \quad \text{or} \quad x = -12$$

The solution set is $\{-12, 2\}$.

Example 3: Solve $(4x + 3)^2 = 17$.

Solution:

$$(4x + 3)^2 = 17$$

$$4x + 3 = \pm\sqrt{17} \qquad \text{Apply the square root property.}$$

$$4x = -3 \pm \sqrt{17} \qquad \text{Subtract 3.}$$

$$x = \frac{-3 \pm \sqrt{17}}{4} \qquad \text{Divide by 4.}$$

The solution set is $\left\{\frac{-3 + \sqrt{17}}{4}, \frac{-3 - \sqrt{17}}{4}\right\}$.

10.1 EXERCISES

Solve the following.

1. $x^2 = \frac{1}{25}$
2. $y^2 = 8$
3. $2k^2 - 98 = 0$
4. $5x^2 - 5 = 0$
5. $n^2 + 16 = 0$
6. $(x + 4)^2 = 16$
7. $(y - 2)^2 = 36$
8. $(m - 7)^2 = 100$
9. $(z + 8)^2 = -100$
10. $p^2 = 28$
11. $(2x + 7)^2 = 48$
12. $(6y - 13)^2 = 12$
13. $(5m + 4)^2 = 18$
14. $(3 + 2y)^2 = 24$
15. $4x^2 = 169$
16. $-3a^2 = 27$

10.2 SOLVING QUADRATIC EQUATIONS BY COMPLETING THE SQUARE

Example 1: Complete the square for the given expression and then factor the resulting perfect square trinomial.

$x^2 + 12x$

Solution: The coefficient of the x-term is 12. Half of 12 is 6, and $6^2 = 36$. Add 36.

$x^2 + 12x + 36 = (x + 6)^2$

Example 2: Complete the square to solve the quadratic equation $y^2 - 8y = -6$.

Solution:

$y^2 - 8y = -6$

$y^2 - 8y + 16 = -6 + 16$ $\qquad$ $\left[\frac{1}{2}(-8)\right]^2 = (-4)^2 = 16.$

$(y - 4)^2 = 10$ $\qquad$ Factor the trinomial and simplify.

$y - 4 = \pm\sqrt{10}$ $\qquad$ Apply the square root property.

$y = 4 \pm \sqrt{10}$ $\qquad$ Add 4 to both sides.

The solution set is $\{4 + \sqrt{10},\ 4 - \sqrt{10}\}$.

Example 3: Solve the quadratic equation $2x^2 - 5x - 3 = 0$ by completing the square.

Solution:

$2x^2 - 5x - 3 = 0$

$x^2 - \frac{5}{2}x - \frac{3}{2} = 0$ $\qquad$ Divide by 2.

$x^2 - \frac{5}{2}x = \frac{3}{2}$ $\qquad$ Isolate variable terms.

$x^2 - \frac{5}{2}x + \frac{25}{16} = \frac{3}{2} + \frac{25}{16}$ $\qquad$ $\left[\frac{1}{2}\left(-\frac{5}{2}\right)\right]^2 = \left(-\frac{5}{4}\right)^2 = \frac{25}{16}.$

$\left(x - \frac{5}{4}\right)^2 = \frac{49}{16}$ $\qquad$ Factor the trinomial and simplify $\frac{3}{2} + \frac{25}{16}$.

$x - \frac{5}{4} = \pm\sqrt{\frac{49}{16}}$ $\qquad$ Apply the square root property.

$x - \frac{5}{4} = \pm\frac{7}{4}$ $\qquad$ Simplify.

$x = \frac{5}{4} \pm \frac{7}{4}$ $\qquad$ Add $\frac{5}{4}$ to both sides.

$x = \frac{5}{4} + \frac{7}{4} = \frac{12}{4} = 3$ or $x = \frac{5}{4} - \frac{7}{4} = -\frac{2}{4} = -\frac{1}{2}$

The solution set is $\left\{3, -\frac{1}{2}\right\}$.

10.2 EXERCISES

Complete the square for each expression and then factor the resulting perfect square trinomial.

1. $x^2 + 20x$
2. $y^2 - 18y$
3. $x^2 - 7x$
4. $m^2 + 13m$

Solve each quadratic equation by completing the square.

5. $x^2 + 4x = 8$
6. $y^2 - 2y = 0$
7. $m^2 - 6m - 0$
8. $x^2 + 14x + 24 = 0$
9. $z^2 + 8z + 1 = 0$
10. $y^2 = 10y + 3$
11. $2x^2 - 14x + 5 = 0$
12. $3x^2 - 17x = -10$
13. $4y^2 = 12y - 9$
14. $5m^2 - 2 = 10m$
15. $z^2 + z + 8 = 0$
16. $x(x - 1) = 6$

10.3 SOLVING QUADRATIC EQUATIONS BY THE QUADRATIC FORMULA

Example 1: Use the quadratic formula to solve $2x^2 + 3x - 5 = 0$.

Solution: The equation is in standard form with

$a = 2,\ b = 3,$ and $c = -5.$

$$x = \frac{-b \pm \sqrt{b^2 - 4ac}}{2a}$$ Quadratic formula.

$$= \frac{-3 \pm \sqrt{3^2 - 4(2)(-5)}}{2(2)}$$ Let $a = 2$, $b = 3$, and $c = -5$.

$$= \frac{-3 \pm \sqrt{9 + 40}}{4}$$ Simplify.

$$= \frac{-3 \pm \sqrt{49}}{4} = \frac{-3 \pm 7}{4}$$

Then, $x = \frac{-3 + 7}{4} = 1$ or $x = \frac{-3 - 7}{4} = -\frac{5}{2}$.

The solution set is $\left\{1, -\frac{5}{2}\right\}$.

Example 2: Use the quadratic formula to solve $5x^2 - 2x = 1$.

Solution: First, write the equation in standard form by subtracting 1 from both sides.

$$5x^2 - 2x = 1$$
$$5x^2 - 2x - 1 = 0$$

Next, $a = 5$, $b = -2$, and $c = -1$. Substitute these values into the quadratic formula.

$$x = \frac{-b \pm \sqrt{b^2 - 4ac}}{2a}$$

$$= \frac{-(-2) \pm \sqrt{(-2)^2 - 4(5)(-1)}}{2(5)}$$

$$= \frac{2 \pm \sqrt{4 + 20}}{10} \quad \text{Simplify.}$$

$$= \frac{2 \pm 2\sqrt{6}}{10} \quad \sqrt{24} = \sqrt{4 \cdot 6} = \sqrt{4}\sqrt{6} = 2\sqrt{6}.$$

$$= \frac{2(1 \pm \sqrt{6})}{10} \quad \text{Factor.}$$

$$= \frac{1 \pm \sqrt{6}}{5} \quad \text{Simplify.}$$

The solution set is $\left\{\frac{1 + \sqrt{6}}{5}, \frac{1 - \sqrt{6}}{5}\right\}$.

Example 3: Use the quadratic formula to solve $8x^2 = 3$.

Solution: Write the equation in standard form by subtracting 3 from both sides.

$$8x^2 = 3$$
$$8x^2 - 3 = 0$$

Next, replace a, b, and c with values: $a = 8$, $b = 0$, and $c = -3$.

$$x = \frac{-b \pm \sqrt{b^2 - 4ac}}{2a}$$

$$= \frac{-0 \pm \sqrt{0^2 - 4(8)(-3)}}{2(8)} \quad \text{Substitute in the formula.}$$

$$= \frac{\pm\sqrt{96}}{16} \qquad \text{Simplify.}$$

$$= \frac{\pm 4\sqrt{6}}{16}$$

$$= \frac{\pm\sqrt{6}}{4}$$

The solution set is $\left\{\frac{\sqrt{6}}{4}, -\frac{\sqrt{6}}{4}\right\}$.

Example 4: Use the quadratic formula to solve $\frac{1}{3}x^2 - 2x + 4 = 0$.

Solution: Clear the equation of fractions by multiplying both sides by the LCD 3.

$$\frac{1}{3}x^2 - 2x + 4 = 0$$

$$x^2 - 6x + 12 = 0$$

Here, $a = 1$, $b = -6$, and $c = 12$.

$$x = \frac{-b \pm \sqrt{b^2 - 4ac}}{2a}$$

$$= \frac{-(-6) \pm \sqrt{(-6)^2 - 4(1)(12)}}{2(1)}$$

$$= \frac{6 \pm \sqrt{36 - 48}}{2} \qquad \text{Simplify.}$$

$$= \frac{6 \pm \sqrt{-12}}{2}$$

There is no real number solution, because the radicand is negative.

10.3 EXERCISES

Use the quadratic formula to solve each quadratic equation.

1. $x^2 - 10x + 24 = 0$
2. $x^2 - 18x + 81 = 0$
3. $49x^2 - 10 = 0$
4. $3x^2 + 7x + 4 = 0$

5. $y^2 + 6y - 8 = 0$

6. $2z^2 + 3z - 1 = 0$

7. $7m^2 + 8m = 2$

8. $5 - x^2 = x$

9. $2x^2 + 3x + 5 = 0$

10. $4y^2 = 5y + 8$

11. $4d^2 - \frac{1}{2}d - 1 = 0$

12. $\frac{x^2}{5} = 2x + 3$

13. $10y^2 + y - 5 = 0$

14. $p^2 - 6p + 18 = 0$

15. $\frac{1}{4}x^2 + 2x - \frac{3}{4} = 0$

16. $1 - 3x - x^2 = 0$

10.4 SUMMARY OF METHODS FOR SOLVING QUADRATIC EQUATIONS

Example 1: Solve $x^2 - 7x + 10 = 0$.

Solution:

$$x^2 - 7x + 10 = 0$$

$$(x - 5)(x - 2) = 0 \qquad \text{Factor.}$$

$$x - 5 = 0 \quad \text{or} \quad x - 2 = 0 \qquad \text{Set the factors equal to 0.}$$

$$x = 5 \quad \text{or} \quad x = 2$$

The solution set is $\{5, 2\}$.

Example 2: Solve $(2x - 3)^2 = 28$.

Solution:

$$(2x - 3)^2 = 28$$

$$2x - 3 = \pm\sqrt{28} \qquad \text{Apply the square root property.}$$

$$2x - 3 = \pm 2\sqrt{7} \qquad \text{Simplify } \sqrt{28}.$$

$$2x = 3 \pm 2\sqrt{7}$$

$$x = \frac{3 \pm 2\sqrt{7}}{2}$$

The solution set is $\left\{\frac{3 + 2\sqrt{7}}{2}, \frac{3 - 2\sqrt{7}}{2}\right\}$.

Example 3: Solve $x^2 - \frac{5}{3}x = \frac{1}{3}$.

Solution: $x^2 - \frac{5}{3} = \frac{1}{3}$

$3x^2 - 5x = 1$ — Multiply both sides by 3.

$3x^2 - 5x - 1 = 0$ — Write in standard form.

$x = \frac{-b \pm \sqrt{b^2 - 4ac}}{2a}$ — The quadratic formula.

$= \frac{-(-5) \pm \sqrt{(-5)^2 - 4(3)(-1)}}{2(3)}$ — Let $a = 3$, $b = -5$, and $c = -1$.

$= \frac{5 \pm \sqrt{25 + 12}}{6}$ — Simplify.

$= \frac{5 \pm \sqrt{37}}{6}$

The solution set is $\left\{\frac{5 + \sqrt{37}}{6}, \frac{5 - \sqrt{37}}{6}\right\}$.

10.4 EXERCISES

Choose and use a method to solve each equation.

1. $6x^2 + 11x + 5 = 0$
2. $b^2 = 44$
3. $y^2 + 5 = 3y$
4. $16x^2 + 24x + 9 = 0$
5. $d^2 - 7d = 2$
6. $14x = x^2 - 4$
7. $(9b - 7)^2 = 48$
8. $x^3 + x^2 - 6x = 0$
9. $y^2 - \frac{7}{4}y - \frac{1}{4} = 0$
10. $(8 - z)^2 = 16$
11. $8x^2 + 7x - 6 = 0$
12. $2b^2 - 200 = 0$
13. $\frac{2}{3}x^2 - \frac{5}{6}x + 1 = 0$
14. $\frac{3}{2}y^2 - \frac{1}{2}y - 1 = 0$
15. $x^2 + 4 = 16x$
16. $m^2 - 10m = -25$

10.5 COMPLEX SOLUTIONS TO QUADRATIC EQUATIONS

Example 1: Write each radical as the product of a real number and i.

a. $\sqrt{-64}$ b. $\sqrt{-17}$

Solution: a. $\sqrt{-64} = \sqrt{-1 \cdot 64} = \sqrt{-1} \cdot \sqrt{64} = i \cdot 8 = 8i$

b. $\sqrt{-17} = \sqrt{-1 \cdot 17} = \sqrt{-1} \cdot \sqrt{17} = i\sqrt{17}$

Example 2: Simplify the sum or difference. Write the result in standard form.

a. $(5 + 2i) + (-8 - 7i)$ b. $(10 + 4i) - (9 - 5i)$

Solution: a. $(5 + 2i) + (-8 - 7i) = [5 + (-8)] + (2i - 7i)$
$= -3 - 5i$

b. $(10 + 4i) - (9 - 5i) = 10 + 4i - 9 + 5i$
$= (10 - 9) + (4i + 5i)$
$= 1 + 9i$

Example 3: Find the following product and write in standard form.

$(2 + 7i)(3 - 5i)$

Solution: $(2 + 7i)(3 - 5i) = 6 - 10i + 21i - 35i^2$ Use FOIL.
$= 6 + 11i - 35(-1)$ Write i^2 as -1.
$= 6 + 11i + 35$
$= 41 + 11i$

Example 4: Write $\dfrac{2 + i}{3 - i}$ in standard form.

Solution: $\dfrac{2 + i}{3 - i} = \dfrac{(2 + i)}{(3 - i)} \cdot \dfrac{(3 + i)}{(3 + i)}$ Mulitply numerator and denominator by $3 + i$.

$= \dfrac{6 + 2i + 3i + i^2}{9 - i^2}$

$= \dfrac{6 + 5i + (-1)}{9 - (-1)}$

$= \dfrac{6 + 5i - 1}{9 + 1}$

$= \dfrac{5 + 5i}{10}$

$= \dfrac{5}{10} + \dfrac{5}{10}i$ Write in standard form.

$= \dfrac{1}{2} + \dfrac{1}{2}i$ Simplify.

Example 5: Solve $x^2 + x + 6 = 0$.

Solution: $x = \dfrac{-b \pm \sqrt{b^2 - 4ac}}{2a}$ The quadratic formula.

$= \dfrac{-1 \pm \sqrt{1^2 - 4(1)(6)}}{2(1)}$ Let $a = 1$, $b = 1$, and $c = 6$.

$= \dfrac{-1 \pm \sqrt{1 - 24}}{2}$

$= \dfrac{-1 \pm \sqrt{-23}}{2}$

$= \dfrac{-1 \pm i\sqrt{23}}{2}$

The solution set is $\left\{\dfrac{-1 + i\sqrt{23}}{2}, \dfrac{-1 - i\sqrt{23}}{2}\right\}$.

10.5 EXERCISES

Write each expression in i notation.

1. $\sqrt{-144}$
2. $\sqrt{-75}$
3. $\sqrt{-19}$

Add or subtract as indicated.

4. $(3 + i) + (-4 + 9i)$
5. $(7 - 6i) - (-10 + 6i)$

Multiply.

6. $3i(5 - 9i)$
7. $(8 + 7i)(5 + 2i)$

Divide. Write each of the following in standard form.

8. $\dfrac{6 - 5i}{2i}$
9. $\dfrac{3 + 4i}{8 - i}$

Solve the following quadratic equations for complex solutions.

10. $(x - 5)^2 = -36$
11. $(3y + 7)^2 = -24$
12. $z^2 + 2z + 6 = 0$
13. $3d^2 - d + 5 = 0$
14. $9x^2 + 49 = 0$
15. $y^2 + 4y + 8 = 0$

10.6 GRAPHING QUADRATIC EQUATIONS

Example 1: Graph $y = -x^2$.

Solution: Select x-values and calculate the corresponding y-values.

x	$y = -x^2$
-2	$-(-2)^2 = -4$
-1	$-(-1)^2 = -1$
0	$-(0)^2 = 0$
1	$-(1)^2 = -1$
2	$-(2)^2 = -4$

Plot these points and connect them with a smooth curve.

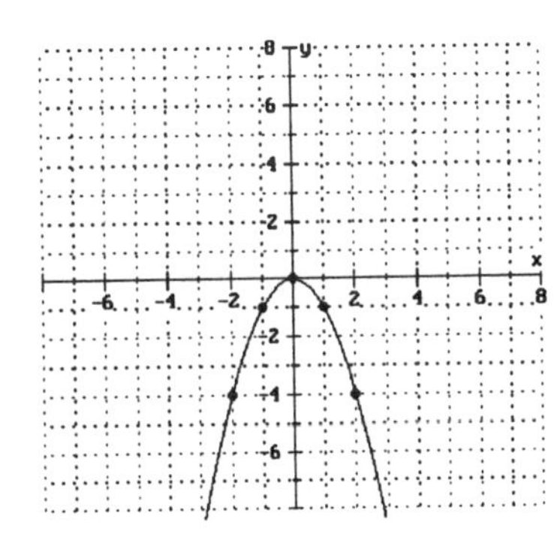

Example 2: Graph $y = 2(x - 1)^2 + 3$.

Solution: The equation is written in the form $y = a(x - h)^2 + k$, with $a = 2$, $h = 1$, and $k = 3$. The vertex is (1, 3), the axis of symmetry is the line $x = 1$, and the graph opens upward since $a = 2 > 0$.

To find the y-intercept, let $x = 0$.

$y = 2(0 - 1)^2 + 3$
$y = 5$

The y-intercept is (0, 5).

Find another point or two on each side of the vertex to determine the shape of the parabola.

x	$y = 2(x - 1)^2 + 3$
-1	$2(-1 - 1)^2 + 3 = 11$
2	$2(2 - 1)^2 + 3 = 5$
3	$2(3 - 1)^2 + 3 = 11$

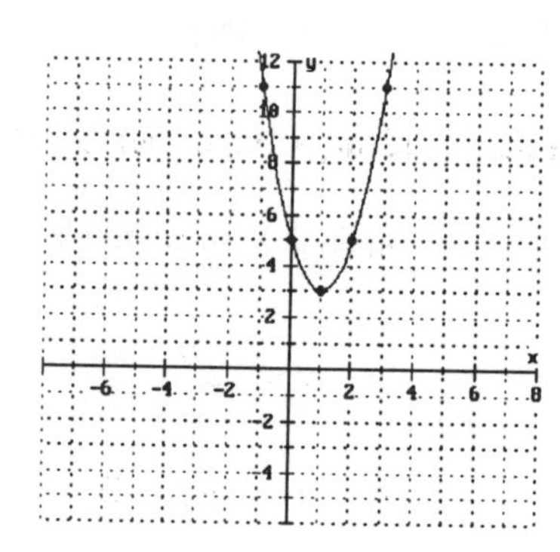

Example 3: Graph $y = -x^2 + 4x + 5$.

Solution: In the equation $y = -x^2 + 4x + 5$, $a = -1$, $b = 4$, and $c = 5$.

The x-coordinate of the vertex is $\dfrac{-b}{2a} = \dfrac{-4}{2(-1)} = 2$.

To find the corresponding y-coordinate, let $x = 2$.

$y = -(2)^2 + 4(2) + 5 = 9$

The vertex is (2, 9), and the axis of symmetry is the line $x = 2$, and the parabola opens downward since $a = -1 < 0$.

To find the x-intercepts, let $y = 0$.

$0 = -x^2 + 4x + 5$
$0 = x^2 - 4x - 5$
$0 = (x - 5)(x + 1)$

The x-intercepts are 5 and -1.

To find the y-intercept, let $x = 0$.
$y = -0^2 + 4(0) + 5 = 5$
The y-intercept is 5.

Plot the vertex and intercepts, then sketch the parabola.

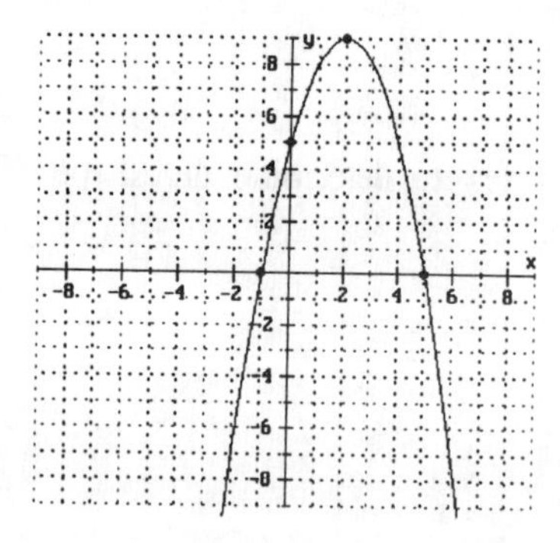

10.6 EXERCISES

Sketch the graph of each equation. Identify the vertex and the intercepts.

1. $y = 3x^2$
2. $y = -3x^2$
3. $y = \frac{1}{2}x^2$
4. $y = (x - 2)^2$
5. $y = -x^2 + 3$
6. $y = 9 - x^2$
7. $y = -2(x + 1)^2$
8. $y = -x^2 - 5$
9. $y = 2(x - 3)^2 - 8$
10. $y = -(x - 3)^2 + 1$
11. $y = \frac{1}{2}(x + 2)^2 + 2$
12. $y = -\frac{1}{9}(x - 6)^2 - 2$
13. $y = x^2 + 4x$
14. $y = x^2 - 6x$
15. $y = -x^2 + 4x - 3$

CHAPTER 10 PRACTICE TEST

Solve by factoring.

1. $3x^2 - x = 10$

2. $x^5 - 7x^4 + 12x^3 = 0$

Solve using the square root property.

3. $6k^2 = 150$

4. $(2m - 7)^2 = 5$

Solve by completing the square.

5. $x^2 - 14x + 48 = 0$

6. $3x^2 + 5x = 4$

Solve using the quadratic formula.

7. $x^2 - 4x - 9 = 0$

8. $p^2 - \frac{7}{6}p - \frac{1}{6} = 0$

Solve by the most appropriate method.

9. $(2x - 5)(x + 3) = -5$

10. $(7x + 6)^2 = 49$

11. $8x^3 = 2x$

12. $x^4 + x^3 - 56x^2 = 0$

Perform the indicated operations. Write the resulting complex number in standard form.

13. $\sqrt{-81}$

14. $(9 - 7i) + (4 - i)$

15. $(8 + 3i) + (8 - 3i)$

16. $(11 - 2i) - (-9 - 7i)$

17. $(1 + 3i)(5 - 4i)$

18. $(2 - 5i)(2 + 5i)$

19. $\dfrac{4 - i}{5 + i}$

20. $\dfrac{3 + 2i}{6 - 2i}$

Graph the quadratic equations. Label the vertex and the intercept points with their coordinates.

21. $y = (x - 2)^2 + 1$

22. $y = x^2 - x - 12$

PRACTICE FINAL EXAMINATION

Simplify the expression.

1. $-17 - (-6)$

2. $-\frac{3}{8} + \frac{2}{5}$

3. $6[2 + 7(5 - 9) - 1]$

4. $-2(-3)^2 - 24$

5. Evaluate $x^2 - 2yz$ if $x = -2$, $y = 5$, and $z = 3$.

Insert <, >, or = in the appropriate space to make each of the following statements true.

6. $-9 \quad -5$

7. $|-10| \quad 10$

8. $|-4| \quad 3 - (-2)$

Identify the property illustrated by each expression.

9. $(-4)(9) = (9)(-4)$

10. $2(3 + 8) = 2 \cdot 3 + 2 \cdot 8$

Simplify each of the following expressions.

11. $2x + 8 - x - 15$

12. $-7(y - 2) - 3(4 - 6y)$

Solve each of the following equations.

13. $-\frac{4}{7}x = 24$

14. $2(y - 3) = -(4 - y)$

15. $\frac{6(a + 1)}{7} = a - 3$

16. $\frac{1}{2} - b + \frac{5}{2} = b + 3$

Solve each of the following applications.

17. A number increased by two-thirds of the number is 25. Find the number.

18. Two trains leave Boston simultaneously traveling on the same track in opposite directions at speeds of 80 and 92 miles per hour. How long will it take before they are 516 miles apart?

19. Solve $6x - 7y = 14$ for y.

20. Solve and graph the inequality $3x - 8 \leq 6x + 16$.

21. Is the ordered pair $(-2, 3)$ a solution of the equation $5x - y = -7$?

22. Complete the ordered pair solution $(4, \quad)$ for the equation $10x - 3y = 55$.

Find the slopes of the following lines.

23.

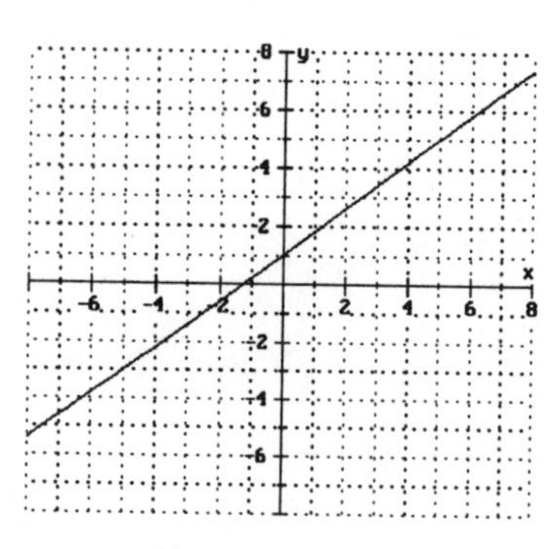

24. Through (7, −9) and (1, 12).

25. $4x - y = 13$

26. Determine whether the lines through the given pairs of points are parallel, perpendicular, or neither.

    (3, 6) and (−1, 4)
    (8, 3) and (9, 1)

Graph the following.

27. $2x + y = 4$

28. $y \le -3x$

29. $x + 6 = 0$

30. Write the following statement as an equation in two variables, and then graph the equation. "The x-value is 5 more than twice the y-value."

31. Evaluate -2^4.

32. Simplify $\left(\dfrac{8x^6y^2}{14xy^7}\right)^3$.

33. Simplify $\dfrac{4^2x^{-10}y^8}{4^5x^{-2}y^{-6}}$. Write the result using only positive exponents.

34. Express 46,500,000 in scientific notation.

35. Write 3.81×10^{-6} in standard form.

36. Find the degree of the polynomial $2x^4y - 10x^2y^5 + x^3y^6$.

Perform the indicated operations.

37. $(2x^4 - 3x^2 + 5x - 11) + (4x^3 + 2x^2 - 6x + 9)$

38. $(4x - 1)(3x + 5)$

39. $(2x - 3y)^2$

40. $(x^2 + 8x - 7) \div (x - 2)$

Factor each polynomial completely. If a polynomial cannot be factored, write "prime".

41. $3a^2b - 6a^3b^2 + 9a^5b^3$

42. $x^2 - 8x + 7$

43. $x^2 - 4x + 12$

44. $3y^2 - 108$

45. $x^2 - 2xy - 15y^2$

46. $8a^3 + 64b^3$

47. $m^2n + 6m^2 - 4n - 24$

48. $x^2 + 6xy + 9y^2$

49. Solve the following equation.

    $2x^2 + 19x = -24$

50. Find the dimensions of a rectangular sign whose width is 5 feet less than its length and whose area is 234 square feet.

51. Simplify $\dfrac{x - 9}{x^2 - 81}$.

Perform the indicated operation and simplify if possible.

52. $\dfrac{x^2 - x - 2}{x^2 + 4x + 3} \cdot \dfrac{x^2 - x - 12}{x^2 + 3x - 10}$

53. $\dfrac{5}{x - 4} \div \dfrac{10}{x^2 - 16}$

54. $\dfrac{7}{x^2 - 36} + \dfrac{4}{x + 6}$

55. $\dfrac{3y}{y^2 + 7y + 6} - \dfrac{4}{y^2 - 2y - 3}$

Solve each equation.

56. $\dfrac{2}{a} - \dfrac{3}{4} = \dfrac{-1}{8}$

57. $\dfrac{4}{x^2 - 9} = \dfrac{3}{x + 3} + \dfrac{6}{x - 3}$

58. Simplify the complex fraction $\dfrac{9 - \dfrac{4}{x^2}}{\dfrac{3}{x} + \dfrac{2}{x^2}}$.

59. One number less 14 times its reciprocal is equal to 5. Find the number.

60. An inlet pipe can fill a tank in 12 hours. A second pipe can fill the tank in 18 hours. If both pipes are used, find how long it takes to fill the tank.

61. Determine the slope and y-intercept of the graph of $3x - 7y = 21$.

Find equations of the following lines. Write the equation in standard form.

62. With slope $-\frac{2}{3}$, through (4, 6).

63. Through (3, 4) and (−1, 5).

64. Through (8, −11) and parallel to $y = 2$.

Graph each equation.

65. $2x - y = 6$

66. $y = |x + 3|$

67. Is $2x - 9y = 4$ a function?

68. Given $f(x) = -2x + 3$, find $f(-4)$.

69. Given $g(x) = -2$, find $g(0)$.

70. Find the domain of $y = \dfrac{2}{x + 1}$.

71. Is the ordered pair (1, −2) a solution to the linear system $\begin{cases} 3x - y = 5 \\ x - 4y = -7 \end{cases}$?

Solve each system of equations.

72. $\begin{cases} x + y = 3 \\ x - y = 5 \end{cases}$

73. $\begin{cases} 2x + 3y = 11 \\ -x + 5y = 14 \end{cases}$

74. $\begin{cases} y = 3x \\ x + 15y = -92 \end{cases}$

75. $\begin{cases} 8x + 5y = 7 \\ 3x + 2y = 2 \end{cases}$

76. $\begin{cases} \frac{1}{2}x + 5y = 32 \\ \frac{1}{4}x - \frac{1}{6}y = 0 \end{cases}$

77. Bob's wallet contains \$5 bills and \$10 bills. There are 21 bills in his wallet. The total value of the bills is \$150. Find the number of \$5 bills and the number of \$10 bills in Bob's wallet.

78. Ricki has invested \$9000, part at 5% simple annual interest and the rest at 7%. Find how much she invested at each rate if the total interest after 1 year is \$590.

Graph the solutions of the following systems of linear inequalities.

79. $\begin{cases} y - 2x \le 4 \\ y \ge 1 \end{cases}$

80. $\begin{cases} 3x + y > -2 \\ x - y < 4 \end{cases}$

Simplify the following. Assume that variables represent positive numbers only.

81. $\sqrt{289}$

82. $-\sqrt[3]{-64}$

83. $81^{3/4}$

84. $25^{-5/2}$

85. $\sqrt{56x^6y^7}$

86. $(x^{4/5})^{10}$

87. $\sqrt{45x} - \sqrt{20x} + 9\sqrt{5x^3}$

88. Rationalize the denominator: $\dfrac{3}{\sqrt{5} + 1}$

89. Solve: $\sqrt{3x + 2} = \sqrt{2x - 5}$

90. Find the distance between $(-8, 4)$ and $(3, 6)$.

Solve.

91. $8x^2 - 19x - 15 = 0$

92. $4y^2 = 144$

93. $(5m - 4)^2 = 12$

94. $2x^2 + 6x - 1 = 0$

95. $8x^3 = x$

96. $x^2 - 5x + 7 = 0$

Perform the indicated operation. Write the resulting complex number in standard form.

97. $\sqrt{-160}$

98. $(4 + 7i) - (-6 + 2i)$

99. $\dfrac{2 - i}{3 + 2i}$

100. Graph $y = x^2 - 8x + 7$. Find the vertex and intercepts.

1.1 SOLUTIONS TO EXERCISES

1. $12 < 21$

2. $8.13 = 8.13$

3. $-18 > -19$; since -18 is to the right of -19 on the number line.

4. $|-22| = 22$; $|-23| = 23$
   Since $22 < 23$, $|-22| < |-23|$.

5. $|-48| = 48$; hence $-48 < |-48|$.

6. $\left|-\frac{4}{3}\right| = \frac{4}{3}$

7. $7 < 10$

8. $-12 \geq -20$

9. $9 \neq -9$

10. True; since $30 = 30$.

11. False; 0 is neither positive nor negative.

12. False; $4(6) = 24$; $4 + 6 = 10$
    $24 > 10$

13. True

14. True; $|-11| = 11$; $|-10| = 10$
    $11 > 10$
    $|-11| > |-10|$

15. True; $|0| = 0$

1.2 SOLUTIONS TO EXERCISES

1. $180 = 2 \cdot 90$
   $= 2 \cdot 2 \cdot 45$
   $= 2 \cdot 2 \cdot 5 \cdot 9$
   $= 2 \cdot 2 \cdot 5 \cdot 3 \cdot 3$

   $180 = 2 \cdot 2 \cdot 3 \cdot 3 \cdot 5$

2. $175 = 5 \cdot 35$
   $= 5 \cdot 5 \cdot 7$

   $175 = 5 \cdot 5 \cdot 7$

3. $48 = 2 \cdot 24$
   $= 2 \cdot 2 \cdot 12$
   $= 2 \cdot 2 \cdot 3 \cdot 4$
   $= 2 \cdot 2 \cdot 3 \cdot 2 \cdot 2$

   $48 = 2 \cdot 2 \cdot 2 \cdot 2 \cdot 3$

4. $882 = 2 \cdot 441$
   $= 2 \cdot 3 \cdot 147$
   $= 2 \cdot 3 \cdot 3 \cdot 49$
   $= 2 \cdot 3 \cdot 3 \cdot 7 \cdot 7$

   $882 = 2 \cdot 3 \cdot 3 \cdot 7 \cdot 7$

5. $\frac{20}{45} = \frac{2 \cdot 2 \cdot 5}{3 \cdot 3 \cdot 5}$

   $= \frac{4}{9}$

6. $\frac{38}{57} = \frac{2 \cdot 19}{3 \cdot 19}$

   $= \frac{2}{3}$

7. $\frac{34}{85} = \frac{2 \cdot 17}{5 \cdot 17}$

   $= \frac{2}{5}$

8. $\frac{138}{161} = \frac{2 \cdot 3 \cdot 23}{7 \cdot 23}$

   $= \frac{2 \cdot 3}{7}$

   $= \frac{6}{7}$

9. $\frac{4}{5} \cdot \frac{20}{14} = \frac{4 \cdot 20}{5 \cdot 14}$

$= \frac{2 \cdot 2 \cdot 2 \cdot 2 \cdot 5}{5 \cdot 2 \cdot 7}$

$= \frac{2 \cdot 2 \cdot 2}{7}$

$= \frac{8}{7}$

10. $\frac{3}{8} \cdot 2\frac{4}{9} = \frac{3}{8} \cdot \frac{22}{9}$

$= \frac{3 \cdot 22}{8 \cdot 9}$

$= \frac{3 \cdot 2 \cdot 11}{2 \cdot 2 \cdot 2 \cdot 3 \cdot 3}$

$= \frac{11}{2 \cdot 2 \cdot 3}$

$= \frac{11}{12}$

11. $\frac{5}{6} \div \frac{15}{24} = \frac{5}{6} \cdot \frac{24}{15}$

$= \frac{5 \cdot 24}{6 \cdot 15}$

$= \frac{5 \cdot 2 \cdot 2 \cdot 2 \cdot 3}{2 \cdot 3 \cdot 3 \cdot 5}$

$= \frac{2 \cdot 2}{3}$

$= \frac{4}{3}$

12. $8 \div \frac{4}{3} = \frac{8}{1} \cdot \frac{3}{4}$

$= \frac{8 \cdot 3}{1 \cdot 4}$

$= \frac{2 \cdot 2 \cdot 2 \cdot 3}{2 \cdot 2}$

$= \frac{2 \cdot 3}{1}$

$= 6$

13. $\frac{19}{26} - \frac{7}{26} = \frac{19 - 7}{26}$

$= \frac{12}{26}$

$= \frac{2 \cdot 2 \cdot 3}{2 \cdot 13}$

$= \frac{2 \cdot 3}{13}$

$= \frac{6}{13}$

14. $\frac{10}{9} + \frac{2}{27} = \frac{10 \cdot 3}{9 \cdot 3} + \frac{2}{27}$

$= \frac{30}{27} + \frac{2}{27}$

$= \frac{30 + 2}{27}$

$= \frac{32}{27}$

15. $9 - \frac{2}{3} = \frac{9}{1} - \frac{2}{3}$

$= \frac{9 \cdot 3}{1 \cdot 3} - \frac{2}{3}$

$= \frac{27}{3} - \frac{2}{3}$

$= \frac{27 - 2}{3}$

$= \frac{25}{3}$

16. $5\frac{3}{7} + 2\frac{1}{2} = 5\frac{3 \cdot 2}{7 \cdot 2} + 2\frac{1 \cdot 7}{2 \cdot 7}$

$= 5\frac{6}{14} + 2\frac{7}{14}$

$= (5 + 2) + \left(\frac{6}{14} + \frac{7}{14}\right)$

$= 7 + \left(\frac{6 + 7}{14}\right)$

$= 7 + \left(\frac{13}{14}\right)$

$= 7\frac{13}{14}$

17. $\frac{3}{4} + \frac{8}{5} - \frac{1}{3} = \frac{3 \cdot 15}{4 \cdot 15} + \frac{8 \cdot 12}{5 \cdot 12} - \frac{1 \cdot 20}{3 \cdot 20}$

$= \frac{45}{60} + \frac{96}{60} - \frac{20}{60}$

$= \frac{45 + 96 - 20}{60}$

$= \frac{141 - 20}{60}$

$= \frac{121}{60}$

18. $6\frac{2}{3} \div \frac{5}{3} = \frac{20}{3} \div \frac{5}{3}$

$= \frac{20}{3} \cdot \frac{3}{5}$

$= \frac{20 \cdot 3}{3 \cdot 5}$

$= \frac{2 \cdot 2 \cdot 5 \cdot 3}{3 \cdot 5}$

$= \frac{2 \cdot 2}{1}$

$= 4$

19. $\frac{8}{9} + \frac{3}{4} = \frac{8 \cdot 4}{9 \cdot 4} + \frac{3 \cdot 9}{4 \cdot 9}$

$= \frac{32}{36} + \frac{27}{36}$

$= \frac{32 + 27}{36}$

$= \frac{59}{36}$

20. $3\frac{1}{8} + 1\frac{2}{7} + \frac{3}{5}$

$= 3\frac{1 \cdot 35}{8 \cdot 35} + 1\frac{2 \cdot 40}{7 \cdot 40} + \frac{3 \cdot 56}{5 \cdot 56}$

$= 3\frac{35}{280} + 1\frac{80}{280} + \frac{168}{280}$

$= (3 + 1) + \left(\frac{35}{280} + \frac{80}{280} + \frac{168}{280}\right)$

$= 4 + \left(\frac{35 + 80 + 168}{280}\right)$

$= 4 + \frac{283}{280}$

$= 4 + 1\frac{3}{280}$

$= 5\frac{3}{280}$

1.3 SOLUTIONS TO EXERCISES

1. $7^3 = 7 \cdot 7 \cdot 7 = 343$

2. $\left(\frac{3}{2}\right)^4 = \left(\frac{3}{2}\right)\left(\frac{3}{2}\right)\left(\frac{3}{2}\right)\left(\frac{3}{2}\right)$

$= \frac{81}{16}$

3. $(0.03)^2 = (0.03)(0.03)$
$= 0.0009$

4. $0^4 = 0 \cdot 0 \cdot 0 \cdot 0 = 0$

5. $1^3 = 1 \cdot 1 \cdot 1 = 1$

6. $4^3 = 4 \cdot 4 \cdot 4 = 64$

7. $9 + 8 \cdot 3 = 9 + 24 = 33$

8. $11 \cdot 4 - 2 \cdot 7 = 44 - 14 = 30$

9. $6 \cdot 2^3 = 6 \cdot 8 = 48$

10. $12(7 - 4) = 12(3) = 36$

11. $\frac{5 - 2}{15 - 4} = \frac{3}{11}$

12. $\frac{2}{5} \cdot \frac{10}{3} - \frac{5}{6} = \frac{4}{3} - \frac{5}{6}$
$= \frac{8}{6} - \frac{5}{6}$
$= \frac{3}{6}$
$= \frac{1}{2}$

13. $6[4 + 3(9 - 5)] = 6[4 + 3(4)]$
$= 6(4 + 12)$
$= 6(16)$
$= 96$

14. $\frac{4 + 4(6 + 3)}{4^2 + 2} = \frac{4 + 4(9)}{16 + 2}$
$= \frac{4 + 36}{18}$
$= \frac{40}{18}$
$= \frac{20}{9}$

15. $\frac{21 + |18 - 4| + 5^2}{23 - 3} = \frac{21 + |14| + 25}{20}$
$= \frac{21 + 14 + 25}{20}$
$= \frac{60}{20}$
$= 3$

16.

The sum of eight and twelve	is greater than	fifteen.
$8 + 12$	$>$	15

17.

The difference of twenty and four	is less than	nineteen.
$20 - 4$	$<$	19

18.

The product of two and six	is greater than	ten.
$2 \cdot 6$	$>$	10

1.4 SOLUTIONS TO EXERCISES

1. $4x - z = 4(2) - 6$
$= 8 - 6$
$= 2$

2. $xz - 2y = (2)(6) - 2(5)$
$= 12 - 10$
$= 2$

3. $3z^2 = 3(6)^2 = 3(36)$
$= 108$

4. $|z - 4| + 8x = |6 - 4| + 8(2)$
$= |2| + 16$
$= 2 + 16$
$= 18$

5. $\frac{z}{x} + y = \frac{6}{2} + 5$
$= 3 + 5$
$= 8$

6. $\frac{x^2 + y}{y^2 - z} = \frac{2^2 + 5}{5^2 - 6}$
$= \frac{4 + 5}{25 - 6}$
$= \frac{9}{19}$

7.

The sum of a number and 4	is	18.
$x + 4$	$=$	18

8.

The quotient of 14 and a number	is	1/7.
$14/x$	$=$	$1/7$

9. This is the same as,
"30 subtract twice a number is 11".

30 subtract twice a number	is	11.
$30 - 2x$	$=$	11

10. Triple a number: $3x$

The sum of triple a number and 5	is	84.
$3x + 5$	$=$	84

11.

Eight minus four times a number	is	31.
$8 - 4x$	$=$	31

12.

The product of 1/5 and	the sum of a number and 12	equals	7.
$1/5$	$\cdot\ (x + 12)$	$=$	7

13. Twice a number: $2x$

The sum of twice a number and one	is	60.
$2x + 1$	$=$	60

14.

The quotient of 15 and a number	is	2/3.
$15/x$	$=$	$2/3$

15. $4x + 8 = 28$
$4(5) + 8 = 28$
$20 + 8 = 28$
$28 = 28$
True

It is a solution.

16.
$$\begin{aligned} 3x - 2 &= 4x \\ 3(2) - 2 &= 4(2) \\ 6 - 2 &= 8 \\ 4 &= 8 \end{aligned}$$
False

It is not a solution.

17.
$$\begin{aligned} 7x + 10 &= 9x - 6 \\ 7(8) + 10 &= 9(8) - 6 \\ 56 + 10 &= 72 - 6 \\ 66 &= 66 \end{aligned}$$
True

It is a solution.

18.
$$\begin{aligned} 5x + 1 &= 5x + 1 \\ 5(50) + 1 &= 5(50) + 1 \\ 250 + 1 &= 250 + 1 \\ 251 &= 251 \end{aligned}$$
True

It is a solution.

19.
$$\begin{aligned} 9 &= 8 - x \\ 9 &= 8 - 1 \\ 9 &= 7 \end{aligned}$$
False

It is not a solution.

20.
$$\begin{aligned} 8 - 3x &= 14 \\ 8 - 3(2) &= 14 \\ 8 - 6 &= 14 \\ 2 &= 14 \end{aligned}$$
False

It is not a solution.

1.5 SOLUTIONS TO EXERCISES

1. $-12 + (-4) = -(12 + 4)$
$= -16$

2. $11 + (-7) = +(11 - 7)$
$= 4$

3. $8 + (+18) = +(8 + 18)$
$= 26$

4. $-13 + 5 = -(13 - 5)$
$= -8$

5. $-\dfrac{8}{15} + \dfrac{3}{5} = -\dfrac{8}{15} + \dfrac{9}{15}$
$= \dfrac{-8 + 9}{15}$
$= \dfrac{1}{15}$

6. $\dfrac{5}{2} + \left(-\dfrac{1}{4}\right) = \dfrac{10}{4} + \left(-\dfrac{1}{4}\right)$
$= \dfrac{10 + (-1)}{4}$
$= \dfrac{9}{4}$

7. $10 + (-30) + 25 = -20 + 25$
$= 5$

8. $-131 + (-64) + 17 = -195 + 17$
$= -178$

9. $|4 + (-19)| = |-15|$
$= 15$

10. $[-2 + 6] + (-12) = (4) + (-12)$
$= -8$

11. $-27 + (-4) + 13 = -31 + 13$
$= -18$

12. $|8 + (-14)| + |-10| = |-6| + |-10|$
$= 6 + 10$
$= 16$

13. 9

14. $\dfrac{2}{3}$

15. $|-17| = 17$

The opposite of 17 is -17.

16. beginning temperature + increase in degrees = high temperature

$-18 + 11 = -7$

The high temperature was -7 degrees.

1.6 SOLUTIONS TO EXERCISES

1. $-8 - 13 = -8 + (-13)$
$= -21$

2. $7 - 12 = 7 + (-12)$
$= -5$

3. $-22 - (-11) = -22 + 11$
$= -11$

4. $\frac{3}{4} - \frac{7}{2} = \frac{3}{4} + \left(-\frac{7}{2}\right)$
$= \frac{3}{4} + \left(-\frac{14}{4}\right)$
$= \frac{3 + (-14)}{4}$
$= -\frac{11}{4}$

5. $-9 - (3 - 12) = -9 - [3 + (-12)]$
$= -9 - (-9)$
$= -9 + 9$
$= 0$

6. $1 - 4(7 - 5) = 1 - 4(2)$
$= 1 - 8$
$= 1 + (-8)$
$= -7$

7. $|-3| + 7^2 + (-5 - 4)$
$= 3 + 49 + [-5 + (-4)]$
$= 3 + 49 + (-9)$
$= 52 + (-9)$
$= 43$

8. $3\frac{5}{9} - \left(-2\frac{1}{3}\right) = 3\frac{5}{9} + 2\frac{1}{3}$
$= 3\frac{5}{9} + 2\frac{3}{9}$
$= 5\frac{8}{9}$

9. $7 - \{2[6 - (-10)] - 13\}$
$= 7 - \{2[6 + 10] - 13\}$
$= 7 - \{2(16) - 13\}$
$7 - (32 - 13)$
$= 7 - (19)$
$= 7 + (-19)$
$= -12$

10. $4 - (2 \cdot 8)^2 = 4 - (16)^2$
$= 4 - 256$
$= 4 + (-256)$
$= -252$

11. $-17 + [(2 - 9) - (-14) - 20]$
$= -17 + \{[2 + (-9)] + 14 + (-20)\}$
$= -17 + [-7 + 14 + (-20)]$
$= -17 + [7 + (-20)]$
$= -17 + (-13)$
$= -30$

12. $5 - 3 \cdot 2^3 = 5 - 3 \cdot 8$
$= 5 - 24$
$= 5 + (-24)$
$= -19$

13. $3t - x = 3(3) - (-5)$
$= 9 - (-5)$
$= 9 + 5$
$= 14$

14. $\frac{x - y}{t} = \frac{-5 - (-4)}{3}$

$= \frac{-5 + 4}{3}$

$= \frac{-1}{3}$

15. $y - t^3 = -4 - (3)^3$
$= -4 - (27)$
$= -4 + (-27)$
$= -31$

16. drops 220 ft climbs 180 ft drops 195 ft
-220 + 180 - 195

(-220 + 180) - 195
= -40 - 195
= -235

The overall vertical change is a drop of 235 ft.

1.7 SOLUTIONS TO EXERCISES

1. $(-12)(10) = -120$

2. $(-8)(5)(0) = 0$ (since 0 is one of the factors.)

3. $\left(\frac{3}{2}\right)\left(-\frac{8}{9}\right) = -\frac{3 \cdot 8}{2 \cdot 9}$

$= -\frac{4}{3}$

4. $(-6)^2 = (-6)(-6) = 36$

5. $\frac{32}{-16} = -2$

6. $\frac{0}{-8} = 0$

7. $\frac{-5^2 + 7}{-9} = \frac{-25 + 7}{-9}$

$= \frac{-18}{-9}$

$= 2$

8. $\frac{12 - (-3)^2}{11 - 14} = \frac{12 - 9}{-3}$

$= \frac{3}{-3}$

$= -1$

9. $\frac{10 - 5(-4)}{7 - 2(-9)} = \frac{10 - (-20)}{7 - (-18)}$

$= \frac{10 + 20}{7 + 18}$

$= \frac{30}{25}$

$= \frac{6}{5}$

10. $-8(4 - 15) = -8(-11)$
$= 88$

11. $-6[(2 - 7) - (3 - 14)]$

$= -6[(-5) - (-11)]$

$= -6(-5 + 11)$

$= -6(6)$
$= -36$

12. $\left(-3\frac{2}{5}\right)\left(-\frac{25}{8}\right) = \left(-\frac{17}{5}\right)\left(-\frac{25}{8}\right)$

$= \frac{85}{8}$

$= 10\frac{5}{8}$

13. $\frac{5 - y}{7x} = \frac{5 - (-2)}{7(-6)}$

$= \frac{5 + 2}{-42}$

$= \frac{7}{-42}$

$= -\frac{1}{6}$

14. $3x^2 - 4y^3 = 3(-6)^2 - 4(-2)^3$
$= 3(36) - 4(-8)$
$= 108 + 32$
$= 140$

15. $-8 + \frac{-20}{10} = -8 + (-2)$

$= -10$

16. $2(10 - 17) - 4 = 2(-7) - 4$
$= -14 - 4$
$= -18$

1.8 SOLUTIONS TO EXERCISES

1. Distributive property; The 7 has been distributed.

2. Commutative property of multiplication; The factors have changed order.

3. Commutative property of multiplication; Inside the parentheses the factors have changed order.

4. Identity property of addition

5. Multiplicative inverse

6. Identity property of multiplication

7. $-7(6x + 3y) = -7(6x) + (-7)(3y)$
$= -42x + (-21y)$
$= -42x - 21y$

8. $-(-a - b + 2c)$
$= -1(-a - b + 2c)$
$= -1(-a) + (-1)(-b) + (-1)(2c)$
$= a + b + (-2c)$
$= a + b - 2c$

9. $3(-5x + 4) = 3(-5x) + 3(4)$
$= -15x + 12$

10. $\frac{10}{13}$

11. $-|-12| = -(12) = -12$

The opposite of -12 is 12.

12. $-2\frac{3}{7} = -\frac{17}{7}$

The reciprocal of $-\frac{17}{7}$ is $-\frac{7}{17}$.

13. $|-6| = 6$

The reciprocal of 6 is $\frac{1}{6}$.

14. $3(a + b) = 3(b + a)$

15. $5 \cdot (3 \cdot 8) = (5 \cdot 3) \cdot 8$

16. $-7 + 0 = -7$

1.9 SOLUTIONS TO EXERCISES

1. Go to the top of the bar corresponding to 1970-79, then move left to the vertical axis. Approximately 75 banks closed.

2. The largest distance up from the top of one bar to the top of the next bar indicates the greatest increase in bank closings, so 1970-79 to 1980-89.

3. The shortest bar indicates the years with the least number of banks failing. The smallest number of banks closed in 1950-59.

4. Look for the highest point on the line graph to determine the year with the greatest number of travelers. The year 1986 had the greatest number of travelers.

5. Go to the year 1994 then up to the graph. From this point on the graph move left to the vertical axis. Approximately 35 million Americans traveled during the Fourth of July holiday in 1994.

6. Between consecutive years look for the least steep line segment to determine when there was the smallest amount of change in the number of travelers. This occurs between 1987 and 1988.

7. Look for the largest pie slice. Most adult shoppers prefer paying with cash.

8. Add the check percentage to the credit card percentage. 30% + 16% = 46% of adult shoppers prefer checks or credit cards.

9. The total pie contains 100%. 100% minus the percent preferring checks yields the percentage of adult shoppers that do not prefer checks. This percentage is 100% - 30% = 70%.

10. Look for the highest point on the top line graph. The year 1988 had the greatest number of total traffic fatalities.

11. Look for the lowest point on the bottom line graph. The year 1994 had the least number of alcohol-related fatalities.

12. Go to the year 1991 and move to the bottom line graph. From this point move left to the vertical axis. Approximately 20,000 alcohol-related fatalities occurred in 1991.

13. Look for the years where the top line graph rises while the bottom line graph falls. This happens between 1986 and 1987, 1987 and 1988, 1992 and 1993, and 1993 and 1994.

14. Go to the year 1960 and move up to the line graph. From this point move left to the vertical axis. This point appears midway between 10 and 20. Hence 15 million people were over the age of 65 in 1960.

15. Look for the least steep line segment. It is projected that the population will increase the least between 2000 and 2010.

16. The difference is given by population in 2020 minus population in 1980.
    50 million - 24 million = 26 million

CHAPTER 1 PRACTICE TEST SOLUTIONS

1. $|-9| > 4$

2. $8 + 3 \geq 7$

3. $-12 + 5 = -7$

4. $-18 - (-4) = -18 + 4$
$= -14$

5. $7 \cdot 2 - 9 \cdot 6 = 14 - 54$
$= 14 + (-54)$
$= -40$

6. $(15)(-4) = -60$

7. $(-4)(-3) = 12$

8. $\dfrac{|-24|}{-6} = \dfrac{24}{-6}$
$= -4$

9. $\dfrac{-10}{0}$ is undefined.

10. $\dfrac{|-9| + 3}{4 - 5} = \dfrac{9 + 3}{4 + (-5)}$
$= \dfrac{12}{-1}$
$= -12$

11. $\dfrac{1}{3} - \dfrac{5}{12} = \dfrac{4}{12} - \dfrac{5}{12}$
$= \dfrac{4 - 5}{12}$
$= \dfrac{4 + (-5)}{12}$
$= -\dfrac{1}{12}$

12. $-2\frac{1}{4} + 6\frac{3}{8} = -2\frac{2}{8} + 6\frac{3}{8}$
$= 4\frac{1}{8}$

13. $-\dfrac{2}{7} + \dfrac{13}{6} = -\dfrac{12}{42} + \dfrac{91}{42}$
$= \dfrac{-12 + 91}{42}$
$= \dfrac{79}{42}$

14. $4(-3)^2 - 60 = 4(9) - 60$
$= 36 - 60$
$= 36 + (-60)$
$= -24$

15. $7[4 + 3(2 - 6) - 5]$
$= 7[4 + 3(-4) - 5]$
$= 7[4 + (-12) + (-5)]$
$= 7[-8 + (-5)]$
$= 7(-13)$
$= -91$

16. $\dfrac{-25 + 2 \cdot 10}{5} = \dfrac{-25 + 20}{5}$
$= \dfrac{-5}{5}$
$= -1$

17. $\dfrac{(-6)(-1)(0)}{-8} = \dfrac{0}{-8}$
$= 0$

18. $-5 > -8$

19. $6 > -9$

20. $|-7| \quad 6$
$7 \quad 6$
$>$

21. $|-3|$ $\quad -2 - (-5)$

$3 \quad -2 + 5$

$3 \quad 3$

$=$

22. a. $\{2, 8\}$

b. $\{0, 2, 8\}$

c. $\{-7, -2, 0, 2, 8\}$

d. $\{-7, -2, \frac{1}{3}, 0, 2, 8, 12.3\}$

e. $\{\sqrt{11}, 5\pi\}$

f. $\left\{-7, -2, \frac{1}{3}, 0, 2, 8, 12.3, \sqrt{11}, 5\pi\right\}$

23. $x^2 - y^2 = (3)^2 - (-4)^2$
$= 9 - 16$
$= 9 + (-16)$
$= -7$

24. $xy + z = (3)(-4) + (-7)$
$= -12 + (-7)$
$= -19$

25. $5 - 2x + 3y$
$= 5 - 2(3) + 3(-4)$
$= 5 - 6 + (-12)$
$= 5 + (-6) + (-12)$
$= -13$

26. $\dfrac{4x + y - 2}{x}$

$= \dfrac{4(3) + (-4) - 2}{3}$

$= \dfrac{12 + (-4) + (-2)}{3}$

$= \dfrac{8 + (-2)}{3}$

$= \dfrac{6}{3}$

$= 2$

27. Associative property of multiplication

28. Commutative property of addition

29. Distributive property

30. Additive inverse property

31. $-(-12) = 12$

32. -8

33. Second down

34. $18 - [4 + (-9) + (-1) + 24]$
$= 18 - [-5 + (-1) + 24]$
$= 18 - [-6 + 24]$
$= 18 - 18$
$= 0$

Yes, a touchdown was scored.

35. $-6 + 22 = 16$

16°

36. $340(-1.80) = -612$

$612 loss

2.1 SOLUTIONS TO EXERCISES

1. $5a - 6b + 9b + 8a$

$= (5 + 8)a + (-6 + 9)b$

$= 13a + 3b$

2. $12 - 5c - 10c + 30$

$= (12 + 30) + (-5 - 10)c$

$= 42 - 15c$

3. $-2(4a - 3b) + 10a = -8a + 6b + 10a$

$= (-8 + 10)a + 6b$

$= 2a + 6b$

4. $-(x + 2y - z) = -x - 2y + z$

5. $20x^2 + 6x - 8 + 4x - x^2$

$= (20 - 1)x^2 + (6 + 4)x - 8$

$= 19x^2 + 10x - 8$

6. $1 + 3y - 5y^2 + 9y - 10$

$= (1 - 10) + (3 + 9)y - 5y^2$

$= -9 + 12y - 5y^2$

7. $4.6m - 5.3n + 2.1m + 7.8n$

$= (4.6 + 2.1)m + (-5.3 + 7.8)n$

$= 6.7m + 2.5n$

8. $-17 - 4(r - 5) = -17 - 4r + 20$

$= (-17 + 20) - 4r$

$= 3 - 4r$

9. $18y - 12 - 13(y + 5)$

$= 18y - 12 - 13y - 65$

$= (18 - 13)y + (-12 - 65)$

$= 5y - 77$

10. $-3(2x + 7) - 8x + 1$

$= -6x - 21 - 8x + 1$

$= (-6 - 8)x + (-21 + 1)$

$= -14x - 20$

11. $-19d - (5 - 3d) = -19d - 5 + 3d$

$= (-19 + 3)d - 5$

$= -16d - 5$

12. $(21 - 3x) - (7 + 6x) + 1$

$= 21 - 3x - 7 - 6x + 1$

$= (21 - 7 + 1) + (-3 - 6)x$

$= 15 - 9x$

13.

Twice a number	added to	the product of 9 and the number
$2x$	$+$	$9x$

14.

The sum of a number and 6	divided by	10
$x + 6$	$\div$	10

or $\dfrac{x + 6}{10}$

15.

The sum of a number and 7	subtract	4
$x + 7$	$-$	4

2.2 SOLUTIONS TO EXERCISES

1.
$$\begin{aligned} x + 17 &= -6 \\ x + 17 - 17 &= -6 - 17 \\ x &= -23 \end{aligned}$$
$\{-23\}$

2.
$$\begin{aligned} 4 + y &= 18 \\ 4 + y - 4 &= 18 - 4 \\ y &= 14 \end{aligned}$$
$\{14\}$

3.
$$\begin{aligned} 3x &= 2x - 8 \\ 3x - 2x &= 2x - 8 - 2x \\ x &= -8 \end{aligned}$$
$\{-8\}$

4.
$$\begin{aligned} 5a - 3 &= 4a + 9 \\ 5a - 3 - 4a &= 4a + 9 - 4a \\ a - 3 &= 9 \\ a - 3 + 3 &= 9 + 3 \\ a &= 12 \end{aligned}$$
$\{12\}$

5.
$$\begin{aligned} 17b - 1 &= 16b + 31 \\ 17b - 1 - 16b &= 16b + 31 - 16b \\ b - 1 &= 31 \\ b - 1 + 1 &= 31 + 1 \\ b &= 32 \end{aligned}$$
$\{32\}$

6.
$$\begin{aligned} 9d + 18 - d &= 7d - 15 \\ 8d + 18 &= 7d - 15 \\ 8d + 18 - 7d &= 7d - 15 - 7d \\ d + 18 &= -15 \\ d + 18 - 18 &= -15 - 18 \\ d &= -33 \end{aligned}$$
$\{-33\}$

7.
$$\begin{aligned} x + 0.9 &= 10.7 \\ x + 0.9 - 0.9 &= 10.7 - 0.9 \\ x &= 9.8 \end{aligned}$$
$\{9.8\}$

8.
$$\begin{aligned} -6(y - 2) &= 5 - 7y \\ -6y + 12 &= 5 - 7y \\ -6y + 12 + 7y &= 5 - 7y + 7y \\ y + 12 &= 5 \\ y + 12 - 12 &= 5 - 12 \\ y &= -7 \end{aligned}$$
$\{-7\}$

9.
$$\begin{aligned} 10 - 5(x + 1) &= 10x + 3 - 16x \\ 10 - 5x - 5 &= -6x + 3 \\ -5x + 5 &= -6x + 3 \\ -5x + 5 + 6x &= -6x + 3 + 6x \\ x + 5 &= 3 \\ x + 5 - 5 &= 3 - 5 \\ x &= -2 \end{aligned}$$
$\{-2\}$

10.
$$\begin{aligned} 2b - (b - 7) &= 3b - 8 - b \\ 2b - b + 7 &= 2b - 8 \\ b + 7 &= 2b - 8 \\ b + 7 - b &= 2b - 8 - b \\ 7 &= b - 8 \\ 7 + 8 &= b - 8 + 8 \\ 15 &= b \end{aligned}$$
$\{15\}$

11.
$$\begin{aligned} -20m + 6(3m + 2) &= 8 - 3(1 + m) \\ -20m + 18m + 12 &= 8 - 3 - 3m \\ -2m + 12 &= 5 - 3m \\ -2m + 12 + 3m &= 5 - 3m + 3m \\ m + 12 &= 5 \\ m + 12 - 12 &= 5 - 12 \\ m &= -7 \end{aligned}$$
$\{-7\}$

12.
$$\begin{aligned} 0.7x + 0.3(0.6 + x) &= 3.61 \\ 0.7x + 0.18 + 0.3x &= 3.61 \\ x + 0.18 &= 3.61 \\ x + 0.18 - 0.18 &= 3.61 - 0.18 \\ x &= 3.43 \end{aligned}$$
$\{3.43\}$

13. $-14(c - 3) - 5(8 - 3c) = -9$
$-14c + 42 - 40 + 15c = -9$
$c + 2 = -9$
$c + 2 - 2 = -9 - 2$
$c = -11$
$\{-11\}$

14. $x + \text{other number} = 15$
$x + \text{other number} - x = 15 - x$
$\text{other number} = 15 - x$

15. $x + \text{other length} = 25$
$x + \text{other length} - x = 25 - x$
$\text{other length} = 25 - x$
$(25 - x)$ feet

2.3 SOLUTIONS TO EXERCISES

1. $-4x = 32$

$\frac{-4x}{4} = \frac{32}{-4}$

$x = -8$

$\{-8\}$

2. $-5y = -125$

$\frac{-5y}{-5} = \frac{-125}{-5}$

$y = 25$

$\{25\}$

3. $-x = -20$

$\frac{-x}{-1} = \frac{-20}{-1}$

$x = 20$

$\{20\}$

4. $2x = 0$

$\frac{2x}{2} = \frac{0}{2}$

$x = 0$

$\{0\}$

5. $\frac{2}{7}x = 30$

$\frac{7}{2} \cdot \frac{2}{7}x = \frac{7}{2} \cdot 30$

$\left(\frac{7}{2} \cdot \frac{2}{7}\right)x = \frac{7}{2} \cdot \frac{30}{1}$

$1x = 105$
$x = 105$

$\{105\}$

6. $\frac{-x}{6} = 8$

$6 \cdot \frac{-x}{6} = 6 \cdot 8$

$\frac{-x}{-1} = \frac{48}{-1}$

$x = -48$

$\{-48\}$

7. $\frac{a}{-2} = 0$

$-2 \cdot \frac{a}{-2} = -2 \cdot 0$

$a = 0$

$\{0\}$

8. $\frac{1}{14}d = \frac{3}{7}$

$14 \cdot \frac{1}{14}d = 14 \cdot \frac{3}{7}$

$d = 6$

$\{6\}$

9.
$$2c - 9 = 5c + 12$$
$$2c - 9 - 5c = 5c + 12 - 5c$$
$$-3c - 9 = 12$$
$$-3c - 9 + 9 = 12 + 9$$
$$-3c = 21$$
$$\frac{-3c}{-3} = \frac{21}{-3}$$
$$c = -7$$
$\{-7\}$

10.
$$14t - 1 = 8t + 35$$
$$14t - 1 - 8t = 8t + 35 - 8t$$
$$6t - 1 = 35$$
$$6t - 1 + 1 = 35 + 1$$
$$6t = 36$$
$$\frac{6t}{6} = \frac{36}{6}$$
$$t = 6$$
$\{6\}$

11.
$$3y - 7 - y = 11y - 20$$
$$2y - 7 = 11y - 20$$
$$2y - 7 - 2y = 11y - 20 - 2y$$
$$-7 = 9y - 20$$
$$-7 + 20 = 9y - 20 + 20$$
$$13 = 9y$$
$$\frac{13}{9} = \frac{9y}{9}$$
$$\frac{13}{9} = y$$
$\left\{\frac{13}{9}\right\}$

12.
$$5 + 0.6m = 5$$
$$5 + 0.6m - 5 = 5 - 5$$
$$0.6m = 0$$
$$\frac{0.6m}{0.6} = \frac{0}{0.6}$$
$$m = 0$$
$\{0\}$

13.
$$\frac{x}{3} + 2 = 13$$
$$\frac{x}{3} + 2 - 2 = 13 - 2$$
$$\frac{x}{3} = 11$$
$$3 \cdot \frac{x}{3} = 3 \cdot 11$$
$$x = 33$$
$\{33\}$

14.
$$y - 7y = -18 + y - 17$$
$$-6y = y - 35$$
$$-6y - y = y - 35 - y$$
$$-7y = -35$$
$$\frac{-7y}{-7} = \frac{-35}{-7}$$
$$y = 5$$
$\{5\}$

15.
$$6y + \frac{1}{5} = \frac{4}{5}$$
$$6y + \frac{1}{5} - \frac{1}{5} = \frac{4}{5} - \frac{1}{5}$$
$$6y = \frac{3}{5}$$
$$\frac{1}{6} \cdot 6y = \frac{1}{6} \cdot \frac{3}{5}$$
$$y = \frac{1}{10}$$
$\left\{\frac{1}{10}\right\}$

2.4 SOLUTIONS TO EXERCISES

1.
$$\begin{aligned} -3(2x + 1) &= 15 \\ -6x - 3 &= 15 \\ -6x - 3 + 3 &= 15 + 3 \\ -6x &= 18 \\ \frac{-6x}{-6} &= \frac{18}{-6} \\ x &= -3 \end{aligned}$$

$\{-3\}$

2.
$$\begin{aligned} -(4x + 3) &= 17 \\ -4x - 3 &= 17 \\ -4x - 3 + 3 &= 17 + 3 \\ -4x &= 20 \\ \frac{-4x}{-4} &= \frac{20}{-4} \\ x &= -5 \end{aligned}$$

$\{-5\}$

3.
$$\begin{aligned} 7(3x - 5) - 2x &= 3 \\ 21x - 35 - 2x &= 3 \\ 19x - 35 &= 3 \\ 19x - 35 + 35 &= 3 + 35 \\ 19x &= 38 \\ \frac{19x}{19} &= \frac{38}{19} \\ x &= 2 \end{aligned}$$

$\{2\}$

4.
$$\begin{aligned} -5(8 + b) + 11 &= 29 \\ -40 - 5b + 11 &= 29 \\ -5b - 29 &= 29 \\ -5b - 29 + 29 &= 29 + 29 \\ -5b &= 58 \\ \frac{-5b}{-5} &= \frac{58}{-5} \\ b &= -\frac{58}{5} \end{aligned}$$

$\left\{-\frac{58}{5}\right\}$

5.
$$\begin{aligned} 10(3y + 2) &= 6(y - 1) + 24y \\ 30y + 20 &= 6y - 6 + 24y \\ 30y + 20 &= 30y - 6 \\ 30y + 20 - 30y &= 30y - 6 - 30y \\ 20 &= -6 \end{aligned}$$

False

$\{\ \}$

6.
$$\begin{aligned} \frac{4}{5}x - \frac{1}{10} &= 2 \\ 10\left(\frac{4}{5}x - \frac{1}{10}\right) &= 10(2) \\ \frac{10}{1} \cdot \frac{4}{5}x - \frac{10}{1} \cdot \frac{1}{10} &= 20 \\ 8x - 1 &= 20 \\ 8x - 1 + 1 &= 20 + 1 \\ 8x &= 21 \\ \frac{8x}{8} &= \frac{21}{8} \\ x &= \frac{21}{8} \end{aligned}$$

$\left\{\frac{21}{8}\right\}$

7.
$$\begin{aligned} \frac{7}{3}x - 4 &= \frac{5}{6} \\ 6\left(\frac{7}{3}x - 4\right) &= 6\left(\frac{5}{6}\right) \\ \frac{6}{1} \cdot \frac{7}{3}x - 6(4) &= 5 \\ 14x - 24 &= 5 \\ 14x - 24 + 24 &= 5 + 24 \\ 14x &= 29 \\ \frac{14x}{14} &= \frac{29}{14} \\ x &= \frac{29}{14} \end{aligned}$$

$\left\{\frac{29}{14}\right\}$

8. $$\frac{8(2-a)}{3} = 2a$$

$$\frac{3}{1} \cdot \frac{8(2-a)}{3} = 3(2a)$$

$$8(2-a) = 6a$$
$$16 - 8a = 6a$$
$$16 - 8a + 8a = 6a + 8a$$
$$16 = 14a$$
$$\frac{16}{14} = \frac{14a}{14}$$
$$\frac{8}{7} = a$$

$\left\{\frac{8}{7}\right\}$

9. $$\frac{-(y+1)}{5} = 2y + 3$$

$$\frac{5}{1} \cdot \frac{-(y+1)}{5} = 5(2y+3)$$

$$-(y+1) = 10y + 15$$
$$-y - 1 = 10y + 15$$
$$-y - 1 + y = 10y + 15 + y$$
$$-1 = 11y + 15$$
$$-1 - 15 = 11y + 15 - 15$$
$$-16 = 11y$$
$$\frac{-16}{11} = \frac{11y}{11}$$
$$-\frac{16}{11} = y$$

$\left\{-\frac{16}{11}\right\}$

10. $$6x - 3 = 3(5x - 4) + 9(1 - x)$$
$$6x - 3 = 15x - 12 + 9 - 9x$$
$$6x - 3 = 6x - 3$$
$$6x - 3 - 6x = 6x - 3 - 6x$$
$$-3 = -3$$

True

$\{x \mid x \text{ is a real number}\}$

11. $$\frac{x-2}{3} = \frac{x+2}{7}$$

$$\frac{21}{1} \cdot \frac{x-2}{3} = \frac{21}{1} \cdot \frac{x+2}{7}$$

$$7(x-2) = 3(x+2)$$
$$7x - 14 = 3x + 6$$
$$7x - 14 - 3x = 3x + 6 - 3x$$
$$4x - 14 = 6$$
$$4x - 14 + 14 = 6 + 14$$
$$4x = 20$$
$$\frac{4x}{4} = \frac{20}{4}$$
$$x = 5$$

$\{5\}$

12. $$13 - 6(1 - b) = 3b + 26$$
$$13 - 6 + 6b = 3b + 26$$
$$7 + 6b = 3b + 26$$
$$7 + 6b - 3b = 3b + 2b - 3b$$
$$7 + 3b = 26$$
$$7 + 3b - 7 = 26 - 7$$
$$3b = 19$$
$$\frac{3b}{3} = \frac{19}{3}$$
$$b = \frac{19}{3}$$

$\left\{\frac{19}{3}\right\}$

13. number: x

$$2x + 8 = x + 11$$
$$2x + 8 - x = x + 11 - x$$
$$x + 8 = 11$$
$$x + 8 - 8 = 11 - 8$$
$$x = 3$$

The number is 3.

14. number: x

$$\frac{1}{4}x = \frac{3}{8}$$

$$4 \cdot \frac{1}{4}x = 4 \cdot \frac{3}{8}$$

$$x = \frac{3}{2}$$

The number is $\frac{3}{2}$.

15. number: x

$$x - 10 = 2x$$
$$x - 10 - x = 2x - x$$
$$-10 = x$$

The number is -10.

2.5 SOLUTIONS TO EXERCISES

1. Bill's salary: x
John's salary: $3x$

$$x + 3x = 92000$$
$$4x = 92000$$

$$\frac{4x}{4} = \frac{92000}{4}$$

$$x = 23000$$
$$3x = 3(23000) = 69000$$

Bill's salary is \$23,000
and John's salary is \$69,000.

2. shorter length: x
longer length: $17 - x$

$$17 - x = 5 + 2x$$
$$17 - x + x = 5 + 2x + x$$
$$17 = 5 + 3x$$
$$17 - 5 = 5 + 3x - 5$$
$$12 = 3x$$

$$\frac{12}{3} = \frac{3x}{3}$$

$$4 = x$$
$$17 - x = 17 - 4 = 13$$

The shorter piece is 4 feet
and the longer piece is 13 feet.

3. number: x

$$(3x)(6) = 2x - \frac{5}{2}$$

$$18x = 2x - \frac{5}{2}$$

$$18x - 2x = 2x - \frac{5}{2} - 2x$$

$$16x = -\frac{5}{2}$$

$$\frac{1}{16} \cdot 16x = \frac{1}{16} \cdot \left(-\frac{5}{2}\right)$$

$$x = -\frac{5}{32}$$

The number is $-\frac{5}{32}$.

4. number: x

$$2(x + 7) = 26 + 3x$$
$$2x + 14 = 26 + 3x$$
$$2x + 14 - 2x = 26 + 3x - 2x$$
$$14 = 26 + x$$
$$14 - 26 = 26 + x - 26$$
$$-12 = x$$

The number is -12.

5. number of miles: x

$$(26.95)(3) + 0.31x = 150$$
$$80.85 + 0.31x = 150$$
$$80.85 + 0.31x - 80.85 = 150 - 80.85$$
$$0.31x = 69.15$$

$$\frac{0.31x}{0.31} = \frac{69.15}{0.31}$$

$$x = 223.1$$

It can be driven 223 miles.

6. smaller angle: x
larger angle: $8x$

$$x + 8x = 180$$
$$9x = 180$$
$$\frac{9x}{9} = \frac{180}{9}$$
$$x = 20$$
$$8x = 8(20) = 160$$

The angle measures are 20° and 160°.

7. 1st integer: x
2nd integer: $x + 1$
3rd integer: $x + 2$

$$x + (x + 1) + (x + 2) = 3(x + 1)$$
$$3x + 3 = 3x + 3$$
$$3x + 3 - 3x = 3x + 3 - 3x$$
$$3 = 3$$
True

This is true for any set of three consecutive integers.

8. 1st integer: x
2nd integer: $x + 2$

$$4(x + 2) = 72 + 2x$$
$$4x + 8 = 72 + 2x$$
$$4x + 8 - 2x = 72 + 2x - 2x$$
$$2x + 8 = 72$$
$$2x + 8 - 8 = 72 - 8$$
$$2x = 64$$
$$\frac{2x}{2} = \frac{64}{2}$$
$$x = 32$$
$$x + 2 = 32 + 2 = 34$$

The integers are 32 and 34.

9. number: x

$$4(x - 11) = 2(x + 12)$$
$$4x - 44 = 2x + 24$$
$$4x - 44 - 2x = 2x + 24 - 2x$$
$$2x - 44 = 24$$
$$2x - 44 + 44 = 24 + 44$$
$$2x = 68$$
$$\frac{2x}{2} = \frac{68}{2}$$
$$x = 34$$

The number is 34.

10. 1st integer: x
2nd integer: $x + 2$
3rd integer: $x + 4$

$$x + (x + 2) + (x + 4) = 273$$
$$3x + 6 = 273$$
$$3x + 6 - 6 = 273 - 6$$
$$3x = 267$$
$$\frac{3x}{3} = \frac{267}{3}$$
$$x = 89$$
$$x + 2 = 89 + 2 = 91$$
$$x + 4 = 89 + 4 = 93$$

The integers are 89, 91 and 93.

11. 1st angle: x
2nd angle: x
3rd angle: $2x$

$$x + x + 2x = 180$$
$$4x = 180$$
$$\frac{4x}{4} = \frac{180}{4}$$
$$x = 45$$
$$2x = 2(45) = 90$$

The angles are 45°, 45° and 90°.

12. daughter's portion: x
husband's portion: $3x$

$$x + 3x = 36000$$
$$4x = 36000$$
$$\frac{4x}{4} = \frac{36000}{4}$$
$$x = 9000$$
$$3x = 3(9000) = 27000$$

The husband will receive $27,000 and the daughter will receive $9000.

13. 1st angle: x
2nd angle: $3x - 6$

$$x + (3x - 6) = 90$$
$$4x - 6 = 90$$
$$4x - 6 + 6 = 90 + 6$$
$$4x = 96$$
$$\frac{4x}{4} = \frac{96}{4}$$
$$x = 24$$
$$3x - 6 = 3(24) - 6 = 66$$

The angles are 24° and 66°.

14. Smith's votes: x
Barne's votes: $x + 260$

$$x + (x + 260) = 4010$$
$$2x + 260 = 4010$$
$$2x + 260 - 260 = 4010 - 260$$
$$2x = 3750$$
$$\frac{2x}{2} = \frac{3750}{2}$$
$$x = 1875$$
$$x + 260 = 1875 + 260 = 2135$$

Smith received 1875 votes and Barnes received 2135 votes.

15. amount in certificate of deposit: x
amount in mutual fund: $3x$

$$x + 3x = 8100$$
$$4x = 8100$$
$$\frac{4x}{4} = \frac{8100}{4}$$
$$x = 2025$$
$$3x = 3(2025) = 6075$$

Maria invested \$2025 in a certificate of deposit and \$6075 in a mutual fund.

2.6 SOLUTIONS TO EXERCISES

1.
$$A = bh$$
$$65 = 5h$$
$$\frac{65}{5} = \frac{5h}{5}$$
$$13 = h$$

2.
$$D = rt$$
$$220 = 55t$$
$$\frac{220}{55} = \frac{55t}{55}$$
$$4 = t$$

3.
$$I = PRT$$
$$1008 = P(0.035)(8)$$
$$1008 = 0.28P$$
$$\frac{1008}{0.28} = \frac{0.28P}{0.28}$$
$$3600 = P$$

4.
$$P = a + b + c$$
$$62 = 12 + b + 21$$
$$62 = b + 33$$
$$62 - 33 = b + 33 - 33$$
$$29 = b$$

5.
$$V = \frac{4}{3}\pi r^3$$
$$V = \frac{4}{3}(3.14)(4)^3$$
$$V = \frac{4}{3}(3.14)(64)$$
$$V = 267.95$$

6.
$$2x - y = 8$$
$$2x - y - 2x = 8 - 2x$$
$$-y = 8 - 2x$$
$$(-1)(-y) = (-1)(8 - 2x)$$
$$y = -8 + 2x$$

7. $A = \frac{1}{2}bh$

$2A = 2\left(\frac{1}{2}bh\right)$

$2A = bh$

$\frac{2A}{b} = \frac{bh}{b}$

$\frac{2A}{b} = h$

8.
$$P = a + b + c$$
$$P - a = a + b + c - a$$
$$P - a = b + c$$
$$P - a - c = b + c - c$$
$$P - a - c = b$$

9.
$$3x + 5y = 15$$
$$3x + 5y - 5y = 15 - 5y$$
$$3x = 15 - 5y$$
$$\frac{3x}{3} = \frac{15 - 5y}{3}$$
$$x = \frac{15}{3} - \frac{5}{3}y$$
$$x = 5 - \frac{5}{3}y$$

10. $C = 2\pi r$

$\frac{C}{2\pi} = \frac{2\pi r}{2\pi}$

$\frac{C}{2\pi} = r$

11.
$$s = 4lw + 2wh$$
$$s - 2wh = 4lw + 2wh - 2wh$$
$$s - 2wh = 4lw$$
$$\frac{s - 2wh}{4w} = \frac{4lw}{4w}$$
$$\frac{s - 2wh}{4w} = l$$

12. $P = 2W + 2L$
$P = 2(7.6) + 2(18)$
$P = 51.2$
51.2 meters of fencing are required.

13. $F = \frac{9}{5}C + 32$

$F = \frac{9}{5}(-15) + 32$
$F = -27 + 32$
$F = 5$

5° F

14.
$$F = \frac{9}{5}C + 32$$
$$41 = \frac{9}{5}C + 32$$
$$41 - 32 = \frac{9}{5}C + 32 - 32$$
$$9 = \frac{9}{5}C$$
$$\frac{5}{9} \cdot 9 = \frac{5}{9} \cdot \frac{9}{5}C$$
$$5 = C$$

5° C

15. $C = 2\pi r$
$C = 2(3.14)(25)$
$C = 157$

157 feet

16.
$$D = rt$$
$$550 = 50t$$
$$\frac{550}{50} = \frac{50t}{50}$$
$$11 = t$$

The trip took 11 hours.

2.7 SOLUTIONS TO EXERCISES

1. $235\% = 2_{\wedge}35.\%$

$= 2.35$

2. $0.32\% = 0_{\wedge}00.32\%$
$= 0.0032$

3. $1.7\% = 0_{\wedge}01.7\%$
$= 0.017$

4. $31.7\% = 0_{\wedge}31.7\%$
$= 0.317$

5. $0.61_{\wedge} = 61\%$

6. $7 = 7.00_{\wedge}$
$= 700\%$

7. $\frac{7}{8} = 0.87_{\wedge}5$
$= 87.5\%$

8. $\frac{1}{5} = 0.20_{\wedge}$
$= 20\%$

9. number: x

$x = 0.15 \cdot 670$
$x = 100.5$
The number is 100.5.

10. unknown percent: x

$13.12 = x \cdot 41$

$\frac{13.12}{41} = \frac{41x}{41}$

$0.32 = x$
$32\% = x$
The percent is 32%.

11. number: x

$117.6 = 0.28 \cdot x$

$\frac{117.6}{0.28} = \frac{0.28x}{0.28}$

$420 = x$
The number is 420.

12. $x = (1.20)(95)$
$x = 114$

13. number: x

$25.2 = 0.42 \cdot x$

$\frac{25.2}{0.42} = \frac{0.42x}{0.42}$

$60 = x$
The number is 60.

14. unknown percent: x

$306 = x \cdot 450$
$\frac{306}{450} = \frac{450x}{450}$

$0.68 = x$
$68\% = x$
The percent is 68%.

15. unknown percent: x

$483 = x \cdot 210$

$\frac{483}{210} = \frac{210x}{210}$

$2.3 = x$
$230\% = x$
The percent is 230%.

16. unknown percent: x

$5 = x \cdot 2505$

$\frac{5}{2505} = \frac{2505x}{2505}$

$0.002 = x$
$0.2\% = x$
The percent is 0.2%.

17. new price: x

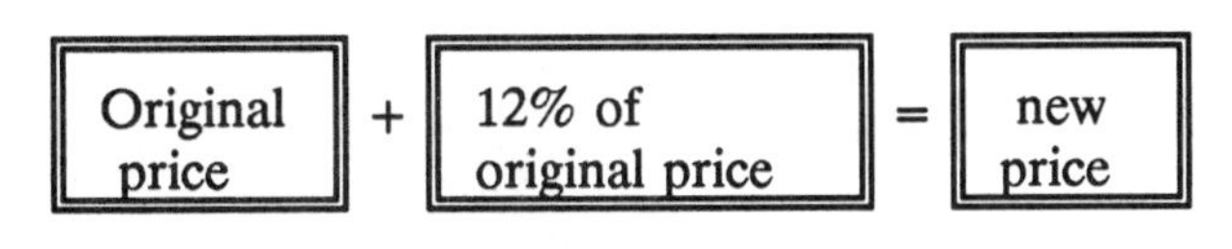

$$1.75 + 0.12 \cdot 1.75 = x$$
$$1.75 + 0.21 = x$$
$$1.96 = x$$

The new price is \$1.96.

18. sale price: x

Original price - 30% of original price = sale price

$$32 - 0.30 \cdot 32 = x$$
$$32 - 9.6 = x$$
$$22.4 = x$$

The sale price was \$22.40.

19. price decrease = 1.17 - 1.07 = 0.10

0.10 is what percent of 1.17?

$$0.10 = x \cdot 1.17$$
$$\frac{0.10}{1.17} = \frac{1.17x}{1.17}$$
$$0.085 = x$$
$$8.5\% = x$$

It is an 8.5% decrease.

20. amount of increase : 1.64 - 1.28 = 0.36

0.36 is what percent of 1.28?

$$0.36 = x \cdot 1.28$$
$$\frac{0.36}{1.28} = \frac{1.28x}{1.28}$$
$$0.281 = x$$
$$28.1\% = x$$

It is a 28.1% increase.

2.8 SOLUTIONS TO EXERCISES

1. width: x
length: $3x - 60$

$$P = 2W + 2L$$
$$712 = 2(x) + 2(3x - 60)$$
$$712 = 2x + 6x - 120$$
$$712 = 8x - 120$$

$$712 + 120 = 8x - 120 + 120$$
$$832 = 8x$$
$$\frac{832}{8} = \frac{8x}{8}$$
$$104 = x$$
$$3x - 60 = 3(104) - 60 = 252$$

The width is 104 feet and the length is 252 feet.

2. shortest length: x
2nd side: $4x$
3rd side: $x + 25$

$$P = a + b + c$$
$$121 = x + (4x) + (x + 25)$$
$$121 = 6x + 25$$
$$121 - 25 = 6x + 25 - 25$$
$$96 = 6x$$
$$\frac{96}{6} = \frac{6x}{6}$$
$$16 = x$$
$$4x = 4(16) = 64$$
$$x + 25 = 16 + 25 = 41$$

The side lengths are 16 feet, 64 feet and 41 feet.

3. longer sides: x
shorter side: $x - 5$

$$P = a + b + c$$
$$94 = x + x + (x - 5)$$
$$94 = 3x - 5$$
$$94 + 5 = 3x - 5 + 5$$
$$99 = 3x$$
$$\frac{99}{3} = \frac{3x}{3}$$
$$33 = x$$
$$x - 5 = 33 - 5 = 28$$

The shorter side is 28 inches.

4.

	rate	· time	= distance
car	65	t	$65t$
truck	52	$t + 3/2$	$52(t + 3/2)$

car's distance = truck's distance

$$65t = 52\left(t + \frac{3}{2}\right)$$

$$\begin{aligned} 65t &= 52t + 78 \\ 65t - 52t &= 52t + 78 - 52t \\ 13t &= 78 \end{aligned}$$

$$\frac{13t}{13} = \frac{78}{13}$$

$$t = 6$$

$D = rt = 65(6) = 390$

They are 390 miles from the starting point.

5.

	rate	· time	= distance
going	60	t	$60t$
returning	50	$8.8 - t$	$50(8.8 - t)$

distance going = distance returning

$$\begin{aligned} 60t &= 50(8.8 - t) \\ 60t &= 440 - 50t \\ 60t + 50t &= 440 - 50t + 50t \\ 110t &= 440 \end{aligned}$$

$$\frac{110t}{110} = \frac{440}{110}$$

$$t = 4$$

$D = rt = (60)(4) = 240$

The distance to New York is 240 miles.

6.

	No. of gal ·	Acid strength =	amount of Acid
pure acid	x	100%	$1x$
35% solution	5	35%	$0.35(5)$
45% solution	$x + 5$	45%	$0.45(x + 5)$

$$\begin{aligned} x + 0.35(5) &= 0.45(x + 5) \\ x + 1.75 &= 0.45x + 2.25 \\ x + 1.75 - 0.45x &= 0.45x + 2.25 - 0.45x \\ 0.55x + 1.75 &= 2.25 \\ 0.55x + 1.75 - 1.75 &= 2.25 - 1.75 \\ 0.55x &= 0.5 \end{aligned}$$

$$\frac{0.55x}{0.55} = \frac{0.5}{0.55}$$

$$x = 0.9$$

0.9 gal. of pure acid should be used.

7.

	No. of lbs ·	price per lb =	price
\$8/lb	x	8	$8x$
\$5/lb	10	5	50
\$6/lb	$x + 10$	6	$6(x + 10)$

$$\begin{aligned} 8x + 50 &= 6(x + 10) \\ 8x + 50 &= 6x + 60 \\ 8x + 50 - 6x &= 6x + 60 - 6x \\ 2x + 50 &= 60 \\ 2x + 50 - 50 &= 60 - 50 \\ 2x &= 10 \end{aligned}$$

$$\frac{2x}{2} = \frac{10}{2}$$

$$x = 5$$

5 lbs of \$8/lb coffee should be added.

8.

	amount invested ·	rate =	interest
7% account	x	0.07	$0.07x$
8.5% account	$x + 400$	0.085	$0.085(x + 400)$

$$\begin{aligned} 0.07x + 0.085(x + 400) &= 127 \\ 0.07x + 0.085x + 34 &= 127 \\ 0.155x + 34 &= 127 \\ 0.155x + 34 - 34 &= 127 - 34 \\ 0.155x &= 93 \\ \frac{0.155x}{0.155} &= \frac{93}{0.155} \\ x &= 600 \\ x + 400 &= 600 + 400 = 1000 \end{aligned}$$

She invested \$600 at 7% and \$1000 at 8.5%.

9.

	amount invested ·	rate =	interest
4% account	x	0.04	$0.04x$
6% account	$68000 - x$	0.06	$0.06(68000 - x)$

$$\begin{aligned} 0.04x &= 0.06(68000 - x) \\ 0.04x &= 4080 - 0.06x \\ 0.04x + 0.06x &= 4080 - 0.06x + 0.06x \\ 0.1x &= 4080 \\ \frac{0.1x}{0.1} &= \frac{4080}{0.1} \\ x &= 40800 \\ 68000 - x &= 68000 - 40800 = 27200 \end{aligned}$$

Invest \$40,800 at 4% and \$27,200 at 6%.

10.

	amt of alloy ·	percent of copper =	amt of copper
25% copper	x	0.25	$0.25x$
55% copper	300	0.55	$0.55(300)$
35% copper	$x + 300$	0.35	$0.35(x + 300)$

$$\begin{aligned} 0.25x + 0.55(300) &= 0.35(x + 300) \\ 0.25x + 165 &= 0.35x + 105 \\ 0.25x + 165 - 0.25x &= 0.35x + 105 - 0.25x \\ 165 &= 0.1x + 105 \\ 165 - 105 &= 0.1x + 105 - 105 \\ 60 &= 0.1x \end{aligned}$$

$$\frac{60}{0.1} = \frac{0.1x}{0.1}$$

$$600 = x$$

600 ounces of 25% copper alloy should be added.

11.

	amt of sol ·	percent antifreeze =	amt of antifreeze
water	x	0	$0x$
80% sol	20	0.80	$0.80(20)$
70% sol	$x + 20$	0.70	$0.70(x + 20)$

$$\begin{aligned} 0x + 0.80(20) &= 0.70(x + 20) \\ 16 &= 0.70x + 14 \\ 16 - 14 &= 0.70x + 14 - 14 \\ 2 &= 0.70x \end{aligned}$$

$$\frac{2}{0.70} = \frac{0.70x}{0.70}$$

$$2.86 = x$$

2.86 gallons of water should be added.

12.

	No. of tickets ·	price per ticket =	Total cost
students	x	3.75	$3.75x$
adults	$620 - x$	6	$6(620 - x)$
total	620		2775

$$
\begin{aligned}
3.75x + 6(620 - x) &= 2775 \\
3.75x + 3720 - 6x &= 2775 \\
-2.25x + 3720 &= 2775 \\
-2.25x + 3720 - 3720 &= 2775 - 3720 \\
-2.25x &= -945
\end{aligned}
$$

$$\frac{-2.25x}{-2.25} = \frac{-945}{-2.25}$$

$$x = 420$$
$$620 - x = 620 - 420 = 200$$

There were 200 adult tickets sold.

13.

	rate ·	time =	distance
upstream	4	x	$4x$
downstream	10	$6 - x$	$10(6 - x)$

Distance upstream = distance downstream

$$
\begin{aligned}
4x &= 10(6 - x) \\
4x &= 60 - 10x \\
4x + 10x &= 60 - 10x + 10x \\
14x &= 60
\end{aligned}
$$

$$\frac{14x}{14} = \frac{60}{14}$$

$$x = 4.3$$

Distance upstream = 4(4.3) = 17.2 miles
Total up and back = 2(17.2) = 34.4 miles

14.

	rate ·	time =	distance
1st hiker	x	1.5	$1.5x$
2nd hiker	$x + 1.4$	1.5	$1.5(x + 1.4)$

$$
\begin{aligned}
1.5x + 1.5(x + 1.4) &= 8 \\
1.5x + 1.5x + 2.1 &= 8 \\
3x + 2.1 &= 8 \\
3x + 2.1 - 2.1 &= 8 - 2.1 \\
3x &= 5.9
\end{aligned}
$$

$$\frac{3x}{3} = \frac{5.9}{3}$$

$$x = 1.97$$
$$x + 1.4 = 1.97 + 1.4 = 3.37$$

One hiker's rate is 1.97 miles per hour while the other's rate is 3.37 miles per hour.

15.

$$
\begin{aligned}
C &= R \\
3720 + 160x &= 470x \\
3720 + 160x - 160x &= 470x - 160x \\
3720 &= 310x
\end{aligned}
$$

$$\frac{3720}{310} = \frac{310x}{310}$$

$$12 = x$$

The break-even quantity is 12 VCRs.

2.9 SOLUTIONS TO EXERCISES

1. $3x > -12$

$$\frac{3x}{3} > \frac{-12}{3}$$

$$x > -4$$

$$\{x \mid x > -4\}$$

-5 -4 -3 -2 -1 0 1 2

2.

$$
\begin{aligned}
x - 1 &\le 5 \\
x - 1 + 1 &\le 5 + 1 \\
x &\le 6
\end{aligned}
$$

$$\{x \mid x \le 6\}$$

-1 0 1 2 3 4 5 6 7 8

3. $-4x \geq -16$

$$\frac{-4x}{-4} \leq \frac{-16}{-4}$$

$$x \leq 4$$

$\{x \mid x \leq 4\}$

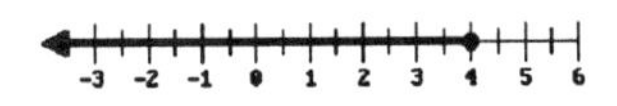

4. $-2x < 8$

$$\frac{-2x}{-2} > \frac{8}{-2}$$

$$x > -4$$

$\{x \mid x > -4\}$

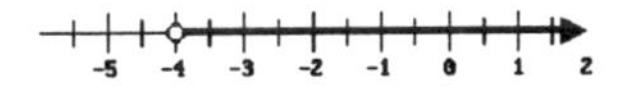

5.
$$\begin{aligned} 4x - 7 &\leq 3x - 5 \\ 4x - 7 - 3x &\leq 3x - 5 - 3x \\ x - 7 &\leq -5 \\ x - 7 + 7 &\leq -5 + 7 \\ x &\leq 2 \end{aligned}$$

$\{x \mid x \leq 2\}$

6.
$$\begin{aligned} 9 - 3x &> 11 - 4x \\ 9 - 3x + 4x &> 11 - 4x + 4x \\ 9 + x &> 11 \\ 9 + x - 9 &> 11 - 9 \\ x &> 2 \end{aligned}$$

$\{x \mid x > 2\}$

7.
$$\begin{aligned} -10x + 3 &\geq 4(12 - x) \\ -10x + 3 &\geq 48 - 4x \\ -10x + 3 + 10x &\geq 48 - 4x + 10x \\ 3 &\geq 48 + 6x \\ 3 - 48 &\geq 48 + 6x - 48 \\ -45 &\geq 6x \end{aligned}$$

$$\frac{-45}{6} \geq \frac{6x}{6}$$

$$-\frac{15}{2} \geq x$$

$$\left\{x \mid x \leq -\frac{15}{2}\right\}$$

8.
$$\begin{aligned} -(2 - 3x) &< 5(x - 4) \\ -2 + 3x &< 5x - 20 \\ -2 + 3x - 3x &< 5x - 20 - 3x \\ -2 &< 2x - 20 \\ -2 + 20 &< 2x - 20 + 20 \\ 18 &< 2x \end{aligned}$$

$$\frac{18}{2} < \frac{2x}{2}$$

$$9 < x$$

$\{x \mid x > 9\}$

9. $7(x + 1) - 6 > -4(x - 3) + 22$
$7x + 7 - 6 > -4x + 12 + 22$
$7x + 1 > -4x + 34$
$7x + 1 + 4x > -4x + 34 + 4x$
$11x + 1 > 34$
$11x + 1 - 1 > 34 - 1$
$11x > 33$

$$\frac{11x}{11} > \frac{33}{11}$$

$$x > 3$$

$\{x \mid x > 3\}$

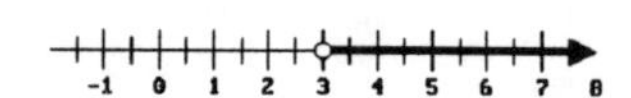

10. $-3(x + 4) - 2x \le -(5x + 2) + 6x$
$-3x - 12 - 2x \le -5x - 2 + 6x$
$-5x - 12 \le x - 2$
$-5x - 12 - x \le x - 2 - x$
$-6x - 12 \le -2$
$-6x - 12 + 12 \le -2 + 12$
$-6x \le 10$

$$\frac{-6x}{-6} \ge \frac{10}{-6}$$

$$x \ge -\frac{5}{3}$$

$\left\{x \mid x \ge -\frac{5}{3}\right\}$

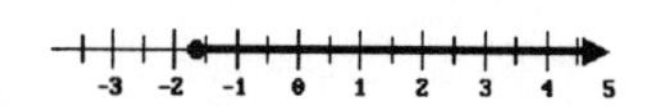

11. $-(x - 8) < 8$
$-x + 8 < 8$
$-x + 8 - 8 < 8 - 8$
$-x < 0$

$$\frac{-x}{-1} > \frac{0}{-1}$$

$$x > 0$$

$\{x \mid x > 0\}$

12. $-9 < 3(x - 1) < 12$
$-9 < 3x - 3 < 12$
$-9 + 3 < 3x - 3 + 3 < 12 + 3$
$-6 < 3x < 15$

$$\frac{-6}{3} < \frac{3x}{3} < \frac{15}{3}$$

$$-2 < x < 5$$

$\{x \mid -2 < x < 5\}$

13. $-6 \le 4(x + 3) < 10$
$-6 \le 4x + 12 < 10$
$-6 - 12 \le 4x + 12 - 12 < 10 - 12$
$-18 \le 4x < -2$

$$\frac{-18}{4} \le \frac{4x}{4} < \frac{-2}{4}$$

$$-\frac{9}{2} \le x < -\frac{1}{2}$$

$\left\{x \mid -\frac{9}{2} \le x < -\frac{1}{2}\right\}$

14. number: x

$9 + 3x > -12$
$9 + 3x - 9 > -12 - 9$
$3x > -21$

$$\frac{3x}{3} > \frac{-21}{3}$$

$$x > -7$$

All numbers greater than -7 make this statement true.

15. length: x

$$
\begin{aligned}
P &\le 150 \\
2W + 2L &\le 150 \\
2(30) + 2x &\le 150 \\
60 + 2x - 60 &\le 150 - 60 \\
2x &\le 90
\end{aligned}
$$

$$\frac{2x}{2} \le \frac{90}{2}$$

$$x \le 45$$

The maximum length is 45 centimeters.

16. number: x

$$
\begin{aligned}
-11 &< 4x - 3 < 17 \\
-11 + 3 &< 4x - 3 + 3 < 17 + 3 \\
-8 &< 4x < 20
\end{aligned}
$$

$$\frac{-8}{4} < \frac{4x}{4} < \frac{20}{4}$$

$$-2 < x < 5$$

All numbers between -2 and 5.

CHAPTER 2 PRACTICE TEST SOLUTIONS

1. $3y - 7 - y + 5$
$= (3 - 1)y + (-7 + 5)$
$= 2y - 2$

2. $3.1x + 4.2 + 5.3x - 6.9$
$= (3.1 + 5.3)x + (4.2 - 6.9)$
$= 8.4x - 2.7$

3. $7(x - 1) - 2(3x - 4)$
$= 7x - 7 - 6x + 8$
$= (7 - 6)x + (-7 + 8)$
$= x + 1$

4. $-8(y + 3) + 4(2 - 7y)$
$= -8y - 24 + 8 - 28y$
$= (-8 - 28)y + (-24 + 8)$
$= -36y - 16$

5. $$-\frac{5}{4}x = 5$$
$$-\frac{4}{5}\left(-\frac{5}{4}x\right) = -\frac{4}{5}(5)$$
$$x = -4$$

6. $$\begin{aligned} 5(n - 4) &= -(6 - 3n) \\ 5n - 20 &= -6 + 3n \\ 5n - 20 - 3n &= -6 + 3n - 3n \\ 2n - 20 &= -6 \\ 2n - 20 + 20 &= -6 + 20 \\ 2n &= 14 \\ \frac{2n}{2} &= \frac{14}{2} \\ n &= 7 \end{aligned}$$

7. $$\begin{aligned} 6y - 8 + y &= -(4y + y) \\ 7y - 8 &= -(5y) \\ 7y - 8 &= -5y \\ 7y - 8 - 7y &= -5y - 7y \\ -8 &= -12y \\ \frac{-8}{-12} &= \frac{-12y}{-12} \\ \frac{2}{3} &= y \end{aligned}$$

8. $$\begin{aligned} 5z + 2 - z &= 2 + 3z \\ 2 + 4z &= 2 + 3z \\ 2 + 4z - 3z &= 2 + 3z - 3z \\ 2 + z &= 2 \\ 2 + z - 2 &= 2 - 2 \\ z &= 0 \end{aligned}$$

9. $$\begin{aligned} \frac{3(x + 5)}{2} &= x - 7 \\ 2 \cdot \frac{3(x + 5)}{2} &= 2(x - 7) \\ 3(x + 5) &= 2(x - 7) \\ 3x + 15 &= 2x - 14 \\ 3x + 15 - 2x &= 2x - 14 - 2x \\ x + 15 &= -14 \\ x + 15 - 15 &= -14 - 15 \\ x &= -29 \end{aligned}$$

10. $$\begin{aligned} \frac{9(y - 2)}{4} &= 3y + 1 \\ 4 \cdot \frac{9(y - 2)}{4} &= 4(3y + 1) \\ 9(y - 2) &= 4(3y + 1) \\ 9y - 18 &= 12y + 4 \\ 9y - 18 - 12y &= 12y + 4 - 12y \\ -3y - 18 &= 4 \\ -3y - 18 + 18 &= 4 + 18 \\ -3y &= 22 \\ \frac{-3y}{-3} &= \frac{22}{-3} \\ y &= -\frac{22}{3} \end{aligned}$$

11. $$\begin{aligned} \frac{1}{3} - x + \frac{5}{3} &= x - 7 \\ 3\left(\frac{1}{3} - x + \frac{5}{3}\right) &= 3(x - 7) \\ 3\left(\frac{1}{3}\right) - 3(x) + 3\left(\frac{5}{3}\right) &= 3x - 21 \end{aligned}$$

$$1 - 3x + 5 = 3x - 21$$
$$-3x + 6 = 3x - 21$$
$$-3x + 6 + 3x = 3x - 21 + 3x$$
$$6 = 6x - 21$$
$$6 + 21 = 6x - 21 + 21$$
$$27 = 6x$$
$$\frac{27}{6} = \frac{6x}{6}$$
$$\frac{9}{2} = x$$

12.
$$\frac{1}{4}(y + 4) = 5y$$
$$4 \cdot \frac{1}{4}(y + 4) = 4(5y)$$
$$y + 4 = 20y$$
$$y + 4 - y = 20y - y$$
$$4 = 19y$$
$$\frac{4}{19} = \frac{19y}{19}$$
$$\frac{4}{19} = y$$

13.
$$-0.2(x - 3) + x = 0.3(5 - x)$$
$$10[-0.2(x - 3) + x] = 10[0.3(5 - x)]$$
$$10[-0.2(x - 3)] + 10x = 3(5 - x)$$
$$-2(x - 3) + 10x = 15 - 3x$$
$$-2x + 6 + 10x = 15 - 3x$$
$$8x + 6 = 15 - 3x$$
$$8x + 6 + 3x = 15 - 3x + 3x$$
$$11x + 6 = 15$$
$$11x + 6 - 6 = 15 - 6$$
$$11x = 9$$
$$\frac{11x}{11} = \frac{9}{11}$$
$$x = \frac{9}{11}$$

14.
$$-6(a + 2) - 5a = -4(3a - 2)$$
$$-6a - 12 - 5a = -12a + 8$$
$$-11a - 12 = -12a + 8$$
$$-11a - 12 + 12a = -12a + 8 + 12a$$
$$a - 12 = 8$$
$$a - 12 + 12 = 8 + 12$$
$$a = 20$$

15. number: x
$$x + \frac{1}{3}(x) = 32$$
$$3\left[x + \frac{1}{3}x\right] = 3(32)$$
$$3x + 3\left(\frac{1}{3}x\right) = 96$$
$$3x + x = 96$$
$$4x = 96$$
$$\frac{4x}{4} = \frac{96}{4}$$
$$x = 24$$
The number is 24.

16. number of gallons: x
$$x = \frac{2(25)(40)}{250}$$
$$= \frac{2000}{250}$$
$$= 8$$
8 gallons

17. amount at 8%: x
amount at 10%: $3x$
$$0.08x + 0.10(3x) = 2280$$
$$100[0.08x + 0.10(3x)] = 100(2280)$$
$$100(0.08x) + 100(0.10)(3x) = 228000$$
$$8x + 30x = 228000$$
$$38x = 228000$$
$$\frac{38x}{38} = \frac{228000}{38}$$
$$x = 6000$$
$$3x = 3(6000) = 18000$$
\$6000 at 8%; \$18,000 at 10%

18.

	rate	·	time	=	distance
1st train	60		t		$60t$
2nd train	72		t		$72t$

$$60t + 72t = 462$$
$$132t = 462$$
$$\frac{132t}{132} = \frac{462}{132}$$
$$t = 3.5$$

3.5 hours

19.
$$y = mx + b$$
$$-19 = -3x - 4$$
$$-19 + 4 = -3x - 4 + 4$$
$$-15 = -3x$$
$$\frac{-15}{-3} = \frac{-3x}{-3}$$
$$5 = x$$

20.
$$C = \pi d$$
$$\frac{C}{\pi} = \frac{\pi d}{\pi}$$
$$\frac{C}{\pi} = d$$

21.
$$2x - 5y = 12$$
$$2x - 5y - 2x = 12 - 2x$$
$$-5y = 12 - 2x$$
$$\frac{-5y}{-5} = \frac{12 - 2x}{-5}$$
$$y = \frac{12}{-5} + \frac{-2x}{-5}$$
$$y = -\frac{12}{5} + \frac{2}{5}x$$

22.
$$2x - 3 > 5x - 9$$
$$2x - 3 - 2x > 5x - 9 - 2x$$
$$-3 > 3x - 9$$
$$-3 + 9 > 3x - 9 + 9$$
$$6 > 3x$$
$$\frac{6}{3} > \frac{3x}{3}$$
$$2 > x$$

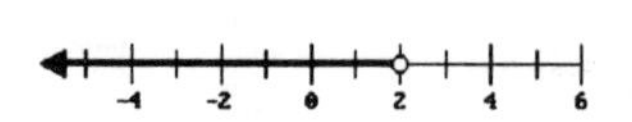

23.
$$x + 4 > 3x - 6$$
$$x + 4 - x > 3x - 6 - x$$
$$4 > 2x - 6$$
$$4 + 6 > 2x - 6 + 6$$
$$10 > 2x$$
$$\frac{10}{2} > \frac{2x}{2}$$
$$5 > x$$

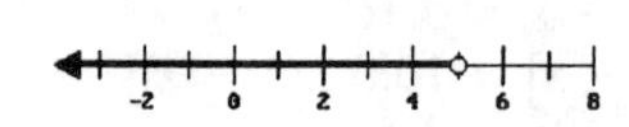

24.
$$-5 < 2x - 1 < 6$$
$$-5 + 1 < 2x - 1 + 1 < 6 + 1$$
$$-4 < 2x < 7$$
$$\frac{-4}{2} < \frac{2x}{2} < \frac{7}{2}$$
$$-2 < x < \frac{7}{2}$$

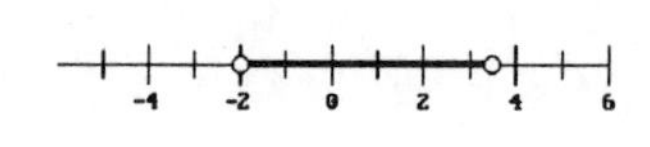

25.

$$0 < 6x - 9 < 15$$
$$0 + 9 < 6x - 9 + 9 < 15 + 9$$
$$9 < 6x < 24$$
$$\frac{9}{6} < \frac{6x}{6} < \frac{24}{6}$$
$$\frac{3}{2} < x < 4$$

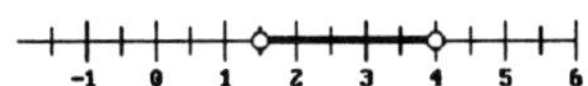

26.

$$\frac{3(4x + 1)}{2} > 5$$
$$2\left[\frac{3(4x + 1)}{2}\right] > 2(5)$$
$$3(4x + 1) > 10$$
$$12x + 3 > 10$$
$$12x + 3 - 3 > 10 - 3$$
$$12x > 7$$
$$\frac{12x}{12} > \frac{7}{12}$$
$$x > \frac{7}{12}$$

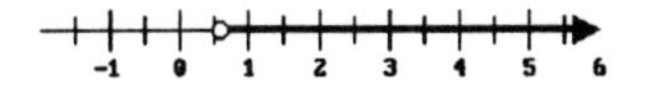

27.

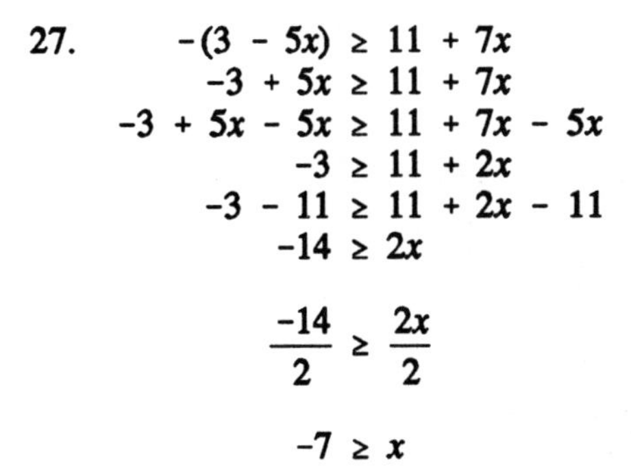

$$-(3 - 5x) \geq 11 + 7x$$
$$-3 + 5x \geq 11 + 7x$$
$$-3 + 5x - 5x \geq 11 + 7x - 5x$$
$$-3 \geq 11 + 2x$$
$$-3 - 11 \geq 11 + 2x - 11$$
$$-14 \geq 2x$$
$$\frac{-14}{2} \geq \frac{2x}{2}$$
$$-7 \geq x$$

3.1 SOLUTIONS TO EXERCISES

1. From the origin, move to the right 5 units and up 1 unit; quadrant I.

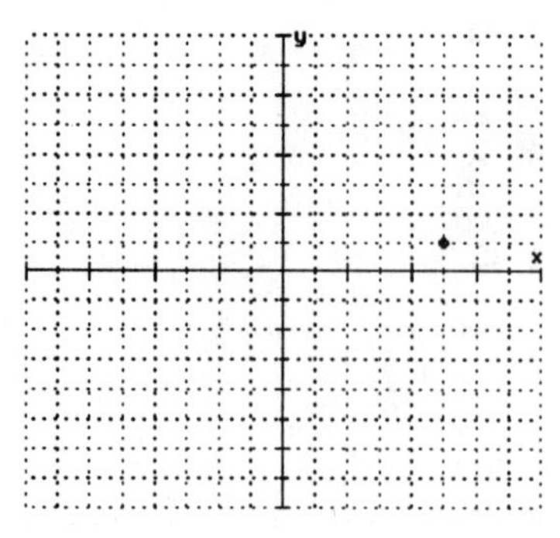

2. From the origin, move to the left 4 units and down 7 units; quadrant III.

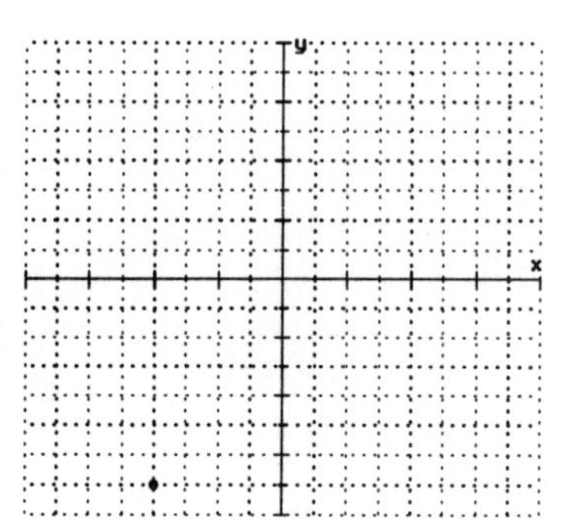

3. From the origin, move down 6 units; y-axis.

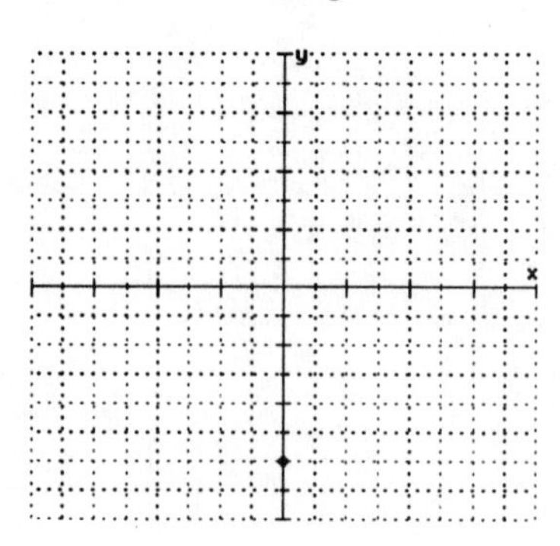

4. From the origin, move to the left $\frac{1}{2}$ unit and up 3 units; quadrant II.

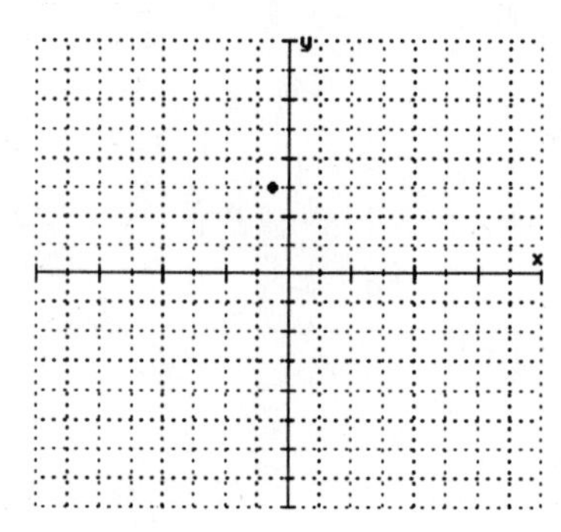

5. From the origin, move to the right 6 units and down 5 units; quadrant IV.

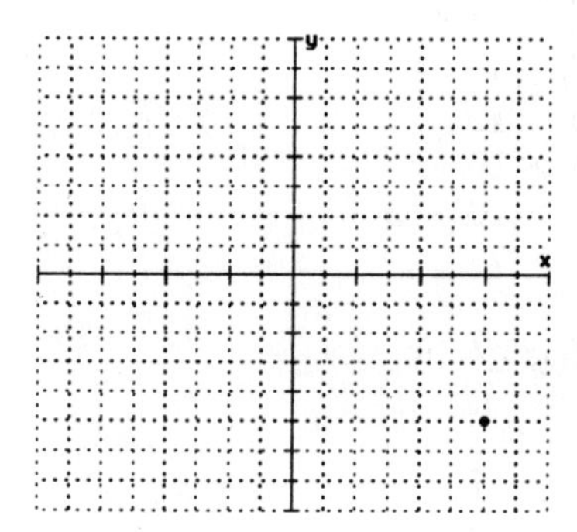

6. From the origin move to the right $2\frac{3}{4}$ units; x-axis.

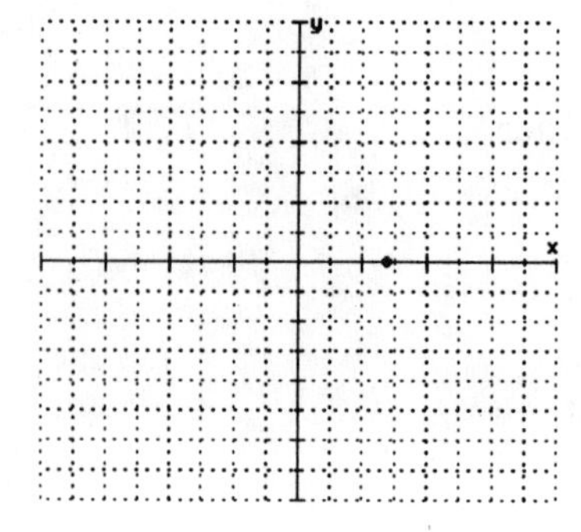

7. $2x - y = 4$
$2(0) - 4 = 4$
$-4 = 4$
False
(0, 4) is not a solution.

$2x - y = 4$
$2(0) - (-4) = 4$
$4 = 4$
True
(0, −4) is a solution.

$2x - y = 4$
$2(1) - (-2) = 4$
$4 = 4$
True
(1, −2) is a solution.

8. $y = 7x$
$1 = 7(-7)$
$1 = -49$
False
(−7, 1) is not a solution.

$y = 7x$
$-7 = 7(1)$
$-7 = 7$
False
(1, −7) is not a solution.

$y = 7x$
$0 = 7(0)$
$0 = 0$
True
(0, 0) is a solution.

9. $x = 3y - 4$
$2 = 3(2) - 4$
$2 = 2$
True
(2, 2) is a solution.

$x = 3y - 4$
$3 = 3(5) - 4$
$3 = 11$
False
(3, 5) is not a solution.

$x = 3y - 4$
$-7 = 3(-1) - 4$
$-7 = -7$
True
(−7, −1) is a solution.

10. $y = -4$
$-4 = -4$
True
(4, −4) is a solution.

$y = -4$
$4 = -4$
False
(−4, 4) is not a solution.

$y = -4$
$-4 = -4$
True
(3, −4) is a solution.

11. $x = \frac{3}{4}y$

$0 = \frac{3}{4}(0)$

$0 = 0$
True
(0, 0) is a solution.

$x = \frac{3}{4}y$

$3 = \frac{3}{4}(4)$

$3 = 3$
True
(3, 4) is a solution.

$x = \frac{3}{4}y$

$1 = \frac{3}{4}\left(\frac{3}{4}\right)$

$1 = \frac{9}{16}$
False

$\left(1, \frac{3}{4}\right)$ is not a solution.

12. $x - 10 = 0$
$0 - 10 = 0$
$-10 = 0$
False
(0, 10) is not a solution.

$x - 10 = 0$
$10 - 10 = 0$
$0 = 0$
True
(10, −10) is a solution.

$x - 10 = 0$
$10 - 10 = 0$
$0 = 0$
True
(10, −100) is a solution.

13. $-x + 2y = 5$
$-0 + 2y = 5$
$2y = 5$
$y = 5/2$
(0, 5/2)

$-x + 2y = 5$
$-x + 2(0) = 5$
$-x = 5$
$x = -5$
(−5, 0)

$-x + 2y = 5$
$-1 + 2y = 5$
$2y = 6$
$y = 3$
(1, 3)

x	y
0	5/2
-5	0
1	3

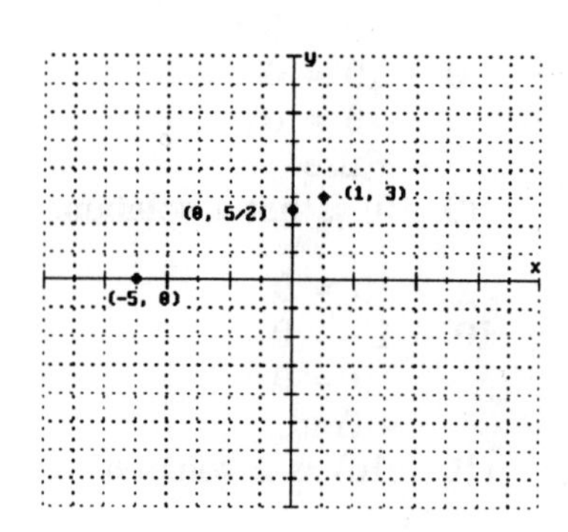

14. $y = -x$
$y = -0$
$y = 0$
(0, 0)

$y = -x$
$-1 = -x$
$1 = x$
(1, -1)

$y = -x$
$y = -2$
(2, -2)

x	y
0	0
1	-1
2	-2

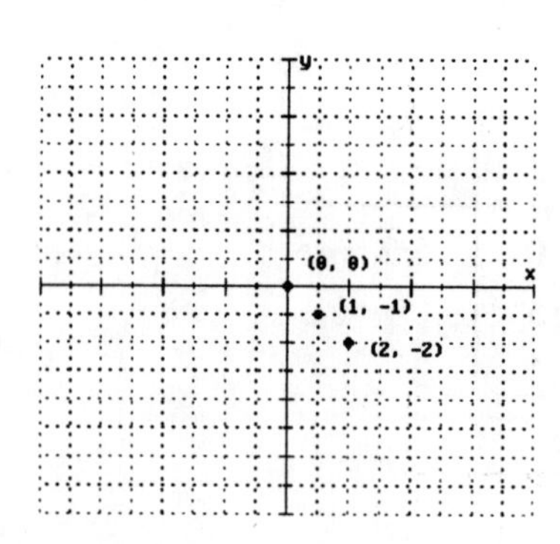

15. $x = -6$
x is -6 for any value of y.

x	y
-6	-2
-6	0.5
-6	-4.75

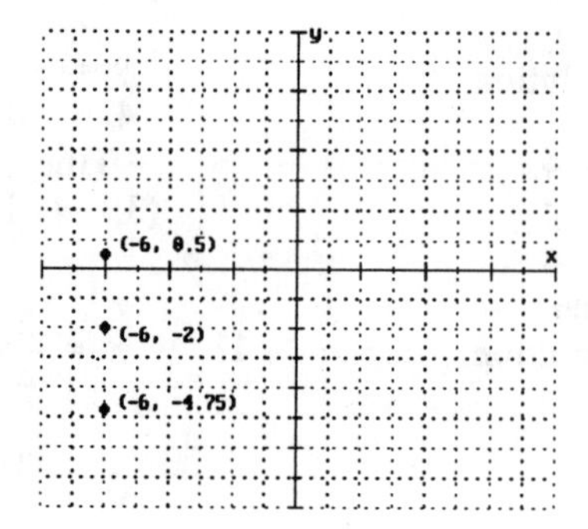

3.2 SOLUTIONS TO EXERCISES

1. $-7x = 3y + 5$
$-7x - 3y = 5$
This is a linear equation in two variables with $A = -7$, $B = -3$ and $C = 5$.

2. This is not a linear equation in two variables since x is squared.

3. This is a linear equation in two variables with $A = 0.6$, $B = -0.8$ and $C = 12$.

4. $y = x - 6$

x	y
-1	-1 - 6 = -7
0	0 - 6 = -6
1	1 - 6 = -5

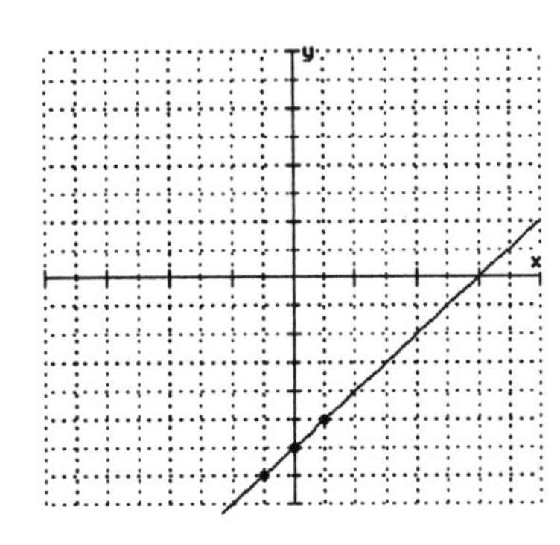

5. $3x + 4y = 12$

Let $x = 0$: $3(0) + 4y = 12$
$4y = 12$
$y = 3$
$(0, 3)$

Let $y = 0$: $3x + 4(0) = 12$
$3x = 12$
$x = 4$
$(4, 0)$

Let $x = 2$: $3(2) + 4y = 12$
$6 + 4y = 12$
$4y = 6$

$y = \frac{3}{2}$

$\left(2, \frac{3}{2}\right)$

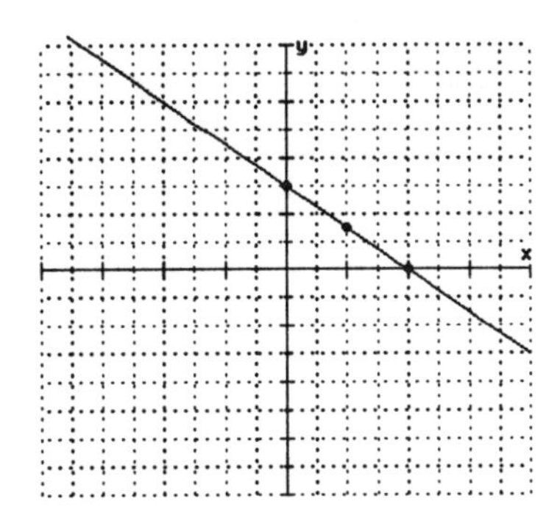

6. $x = \frac{1}{2}y$

x	y
$(1/2)(0) = 0$	0
$(1/2)(-2) = -1$	-2
$(1/2)(2) = 1$	2

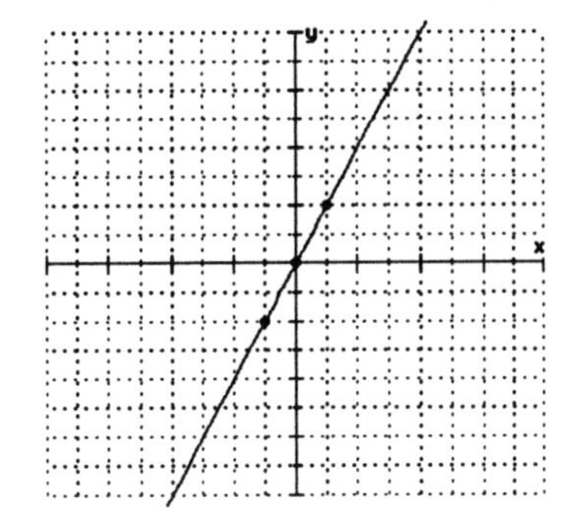

7. Let $x = 0$: $x - y = 3$
$0 - y = 3$
$-y = 3$
$y = -3$
$(0, -3)$

Let $y = 0$: $x - y = 3$
$x - 0 = 3$
$x = 3$
$(3, 0)$

Let $x = 1$: $x - y = 3$
$1 - y = 3$
$-y = 2$
$y = -2$
$(1, -2)$

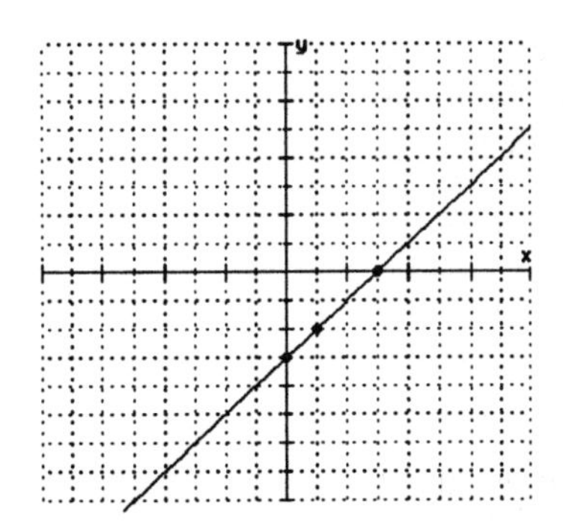

8. Let $x = 0$:
$$\begin{aligned} x + 2y &= 6 \\ 0 + 2y &= 6 \\ 2y &= 6 \\ y &= 3 \end{aligned}$$
(0, 3)

Let $y = 0$:
$$\begin{aligned} x + 2y &= 6 \\ x + 2(0) &= 6 \\ x &= 6 \end{aligned}$$
(6, 0)

Let $x = 2$:
$$\begin{aligned} x + 2y &= 6 \\ 2 + 2y &= 6 \\ 2y &= 4 \\ y &= 2 \end{aligned}$$
(2, 2)

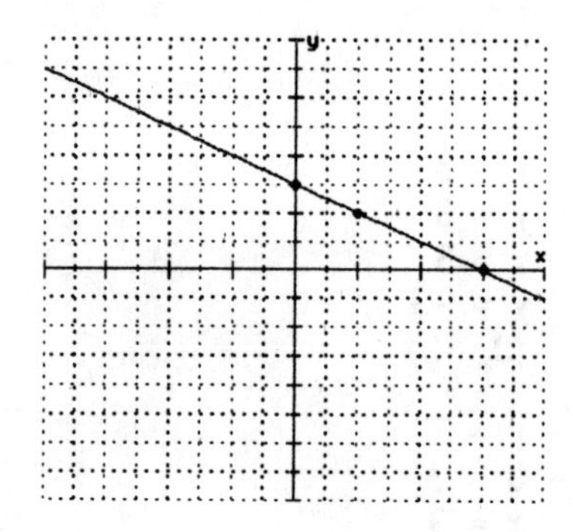

9.

x	$y = -4x + 5$
0	$-4(0) + 5 = 5$
1	$-4(1) + 5 = 1$
2	$-4(2) + 5 = -3$

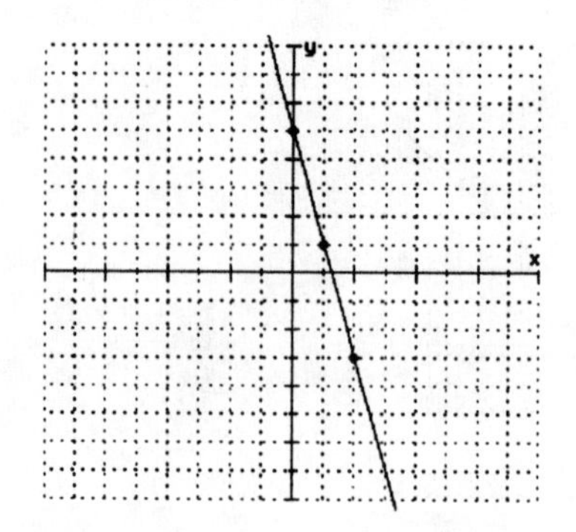

10.

x	$y = -3x - 1$
-1	$-3(-1) - 1 = 2$
0	$-3(0) - 1 = -1$
1	$-3(1) - 1 = -4$

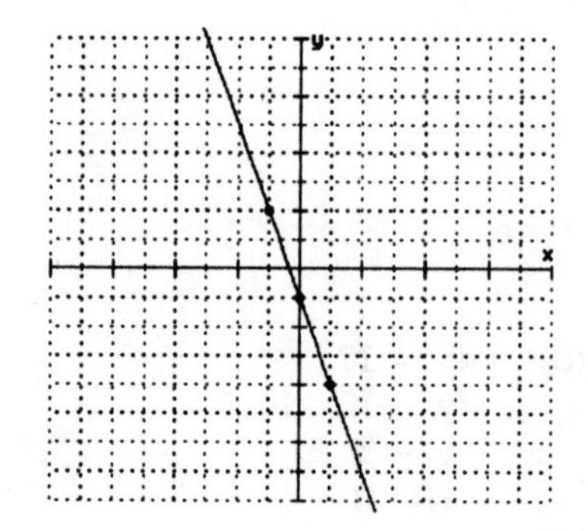

11.

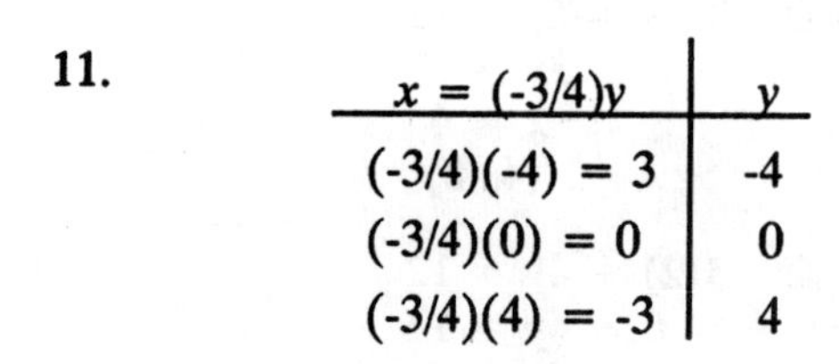

$x = (-3/4)y$	y
$(-3/4)(-4) = 3$	-4
$(-3/4)(0) = 0$	0
$(-3/4)(4) = -3$	4

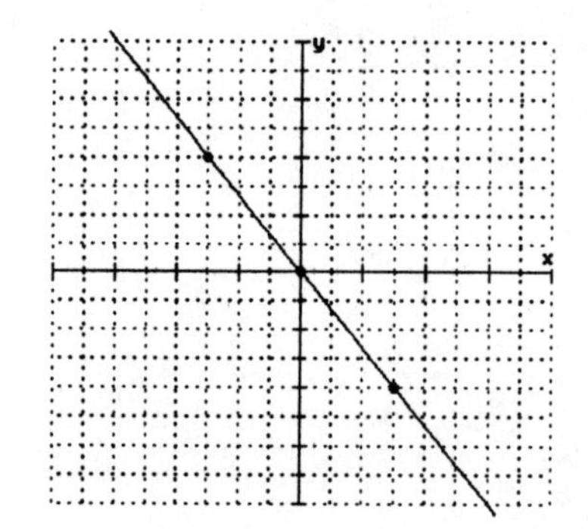

12.

x	$y = (-1/2)x + 2$
-2	$(-1/2)(-2) + 2 = 3$
0	$(-1/2)(0) + 2 = 2$
2	$(-1/2)(2) + 2 = 1$

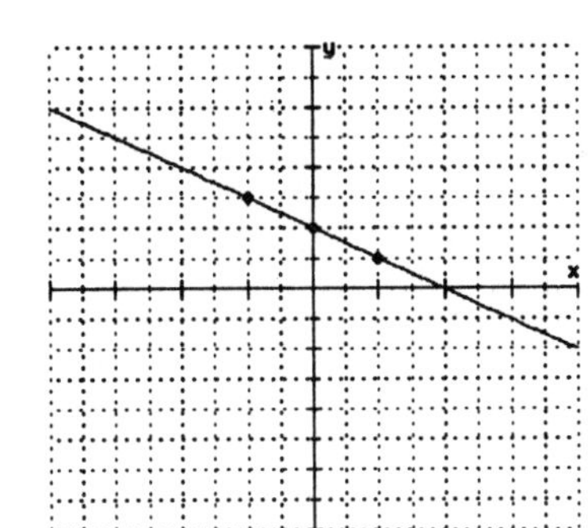

13. Let $x = 0$:

$$\begin{aligned} 2y - 6 &= 3x \\ 2y - 6 &= 3(0) \\ 2y - 6 &= 0 \\ 2y &= 6 \\ y &= 3 \end{aligned}$$
$(0, 3)$

Let $y = 0$:

$$\begin{aligned} 2y - 6 &= 3x \\ 2(0) - 6 &= 3x \\ -6 &= 3x \\ -2 &= x \end{aligned}$$
$(-2, 0)$

Let $x = -4$:

$$\begin{aligned} 2y - 6 &= 3x \\ 2y - 6 &= 3(-4) \\ 2y - 6 &= -12 \\ 2y &= -6 \\ y &= -3 \end{aligned}$$
$(-4, -3)$

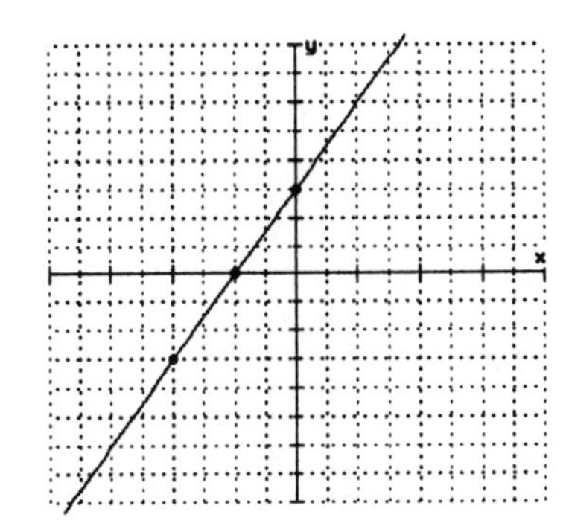

14. Let $x = 0$:

$$\begin{aligned} 7x + 1 &= 2y \\ 7(0) + 1 &= 2y \\ 1 &= 2y \\ \frac{1}{2} &= y \end{aligned}$$
$\left(0, \frac{1}{2}\right)$

Let $y = 0$:

$$\begin{aligned} 7x + 1 &= 2y \\ 7x + 1 &= 2(0) \\ 7x + 1 &= 0 \\ 7x &= -1 \\ x &= -\frac{1}{7} \end{aligned}$$
$\left(-\frac{1}{7}, 0\right)$

Let $x = 1$:

$$\begin{aligned} 7x + 1 &= 2y \\ 7(1) + 1 &= 2y \\ 8 &= 2y \\ 4 &= y \end{aligned}$$
$(1, 4)$

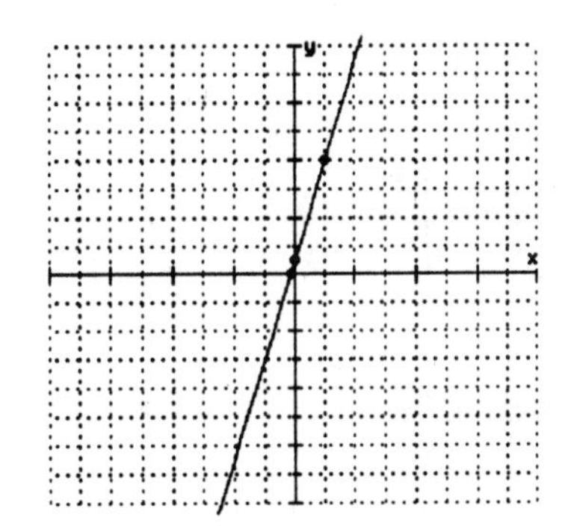

15.

x	$y = 0.2x$
-5	-1
0	0
5	1

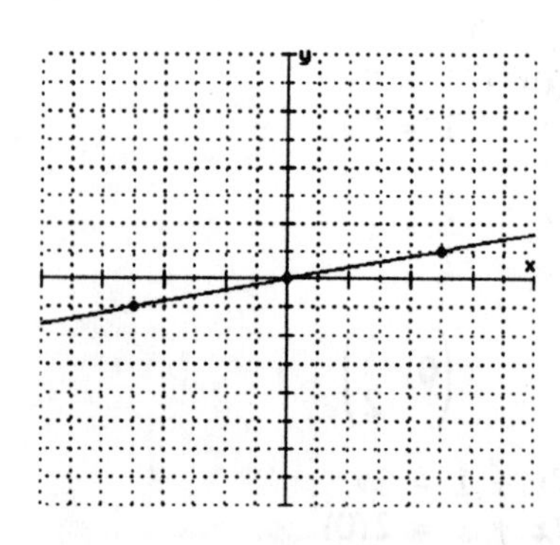

3.3 SOLUTIONS TO EXERCISES

1. $x = -2$; $y = 6$; $(-2, 0)$; $(0, 6)$

2. $x = -2$; $x = 2$; $y = 2$; $(-2, 0)$; $(2, 0)$; $(0, 2)$

3. $x = -2$; $x = 2$; $y = -3$; $y = 3$; $(-2, 0)$; $(2, 0)$; $(0, -3)$; $(0, 3)$

4.

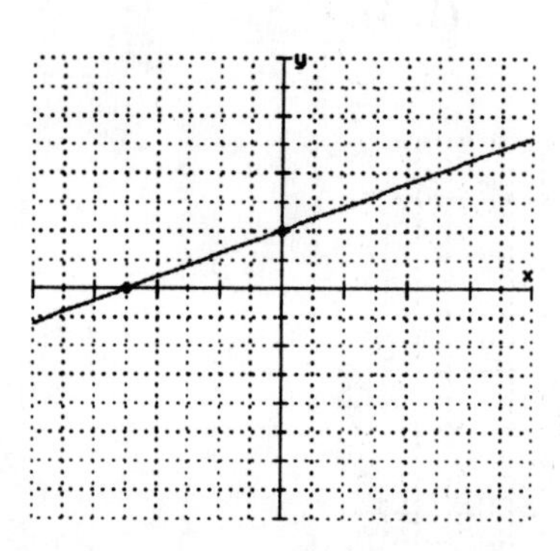

5.

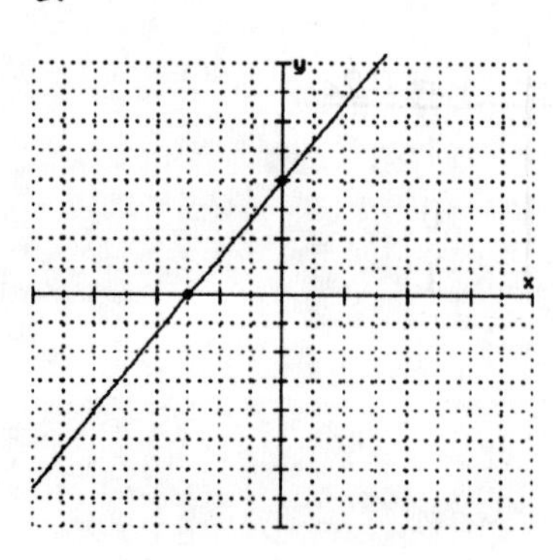

6.

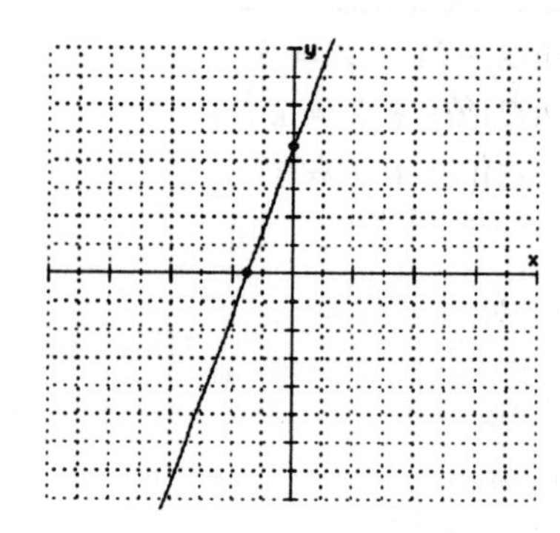

7.

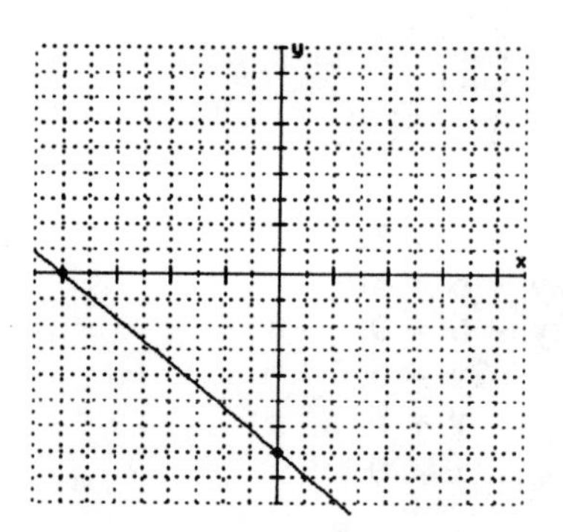

8. Let $x = 0$:

$$\begin{aligned} x + y &= 5 \\ 0 + y &= 5 \\ y &= 5 \\ &(0, 5) \end{aligned}$$

Let $y = 0$:

$$\begin{aligned} x + y &= 5 \\ x + 0 &= 5 \\ x &= 5 \\ &(5, 0) \end{aligned}$$

Let $x = 1$:

$$\begin{aligned} x + y &= 5 \\ 1 + y &= 5 \\ y &= 4 \\ &(1, 4) \end{aligned}$$

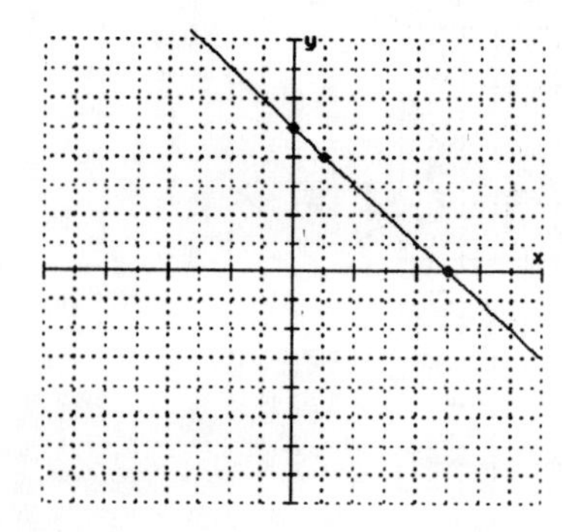

9. Let $x = 0$:
$$3x = -y$$
$$3(0) = -y$$
$$0 = -y$$
$$0 = y$$
$$(0, 0)$$

Let $y = 0$:
$$3x = -y$$
$$3x = -0$$
$$3x = 0$$
$$x = 0$$
$$(0, 0)$$

Let $x = 1$:
$$3x = -y$$
$$3(1) = -y$$
$$3 = -y$$
$$-3 = y$$
$$(1, -3)$$

Let $x = -1$:
$$3x = -y$$
$$3(-1) = -y$$
$$-3 = -y$$
$$3 = y$$
$$(-1, 3)$$

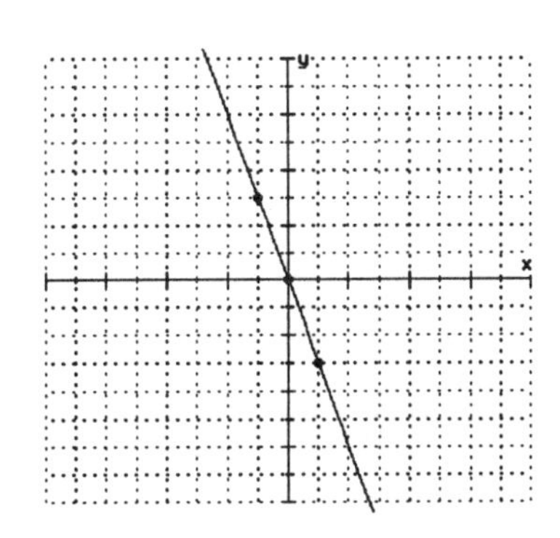

10. Let $x = 0$:
$$2x - y = 7$$
$$2(0) - y = 7$$
$$-y = 7$$
$$y = -7$$
$$(0, -7)$$

Let $y = 0$:
$$2x - y = 7$$
$$2x - 0 = 7$$
$$2x = 7$$
$$x = 7/2$$
$$\left(\frac{7}{2}, 0\right)$$

Let $x = 2$:
$$2x - y = 7$$
$$2(2) - y = 7$$
$$4 - y = 7$$
$$-y = 3$$
$$y = -3$$
$$(2, -3)$$

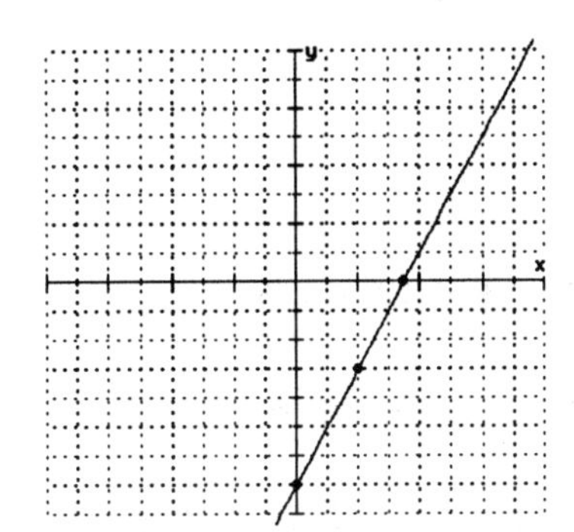

11. Let $x = 0$:
$$3x - 6y = 9$$
$$3(0) - 6y = 9$$
$$-6y = 9$$
$$y = -\frac{3}{2}$$
$$\left(0, -\frac{3}{2}\right)$$

Let $y = 0$;
$$3x - 6y = 9$$
$$3x - 6(0) = 9$$
$$3x = 9$$
$$x = 3$$
$$(3, 0)$$

Let $x = 1$:
$$3x - 6y = 9$$
$$3(1) - 6y = 9$$
$$3 - 6y = 9$$
$$-6y = 6$$
$$y = -1$$
$$(1, -1)$$

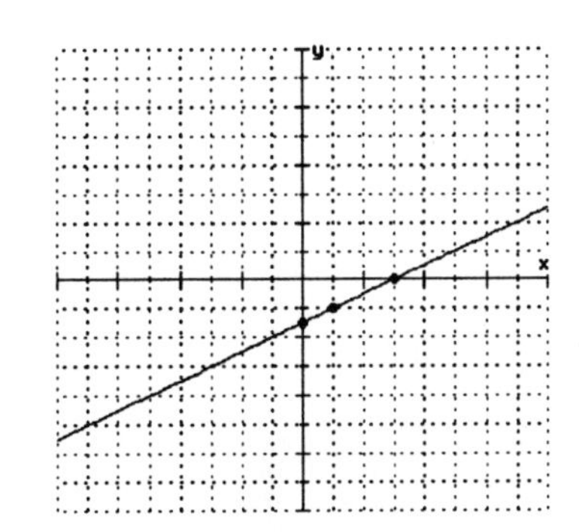

12. $x = -7$ is the vertical line with intercept point $(-7, 0)$.

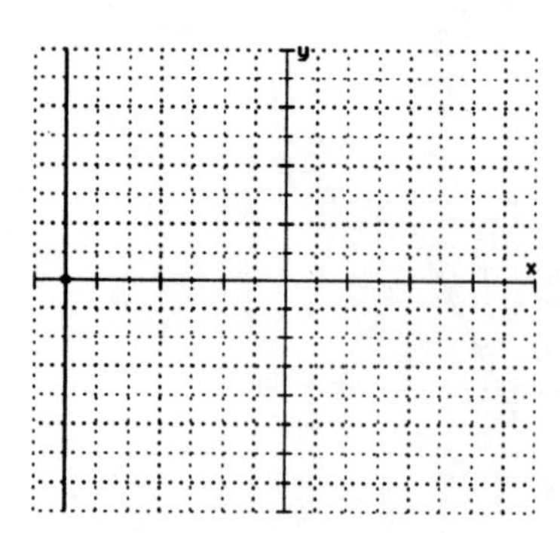

13. $y = 5$ is the horizontal line with intercept point $(0, 5)$.

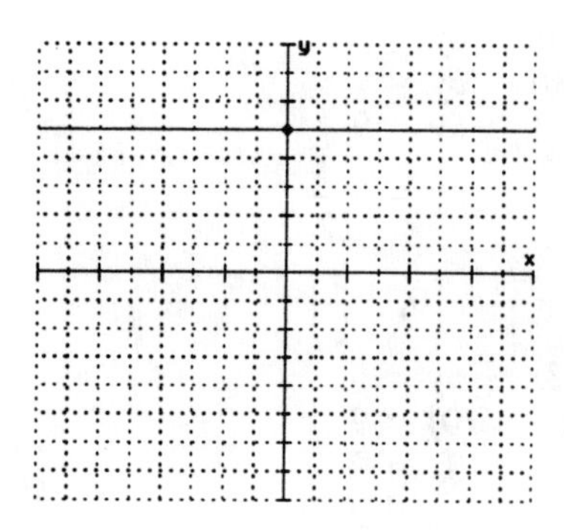

14. $x - 2 = 0$
$x = 2$

This is the vertical line with intercept point $(2, 0)$.

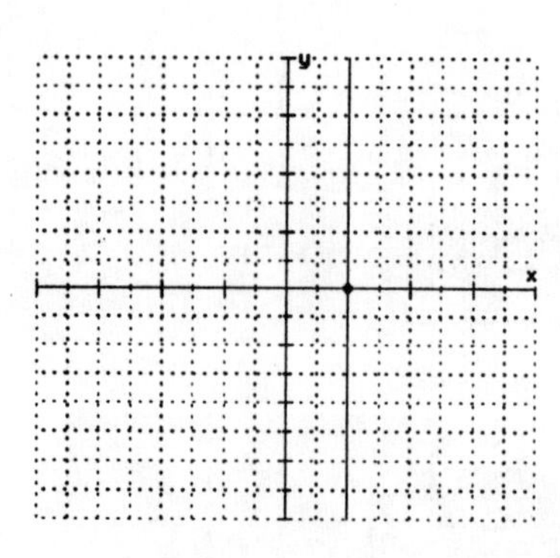

15.

x	$y = (-3/4)x + 1/2$
0	$(-3/4)(0) + 1/2 = 1/2$
-2	$(-3/4)(-2) + 1/2 = 2$
2	$(-3/4)(2) + 1/2 = -1$

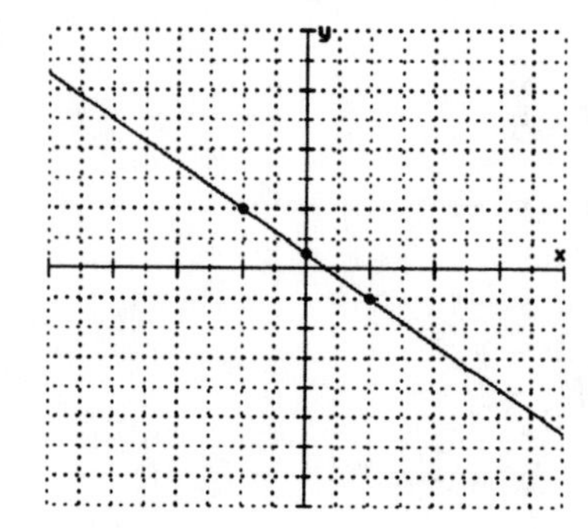

3.4 SOLUTIONS TO EXERCISES

1. $m = \dfrac{y_2 - y_1}{x_2 - x_1}$

$= \dfrac{-6 - 1}{0 - 8}$

$= \dfrac{-7}{-8}$

$= \dfrac{7}{8}$

2. $m = \dfrac{y_2 - y_1}{x_2 - x_1}$

$= \dfrac{-9 - 9}{-4 - 3}$

$= \dfrac{-18}{-7}$

$= \dfrac{18}{7}$

3. $m = \dfrac{y_2 - y_1}{x_2 - x_1}$

$= \dfrac{-6 - 1}{-6 - 1}$

$= \dfrac{-7}{-7}$

$= 1$

4. $m = \dfrac{y_2 - y_1}{x_2 - x_1}$

$= \dfrac{0 - (-6)}{12 - 0}$

$= \dfrac{6}{12}$

$= \dfrac{1}{2}$

5. $m = \dfrac{y_2 - y_1}{x_2 - x_1}$

$= \dfrac{8 - (-1)}{7 - 7}$

$= \dfrac{9}{0}$

undefined slope

6. $m = \dfrac{y_2 - y_1}{x_2 - x_1}$

$= \dfrac{3 - 3}{-2 - (-6)}$

$= \dfrac{0}{4}$

$= 0$

7. Locate (0, -5), then move up 3 units and to the right 4 units to a second point on the line.

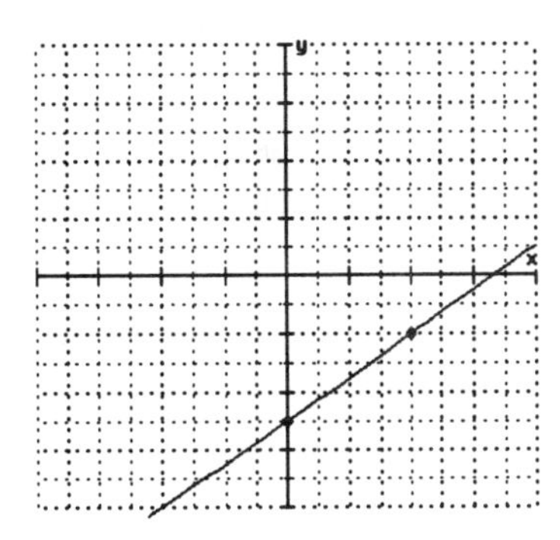

8. $m = -2 = \dfrac{-2}{1}$

Locate (1, 2), then move down 2 units and to the right 1 unit to a second point on the line.

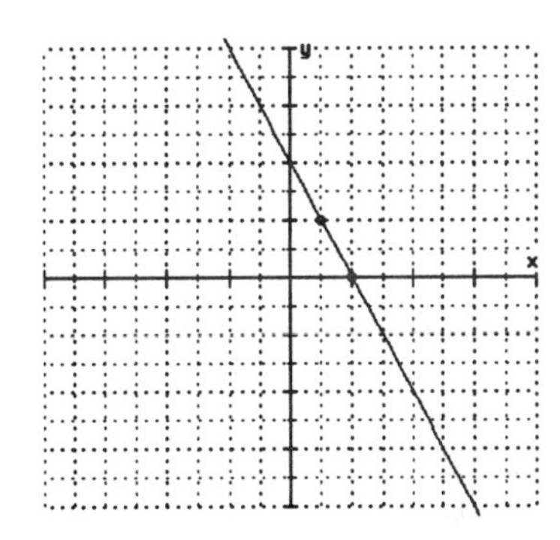

9. $m = -\dfrac{2}{5} = \dfrac{2}{-5}$

Locate (4, -3), then move up 2 units and to the left 5 units to a second point on the line.

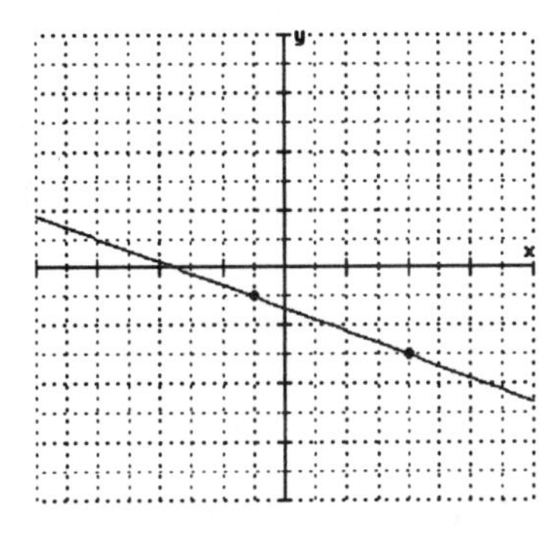

10. $m = 1 = \dfrac{1}{1}$

Locate (-2, 0), then move up 1 unit and to the right 1 unit to a second point on the line.

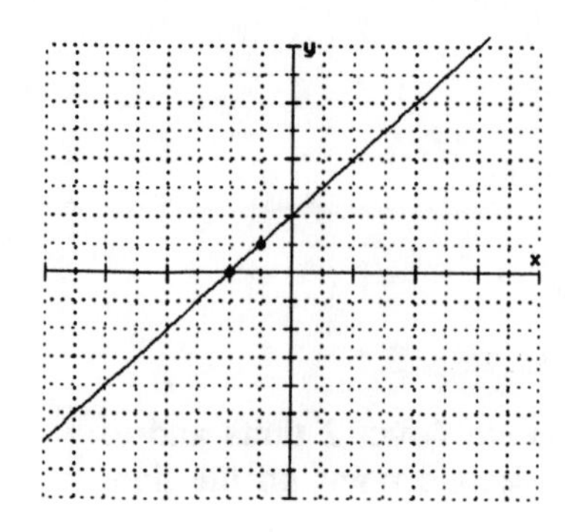

11. Let $x = 0$:

$$\begin{aligned} 2x - 5y &= 10 \\ 2(0) - 5y &= 10 \\ -5y &= 10 \\ y &= -2 \end{aligned}$$

(0, -2)

Let $y = 0$:

$$\begin{aligned} 2x - 5y &= 10 \\ 2x - 5(0) &= 10 \\ 2x &= 10 \\ x &= 5 \end{aligned}$$

(5, 0)

$$\begin{aligned} m &= \frac{y_2 - y_1}{x_2 - x_1} \\ &= \frac{-2 - 0}{0 - 5} \\ &= \frac{2}{5} \end{aligned}$$

12. Let $x = 0$:

$$\begin{aligned} -x - y &= 7 \\ -0 - y &= 7 \\ -y &= 7 \\ y &= -7 \end{aligned}$$

(0, -7)

Let $y = 0$:

$$\begin{aligned} -x - y &= 7 \\ -x - 0 &= 7 \\ -x &= 7 \\ x &= -7 \end{aligned}$$

(-7, 0)

$$\begin{aligned} m &= \frac{y_2 - y_1}{x_2 - x_1} \\ &= \frac{-7 - 0}{0 - (-7)} \\ &= \frac{-7}{7} \\ &= -1 \end{aligned}$$

13. $x - 9 = 0$
$x = 9$

This is a vertical line and all vertical lines have undefined slope.

14. $y = -12$

This is a horizontal line and all horizontal lines have slope 0.

15. (0, -3) and (3, 1) $m_1 = \dfrac{y_2 - y_1}{x_2 - x_1} = \dfrac{-3 - 1}{0 - 3} = \dfrac{4}{3}$

(7, 2) and (3, -1) $m_2 = \dfrac{y_2 - y_1}{x_2 - x_1} = \dfrac{2 - (-1)}{7 - 3} = \dfrac{3}{4}$

Neither, since $m_1 \neq m_2$ and $m_1 \neq -\dfrac{1}{m_2}$.

16. (-1, 2) and (4, 8) $m_1 = \dfrac{y_2 - y_1}{x_2 - x_1} = \dfrac{2 - 8}{-1 - 4} = \dfrac{6}{5}$

(11, 20) and (17, 15) $m_2 = \frac{y_2 - y_1}{x_2 - x_1}$

$= \frac{20 - 15}{11 - 17}$

$= -\frac{5}{6}$

Perpendicular, since $m_1 = -\frac{1}{m_2}$.

3.5 SOLUTIONS TO EXERCISES

1. $x + y < 2$
   $1 - 2 < 2$
   $-1 < 2$
   True
   (1, -2) is a solution.

   $x + y < 2$
   $3 - 6 < 2$
   $-3 < 2$
   True
   (3, -6) is a solution.

   $x + y < 2$
   $-2 + 7 < 2$
   $5 < 2$
   False
   (-2, 7) is not a solution.

2. $x \geq -y$
   $0 \geq -0$
   $0 \geq 0$
   True
   (0, 0) is a solution.

   $x \geq -y$
   $1 \geq -(-1)$
   $1 \geq 1$
   True
   (1, -1) is a solution.

   $x \geq -y$
   $5 \geq -3$
   True
   (5, 3) is a solution.

3. $y > \frac{1}{2}x$
   $0 > \frac{1}{2}(0)$
   $0 > 0$
   False
   (0, 0) is not a solution.

   $y > \frac{1}{2}x$
   $1 > \frac{1}{2}(4)$
   $1 > 2$
   False
   (4, 1) is not a solution.

   $y > \frac{1}{2}x$
   $8 > \frac{1}{2}(6)$
   $8 > 3$
   True
   (6, 8) is a solution.

4. $-x - 3y \leq 12$
   $-4 - 3(4) \leq 12$
   $-16 \leq 12$
   True
   (4, 4) is a solution.

   $-x - 3y \leq 12$
   $-(-2) - 3(-5) \leq 12$
   $17 \leq 12$
   False
   (-2, -5) is not a solution.

   $-x - 3y \leq 12$
   $-0 - 3(4) \leq 12$
   $-12 \leq 12$
   True
   (0, 4) is a solution.

5. $2x + 7y < 9$
   $2(0) + 7(1) < 9$
   $7 < 9$
   True
   (0, 1) is a solution.

$$2x + 7y < 9$$
$$2(1) + 7(1) < 9$$
$$9 < 9 \quad \text{False}$$
(1, 1) is not a solution.

$$2x + 7y < 9$$
$$2(3) + 7(5) < 9$$
$$41 < 9 \quad \text{False}$$
(3, 5) is not a solution.

6. $x + y \geq -1$ (Solid line)

$x + y = -1$

x	y
0	-1
-1	0
1	-2

Test point: (0, 0)
$$x + y \geq -1$$
$$0 + 0 \geq -1$$
$$0 \geq -1 \quad \text{True}$$
Shade the half-plane containing the test point (0, 0).

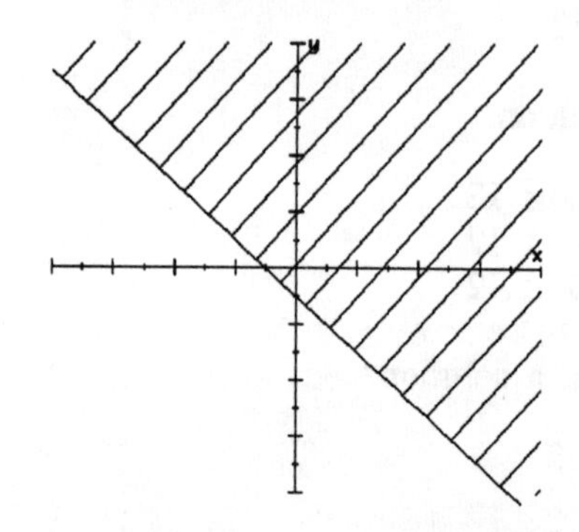

7. $2x - y < -5$ (Dashed line)

$2x - y = -5$

x	y
0	5
-5/2	0
-1	3

Test point: (0, 0)
$$2x - y < -5$$
$$2(0) - 0 < -5$$
$$0 < -5 \quad \text{False}$$
Shade the half plane not containing the test point (0, 0).

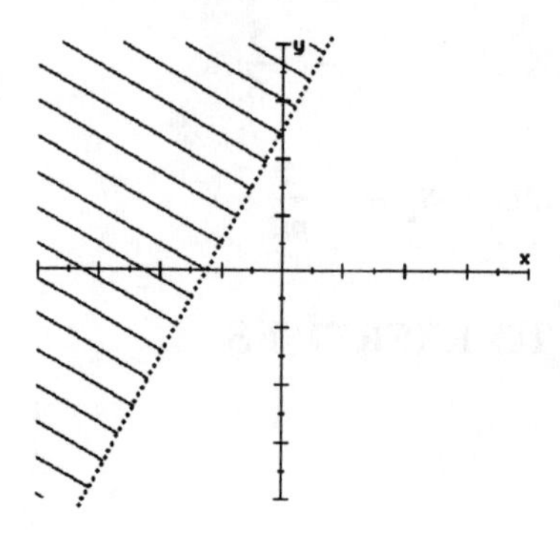

8. $7x + 2y > -14$ (Dashed line)

$7x + 2y = -14$

x	y
0	-7
-2	0
2	-14

Test point: (0, 0)
$$7x + 2y > -14$$
$$7(0) + 2(0) > -14$$
$$0 > -14 \quad \text{True}$$
Shade the half-plane containing the test point (0, 0).

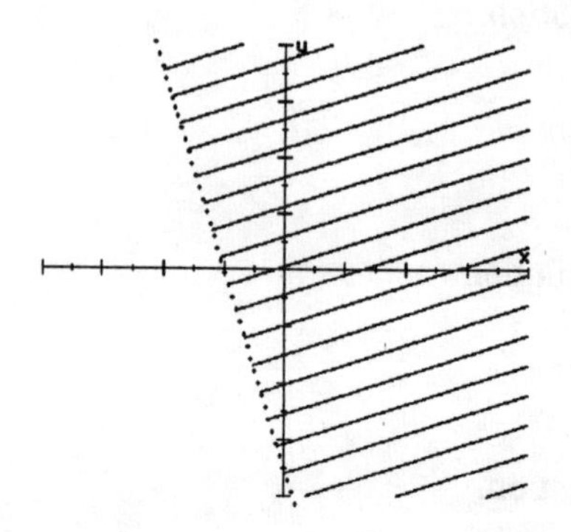

9. $x \ge 6$ (Solid line)

$x = 6$ vertical line with x-intercept (6, 0).
Test point: (0, 0)
$x \ge 6$
$0 \ge 6$ False
Shade the half-plane not containing the test point (0, 0).

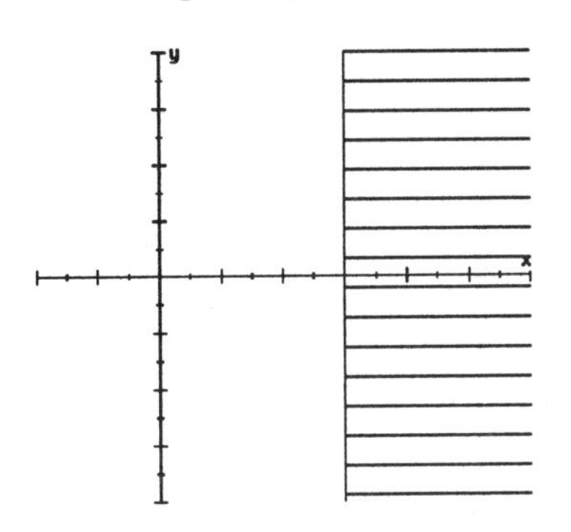

10. $y < -4$ (Dashed line)

$y = -4$ horizontal line with y-intercept (0, −4)
Test point: (0, 0)
$y < -4$
$0 < -4$ False
Shade the half-plane not containing the test point (0, 0).

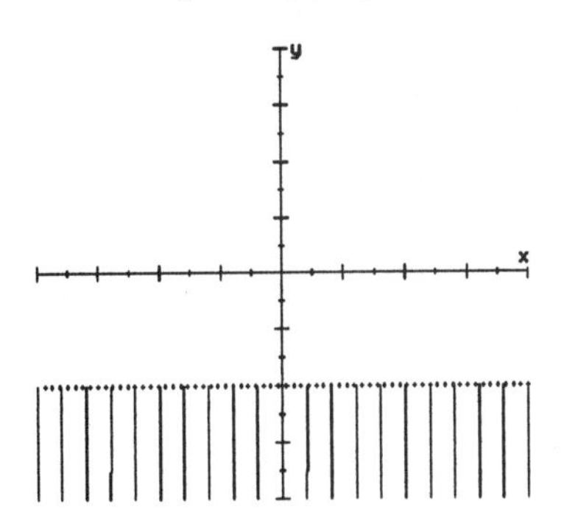

11. $2x + 5y \ge 0$ (Solid line)

$2x + 5y = 0$

x	y
0	0
-5	2
5	-2

Test point: (1, 1) (Note: (0, 0) is on the line so you may not use it as a test point.)

$2x + 5y \ge 0$
$2(1) + 5(1) \ge 0$
$7 \ge 0$ True

Shade the half-plane containing the test point (1, 1).

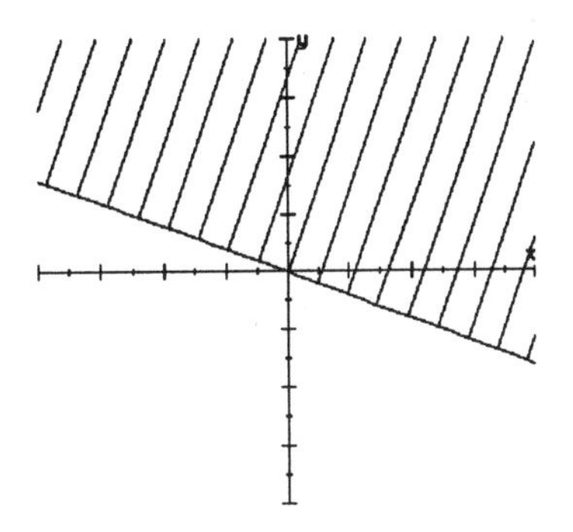

12. $-\frac{1}{5}x + \frac{1}{2}y \le 1$ (Solid line)

$$10\left(-\frac{1}{5}x + \frac{1}{2}y\right) \le 10(1)$$

$$-2x + 5y \le 10$$

$$-2x + 5y = 10$$

x	y
0	2
-5	0
5	4

Test point: (0, 0)
$-2x + 5y \le 10$
$-2(0) + 5(0) \le 10$
$0 \le 10$ True
Shade the half-plane containing the the test point (0, 0).

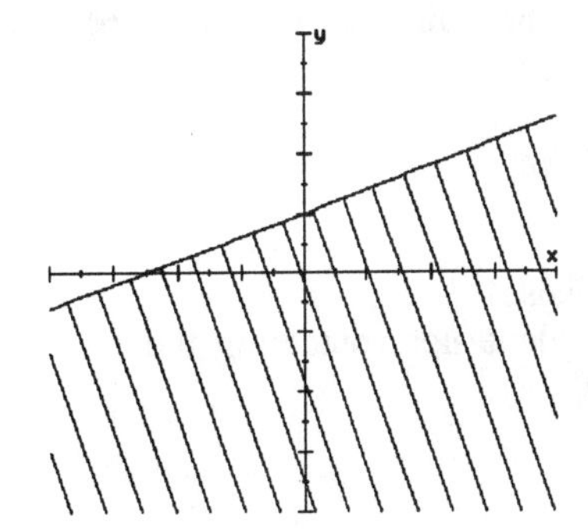

13. $\frac{3}{4}x + \frac{1}{8}y > \frac{1}{6}$ (Dashed line)

$$24\left(\frac{3}{4}x + \frac{1}{8}y\right) > 24\left(\frac{1}{6}\right)$$

$$18x + 3y > 4$$

$$18x + 3y = 4$$

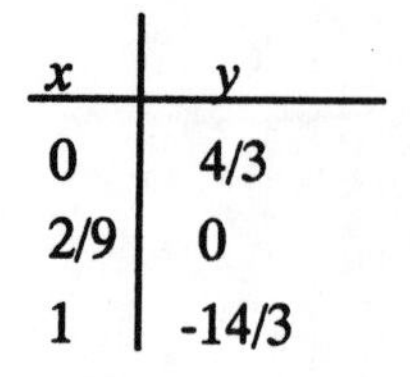

x	y
0	4/3
2/9	0
1	-14/3

Test point: (0, 0)
$18x + 3y > 4$
$18(0) + 3(0) > 4$
$0 > 4$ False
Shade the half-plane not containing the test point (0, 0).

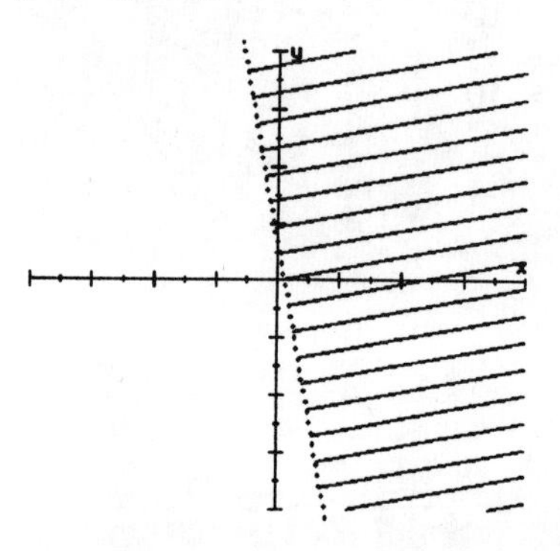

14. $x - 1 \geq 0$ (Solid line)

$x - 1 = 0$
$x = 1$ vertical line with x-intercept (1, 0).

Test point: (0, 0)
$x - 1 \geq 0$
$0 - 1 \geq 0$
$-1 \geq 0$ False
Shade the half-plane not containing the test point (0, 0).

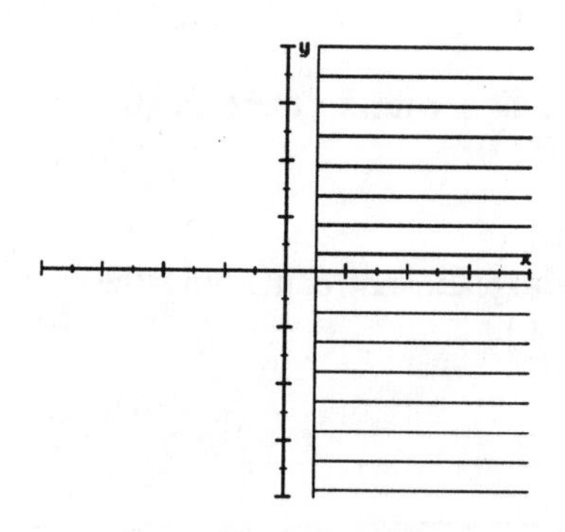

15. $-y < -7$ (Dashed line)

$-y = -7$
$y = 7$ horizontal line with y-intercept (0, 7).

Test point: (0, 0)
$-y < -7$
$-0 < -7$
$0 < -7$ False
Shade the half-plane not containing the test point (0, 0).

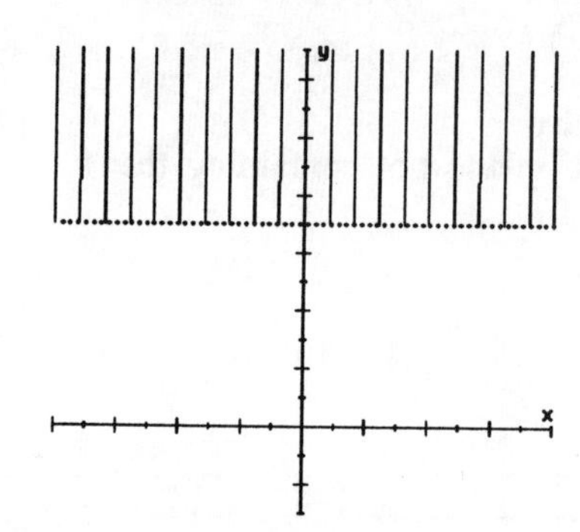

CHAPTER 3 PRACTICE TEST SOLUTIONS

1. $2x - y = 5$
$2(1) - (-3) = 5$
$2 + 3 = 5$
$5 = 5$
True

(1, -3) is a solution.

2. $3x + 2y = 12$
$3(0) + 2(-6) = 12$
$0 - 12 = 12$
$-12 = 12$
False

(0, -6) is not a solution.

3. $17y - 12x = 5$
$17(1) - 12x = 5$
$17 - 12x = 5$
$-12x = -12$
$x = 1$

(1, 1)

4. $y = 6$
$0x + y = 6$
$0(3) + y = 6$
$0 + y = 6$
$y = 6$

(3, 6)

5. (-6, 0) and (0, 2) are on the graph.

Vertical rise $= 2 - 0 = 2$
Horizontal run $= 0 - (-6) = 6$

$$m = \frac{\text{rise}}{\text{run}} = \frac{2}{6} = \frac{1}{3}$$

6. (-2, 0) and (0, -4) are on the graph.

Vertical rise $= 0 - (-4) = 4$
Horizontal run $= -2 - 0 = -2$

$$m = \frac{\text{rise}}{\text{run}} = \frac{4}{-2} = -2$$

7. $m = \dfrac{y_2 - y_1}{x_2 - x_1}$

$= \dfrac{15 - 3}{-2 - (-6)}$

$= \dfrac{12}{4}$

$= 3$

8. $m = \dfrac{y_2 - y_1}{x_2 - x_1}$

$= \dfrac{5 - (-4)}{-3 - 0}$

$= \dfrac{9}{-3}$

$= -3$

9. $-7x + y = 4$
$y = 7x + 4$
$m = 7$

10. $x = -4$ vertical line

undefined slope

11. (-2, 4) and (2, -4):

$m_1 = \dfrac{4 - (-4)}{-2 - 2}$

$= \dfrac{8}{-4}$

$= -2$

(1, 3) and (-1, 2):

$m_2 = \dfrac{3 - 2}{1 - (-1)}$

$= \dfrac{1}{2}$

Perpendicular, since $m_1 = -\dfrac{1}{m_2}$

12. (5, 7) and (-3, 6):

$m_1 = \dfrac{7 - 6}{5 - (-3)}$

$= \dfrac{1}{8}$

(-4, 8) and (4, 9):

$m_2 = \dfrac{9 - 8}{4 - (-4)}$

$= \dfrac{1}{8}$

Parallel, since $m_1 = m_2$

13. $x + 3y = 6$

Let $x = 0$: $0 + 3y = 6$
$3y = 6$
$y = 2$
(0, 2)

Let $y = 0$: $x + 3(0) = 6$
$x + 0 = 6$
$x = 6$
(6, 0)

Let $x = 3$: $3 + 3y = 6$
$3y = 3$
$y = 1$
(3, 1)

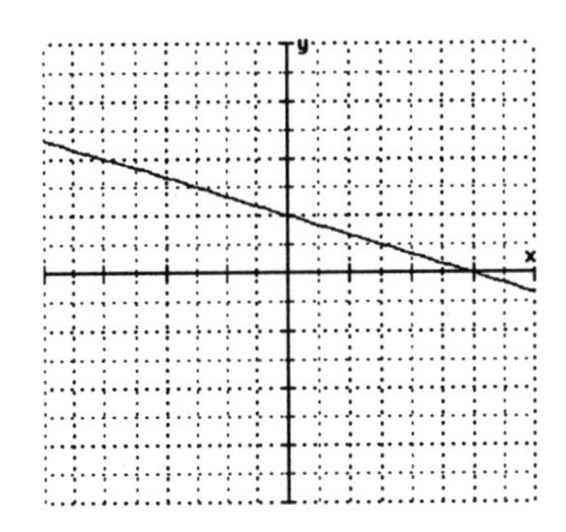

14. $-x + 2y = 7$

Let $x = 0$: $-0 + 2y = 7$
$2y = 7$
$y = \frac{7}{2}$
$\left(0, \frac{7}{2}\right)$

Let $y = 0$: $-x + 2(0) = 7$
$-x = 7$
$x = -7$
$(-7, 0)$

Let $x = 1$: $-1 + 2y = 7$
$2y = 8$
$y = 4$
$(1, 4)$

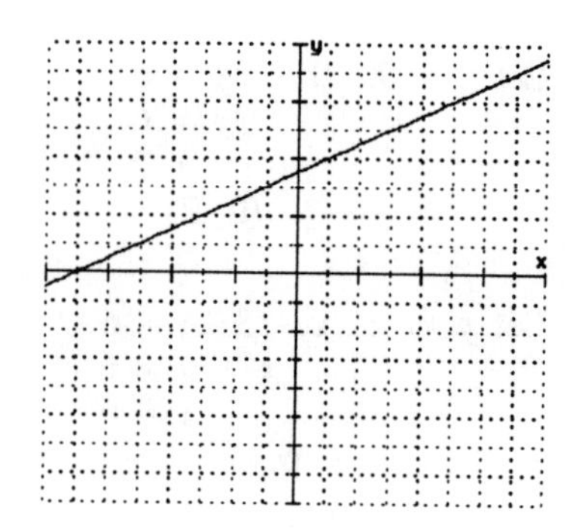

15. $x - y \leq -3$

$x - y = -3$ (Solid line)

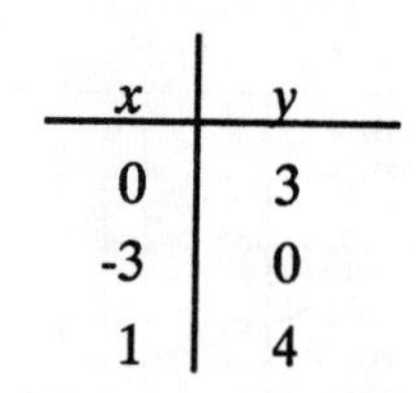

x	y
0	3
-3	0
1	4

Test point: $(0, 0)$
$x - y \leq -3$
$0 - 0 \leq -3$
$0 \leq -3$ False

Shade the half-plane not containing $(0, 0)$.

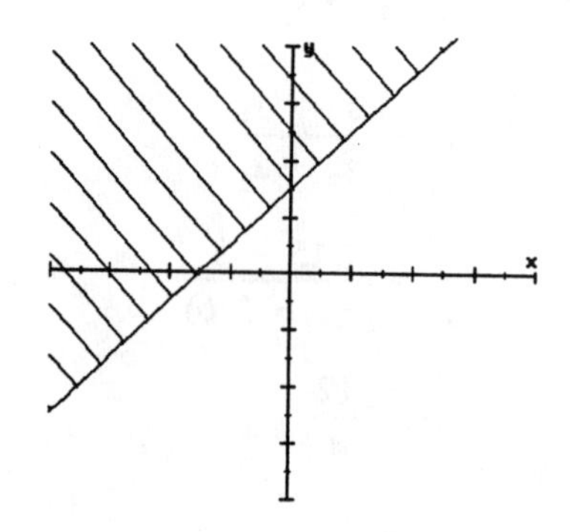

16. $y \geq -2x$

$y = -2x$ (Solid line)

x	y
0	0
1	-2
2	-4

Test point: $(1, 1)$

$y \geq -2x$
$1 \geq -2(1)$
$1 \geq -2$ True

Shade the half-plane containing $(1, 1)$.

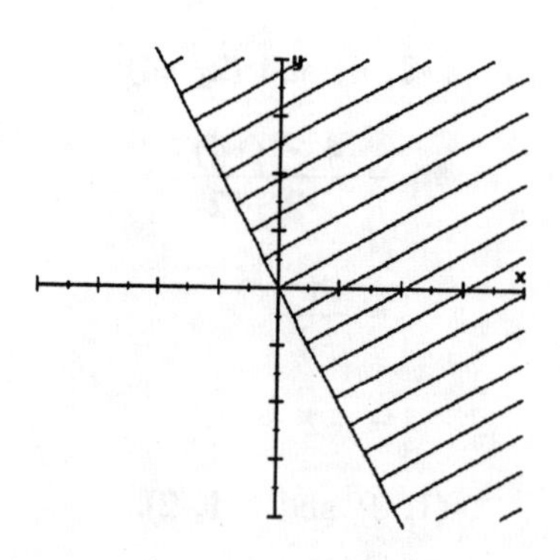

17. $6x - 5y = 12$

Let $x = 0$:

$6(0) - 5y = 12$
$-5y = 12$
$y = -\frac{12}{5}$
$\left(0, -\frac{12}{5}\right)$

Let $y = 0$:

$6x - 5(0) = 12$
$6x = 12$
$x = 2$
$(2, 0)$

Let $y = -2$:

$6x - 5(-2) = 12$
$6x + 10 = 12$
$6x = 2$
$x = \frac{1}{3}$
$\left(\frac{1}{3}, -2\right)$

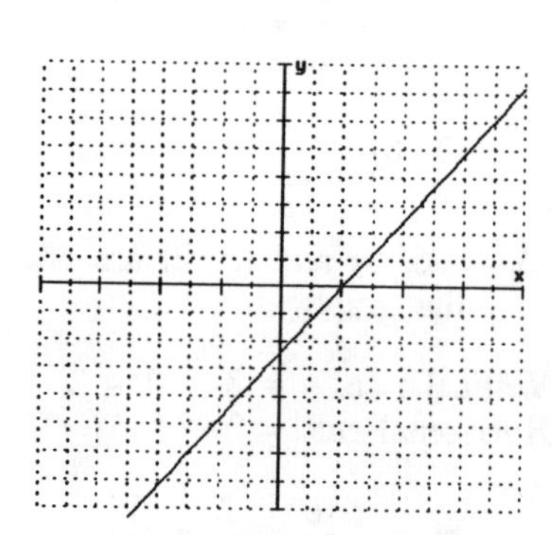

18. $3x - 4y < -8$

$3x - 4y = -8$ (Dashed line)

x	y
0	2
-8/3	0
1	11/4

Test point: (0, 0)

$$3x - 4y < -8$$
$$3(0) - 4(0) < -8$$
$$0 < -8 \quad \text{False}$$

Shade the half-plane not containing (0, 0).

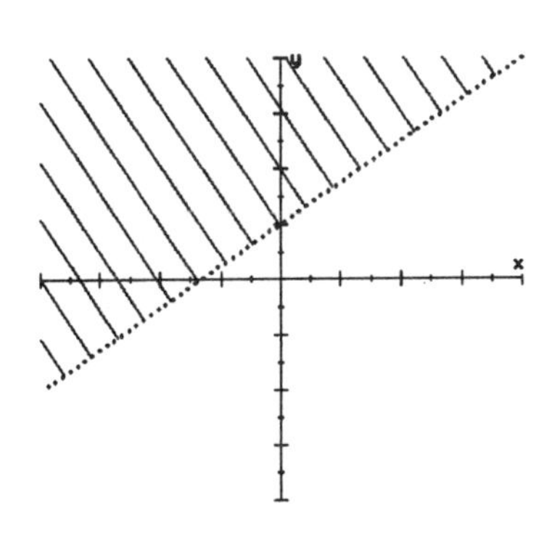

19. $5x + y > -2$

$5x + y = -2$ (Dashed line)

x	y
0	-2
-2/5	0
1	-7

Test point: (0, 0)

$$5x + y > -2$$
$$5(0) + 0 > -2$$
$$0 > -2 \quad \text{True}$$

Shade the half-plane containing (0, 0).

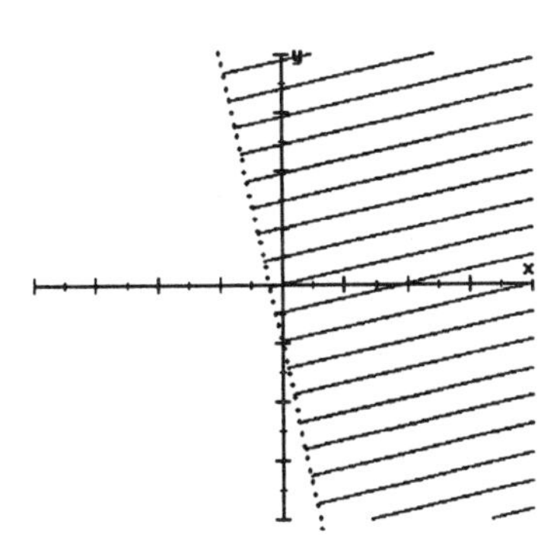

20. $y = -4$

For any x-value chosen, $y = -4$. For example, (0, -4), (1, -4), (2, -4)

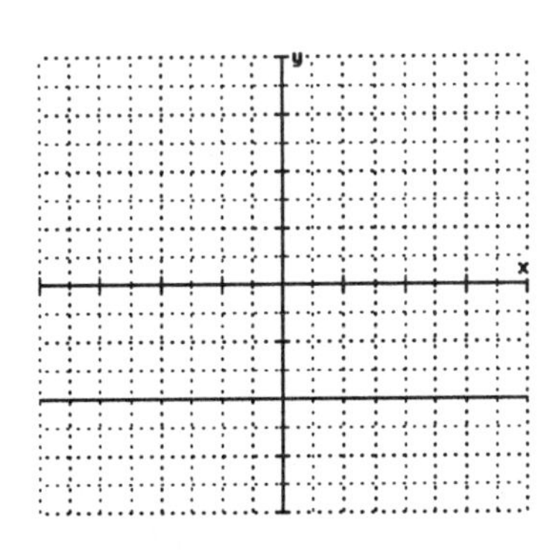

21. $x - 4 = 0$
$x = 4$

For any y-value chosen $x = 4$. For example, (4, -1), (4, 0), (4, 1)

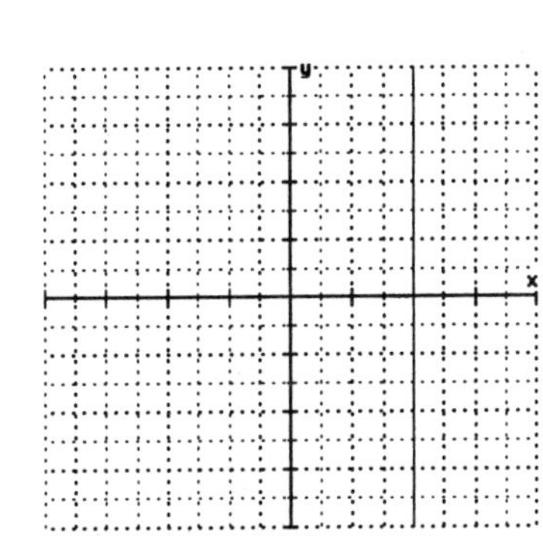

22. $4x - 5y = 20$

Let $x = 0$: $4(0) - 5y = 20$
$-5y = 20$
$y = -4$
(0, -4)

Let $y = 0$: $4x - 5(0) = 20$
$4x = 20$
$x = 5$
(5, 0)

Let $x = 1$: $4(1) - 5y = 20$
$4 - 5y = 20$
$-5y = 16$
$y = -\frac{16}{5}$

$\left(1, -\frac{16}{5}\right)$

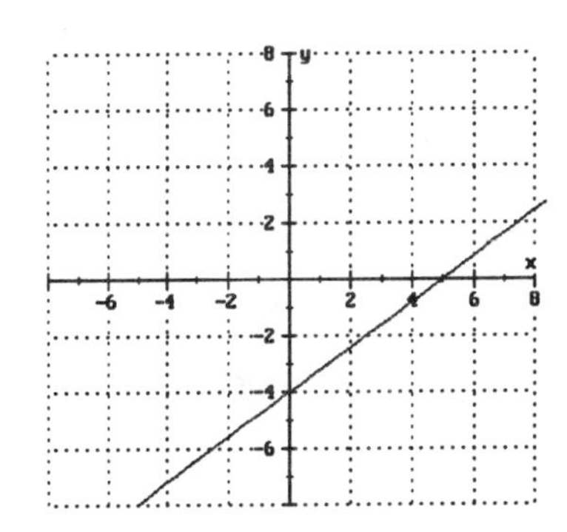

23. $y = 2 + 3x$

x	y
-1	-1
0	2
1	5

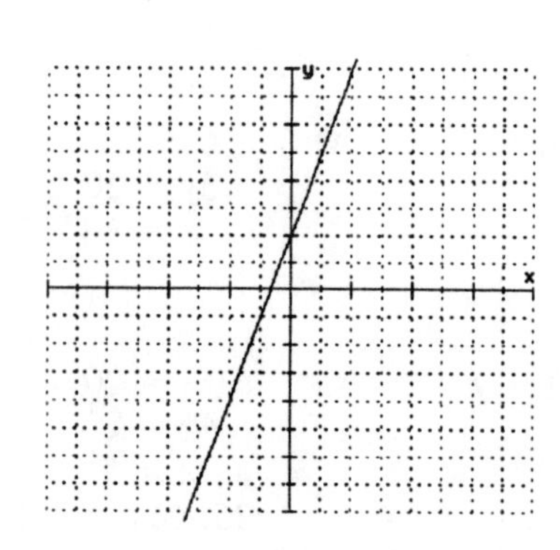

25. $P = 2(x) + 2(3y)$
$P = 2x + 6y$
$48 = 2x + 6(4)$
$48 = 2x + 24$
$24 = 2x$
$12 = x$

24. $x + 2y < -6$

$x + 2y = -6$ (Dashed line)

x	y
0	-3
-6	0
-2	-2

Test point: (0, 0)

$x + 2y < -6$
$0 + 2(0) < -6$
$0 < -6$ False

Shade the half-plane not containing (0, 0).

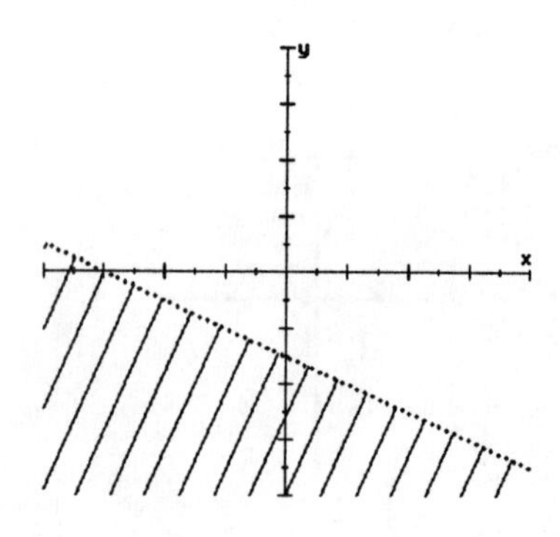

4.1 SOLUTIONS TO EXERCISES

1. $-3^4 = -1 \cdot 3 \cdot 3 \cdot 3 \cdot 3 = -81$

2. $\left(\frac{2}{5}\right)^2 = \frac{2}{5} \cdot \frac{2}{5} = \frac{4}{25}$

3. $4 \cdot 2^3 = 4 \cdot 2 \cdot 2 \cdot 2 = 32$

4. $2x^3y = 2(-1)^3 \cdot (7)$
$= 2(-1) \cdot 7$
$= -2 \cdot 7$
$= -14$

5. $\frac{-2a}{b^4} = \frac{-2(12)}{(-2)^4}$
$= \frac{-24}{16}$
$= -\frac{3}{2}$

6. $(5t)^3 = 5^3t^3$
$= 125t^3$

7. $(2a^2b^5)^4 = 2^4(a^2)^4(b^5)^4$
$= 16a^{2\cdot4}b^{5\cdot4}$
$= 16a^8b^{20}$

8. $\left(\frac{y^3}{-2z^2}\right)^5 = \frac{(y^3)^5}{(-2z^2)^5}$
$= \frac{y^{3\cdot5}}{(-2)^5(z^2)^5}$
$= \frac{y^{15}}{-32z^{2\cdot5}}$
$= \frac{y^{15}}{-32z^{10}}$
$= -\frac{y^{15}}{32z^{10}}$

9. $\frac{8x^3y^9}{32xy^6} = \frac{8}{32}x^{3-1}y^{9-6}$
$= \frac{1}{4}x^2y^3$
or $\frac{x^2y^3}{4}$

10. $6^0 + 3y^0 = 1 + 3(1) = 4$

11. $(12x^2y)^2 = 12^2(x^2)^2y^2$
$= 144x^{2\cdot2}y^2$
$= 144x^4y^2$

12. $\frac{(2xy^2z^6)^3}{4xyz^4} = \frac{2^3x^3(y^2)^3(z^6)^3}{4xyz^4}$
$= \frac{8x^3y^{2\cdot3}z^{6\cdot3}}{4xyz^4}$
$= \frac{8x^3y^6z^{18}}{4xyz^4}$
$= \frac{8}{4}x^{3-1}y^{6-1}z^{18-4}$
$= 2x^2y^5z^{14}$

13. $\frac{a^{19} \cdot a^6}{a^3} = \frac{a^{19+6}}{a^3}$
$= \frac{a^{25}}{a^3}$
$= a^{25-3}$
$= a^{22}$

14. $\frac{(5cd^2)^3}{-25c^2d} = \frac{5^3c^3(d^2)^3}{-25c^2d}$
$= \frac{125c^3d^{2\cdot3}}{-25c^2d}$

$= \dfrac{125c^3d^6}{-25c^2d}$

$= \dfrac{125}{-25}c^{3-2}d^{6-1}$

$= -5cd^5$

15. $\dfrac{(x^9)^8}{(4x)^3} = \dfrac{x^{9\cdot 8}}{4^3x^3}$

$= \dfrac{x^{72}}{64x^3}$

$= \dfrac{1}{64}x^{72-3}$

$= \dfrac{1}{64}x^{69}$

or $\dfrac{x^{69}}{64}$

16. $V = x^3$
$V = 8^3$
$V = 512$

512 cm^3

4.2 SOLUTIONS TO EXERCISES

1. Degree: 2 (largest exponent)
Binomial; since it has 2 terms

2. Degree: 5 (largest exponent)
Trinomial; since it has 3 terms

3. Degree: 9 (add the exponents 3 + 1 + 5 = 9)
Monomial; since it has 1 term

4. (a) $5x - 12 = 5(0) - 12 = -12$

(b) $5x - 12 = 5(-2) - 12$
$= -10 - 12$
$= -22$

5. (a) $-x^2 + x + 5 = -0^2 + 0 + 5 = 5$

(b) $-x^2 + x + 5 = -(-2)^2 + (-2) + 5$
$= -4 - 2 + 5$
$= -1$

6. $(-8x + 11) + (5x - 4)$
$= -8x + 11 + 5x - 4$
$= (-8x + 5x) + (11 - 4)$
$= -3x + 7$

7. $(x^2 - 3x + 4) - (6x^2 + 2x - 1)$
$= (x^2 - 3x + 4) + (-6x^2 - 2x + 1)$
$= x^2 - 3x + 4 - 6x^2 - 2x + 1$
$= (x^2 - 6x^2) + (-3x - 2x) + (4 + 1)$
$= -5x^2 - 5x + 5$

8. $(31x^2 - 9) - (15x^2 + 4)$
$= (31x^2 - 9) + (-15x^2 - 4)$
$= 31x^2 - 9 - 15x^2 - 4$
$= (31x^2 - 15x^2) + (-9 - 4)$
$= 16x^2 - 13$

9. $(17x - 8) - (-3x^2 - 5x + 11)$
$= (17x - 8) + (3x^2 + 5x - 11)$
$= 17x - 8 + 3x^2 + 5x - 11$
$= 3x^2 + (17x + 5x) + (-8 - 11)$
$= 3x^2 + 22x - 19$

10. $(6x + 1) - 8x = 6x + 1 - 8x$
$= (6x - 8x) + 1$
$= -2x + 1$

11. $[(9y - 2x) + (12 + 4x^2)] - (15y + 8x^2)$
$= (9y + 4x^2 - 2x + 12) - (15y + 8x^2)$
$= (9y + 4x^2 - 2x + 12) + (-15y - 8x^2)$
$= 9y + 4x^2 - 2x + 12 - 15y - 8x^2$
$= (9y - 15y) + (4x^2 - 8x^2) - 2x + 12$
$= -6y - 4x^2 - 2x + 12$

12. $(14 + 6a) - (-a - 9)$
$= (14 + 6a) + (a + 9)$
$= 14 + 6a + a + 9$
$= (6a + a) + (14 + 9)$
$= 7a + 23$

13. $(-13x^2 + 5x + 1) + (12x^2 + 7x - 1)$
$= -13x^2 + 5x + 1 + 12x^2 + 7x - 1$
$= (-13x^2 + 12x^2) + (5x + 7x) + (1 - 1)$
$= -x^2 + 12x$

14. $(-9x^5 + 6x^2) + (-9x^5 + 5x^2 + 10)$
$= -9x^5 + 6x^2 - 9x^5 + 5x^2 + 10$
$= (-9x^5 - 9x^5) + (6x^2 + 5x^2) + 10$
$= -18x^5 + 11x^2 + 10$

15. $[(9y^2 + 6) + (3y - 7)] - (3y^2 - 5y - 2)$
$= (9y^2 + 6 + 3y - 7) - (3y^2 - 5y - 2)$
$= (9y^2 + 3y - 1) + (-3y^2 + 5y + 2)$
$= 9y^2 + 3y - 1 - 3y^2 + 5y + 2$
$= (9y^2 - 3y^2) + (3y + 5y) + (-1 + 2)$
$= 6y^2 + 8y + 1$

4.3 SOLUTIONS TO EXERCISES

1. $7a(3a - 5) = (7a)(3a) + (7a)(-5)$
$= 21a^2 - 35a$

2. $-2x^2(9x - 11) = (-2x^2)(9x) + (-2x^2)(-11)$
$= -18x^3 + 22x^2$

3. $(a + 8)(a - 3)$
$= a(a) + a(-3) + 8(a) + 8(-3)$
$= a^2 - 3a + 8a - 24$
$= a^2 + 5a - 24$

4. $(4x + y)(x - 5y)$
$= 4x(x) + 4x(-5y) + y(x) + y(-5y)$
$= 4x^2 - 20xy + xy - 5y^2$
$= 4x^2 - 19xy - 5y^2$

5. $(3x + 2y)(2x + 3y)$
$= 3x(2x) + 3x(3y) + 2y(2x) + 2y(3y)$
$= 6x^2 + 9xy + 4xy + 6y^2$
$= 6x^2 + 13xy + 6y^2$

6. $(5a - 7)^2$
$= (5a - 7)(5a - 7)$
$= 5a(5a) + 5a(-7) + (-7)(5a) + (-7)(-7)$
$= 25a^2 - 35a - 35a + 49$
$= 25a^2 - 70a + 49$

7. $(x^2 - 3)^2$
$= (x^2 - 3)(x^2 - 3)$
$= x^2(x^2) + x^2(-3) + (-3)(x^2) + (-3)(-3)$
$= x^4 - 3x^2 - 3x^2 + 9$
$= x^4 - 6x^2 + 9$

8. $(x + 3)^3$
$= (x + 3)(x + 3)(x + 3)$
$= [(x + 3)(x + 3)](x + 3)$
$= [x(x) + 3x + 3x + (3)(3)](x + 3)$
$= (x^2 + 6x + 9)(x + 3)$
$= x^2(x) + x^2(3) + 6x(x) + 6x(3)$
$+ 9(x) + 9(3)$
$= x^3 + 3x^2 + 6x^2 + 18x + 9x + 27$
$= x^3 + 9x^2 + 27x + 27$

9. $(3y - 1)^3$
$= (3y - 1)(3y - 1)(3y - 1)$
$= [(3y - 1)(3y - 1)](3y - 1)$
$= [3y(3y) + 3y(-1) + (-1)(3y) + (-1)(-1)]$
$(3y - 1)$
$= (9y^2 - 3y - 3y + 1)(3y - 1)$
$= (9y^2 - 6y + 1)(3y - 1)$
$= 9y^2(3y) + 9y^2(-1) + (-6y)(3y) + (-6y)(-1)$
$+ 1(3y) + 1(-1)$
$= 27y^3 - 9y^2 - 18y^2 + 6y + 3y - 1$
$= 27y^3 - 27y^2 + 9y - 1$

10. $(x + 2)(x^2 + 7x - 3)$
$= x(x^2) + x(7x) + x(-3) + 2(x^2) + 2(7x)$
$+ 2(-3)$
$= x^3 + 7x^2 - 3x + 2x^2 + 14x - 6$
$= x^3 + 9x^2 + 11x - 6$

11. $(b + 5)(b^2 + 8b + 5)$
$= b(b^2) + b(8b) + b(5) + 5(b^2)$
$+ 5(8b) + 5(5)$
$= b^3 + 8b^2 + 5b + 5b^2 + 40b + 25$
$= b^3 + 13b^2 + 45b + 25$

12. $(9 + d)(1 - 4d - 2d^2)$
$= 9(1) + 9(-4d) + 9(-2d^2) + d(1) + d(-4d)$
$+ d(-2d^2)$
$= 9 - 36d - 18d^2 + d - 4d^2 - 2d^3$
$= 9 - 35d - 22d^2 - 2d^3$

13. $(x + 6)(5x^2 - x - 1)$
$= x(5x^2) + x(-x) + x(-1) + 6(5x^2) + 6(-x) + 6(-1)$
$= 5x^3 - x^2 - x + 30x^2 - 6x - 6$
$= 5x^3 + 29x^2 - 7x - 6$

14. $(4x + 3)(4x - 3)$
$= 4x(4x) + 4x(-3) + 3(4x) + 3(-3)$
$= 16x^2 - 12x + 12x - 9$
$= 16x^2 - 9$

15. $(y - 3)(y - 2)$
$= y(y) + y(-2) + (-3)(y) + (-3)(-2)$
$= y^2 - 2y - 3y + 6$
$= y^2 - 5y + 6$

16. $(x^2 + x + 2)(3x^2 - x + 4)$
$= x^2(3x^2) + x^2(-x) + x^2(4) + x(3x^2) + x(-x) + x(4) + 2(3x^2) + 2(-x) + 2(4)$
$= 3x^4 - x^3 + 4x^2 + 3x^3 - x^2 + 4x + 6x^2 - 2x + 8$
$= 3x^4 + 2x^3 + 9x^2 + 2x + 8$

4.4 SOLUTIONS TO EXERCISES

1. $(x - 11)^2 = x^2 + 2(x)(-11) + 11^2$
$= x^2 - 22x + 121$

2. $(3x + 10)^2 = (3x)^2 + 2(3x)(10) + 10^2$
$= 9x^2 + 60x + 100$

3. $(y + 8)^2 = y^2 + 2(y)(8) + 8^2$
$= y^2 + 16y + 64$

4. $(2a - 1)^2 = (2a)^2 + 2(2a)(-1) + 1^2$
$= 4a^2 - 4a + 1$

5. $(y + 5)(y - 8) = y(y) - 8(y) + 5(y) + 5(-8)$
$= y^2 - 8y + 5y - 40$
$= y^2 - 3y - 40$

6. $(b - 2)(b - 9)$
$= b(b) + b(-9) + (-2)(b) + (-2)(-9)$
$= b^2 - 9b - 2b + 18$
$= b^2 - 11b + 18$

7. $(x + 9)(x - 9) = x^2 - 9^2$
$= x^2 - 81$

8. $(7y - 6)(7y + 6) = (7y)^2 - 6^2$
$= 49y^2 - 36$

9. $\left(x + \frac{1}{4}\right)\left(x - \frac{1}{4}\right) = x^2 - \left(\frac{1}{4}\right)^2$
$= x^2 - \frac{1}{16}$

10. $\left(\frac{2}{5}a^2 + b\right)\left(\frac{2}{5}a^2 - b\right) = \left(\frac{2}{5}a^2\right)^2 - b^2$
$= \frac{4}{25}a^4 - b^2$

11. $(5b + 2)^2 = (5b)^2 + 2(5b)(2) + 2^2$
$= 25b^2 + 20b + 4$

12. $(2a + 3)(a - 5)$
$= 2a(a) + 2a(-5) + 3(a) + 3(-5)$
$= 2a^2 - 10a + 3a - 15$
$= 2a^2 - 7a - 15$

13. $(x + 10)(x + 10) = (x + 10)^2$
$= x^2 + 2(x)(10) + 10^2$
$= x^2 + 20x + 100$

14. $(5a + 1)(a - 5)$
$= 5a(a) + 5a(-5) + 1(a) + 1(-5)$
$= 5a^2 - 25a + a - 5$
$= 5a^2 - 24a - 5$

15. $(4x + 5y)^2 = (4x)^2 + 2(4x)(5y) + (5y)^2$
$= 16x^2 + 40xy + 25y^2$

4.5 SOLUTIONS TO EXERCISES

1. $7^{-3} = \frac{1}{7^3} = \frac{1}{343}$

2. $\dfrac{1}{a^{-4}} = a^4$

3. $5^{-1} + 6^{-1} = \dfrac{1}{5^1} + \dfrac{1}{6^1}$

$= \dfrac{6}{30} + \dfrac{5}{30}$

$= \dfrac{11}{30}$

4. $\dfrac{a^{-3}}{a^{-8}} = a^{-3-(-8)}$

$= a^5$

5. $(x^{-5})^6 = x^{(-5)(6)}$

$= x^{-30}$

$= \dfrac{1}{x^{30}}$

6. $(2x^{-4})^{-3} = 2^{-3}(x^{-4})^{-3}$

$= \dfrac{x^{-4(-3)}}{2^3}$

$= \dfrac{x^{12}}{8}$

7. $-8^0 - 4x^0 = -1 - 4(1)$

$= -5$

8. $(2x^{-3}y^5)^{-4} = 2^{-4}(x^{-3})^{-4}(y^5)^{-4}$

$= \dfrac{x^{(-3)(-4)}y^{(5)(-4)}}{2^4}$

$= \dfrac{x^{12}y^{-20}}{16}$

$= \dfrac{x^{12}}{16y^{20}}$

9. $\dfrac{9c^3d^{-5}}{9^{-1}c^{-6}d^8} = 9^{1-(-1)}c^{3-(-6)}d^{-5-8}$

$= 9^2c^9d^{-13}$

$= \dfrac{81c^9}{d^{13}}$

10. $\dfrac{(x^2y^4)^6}{(xy^3)^{-5}} = \dfrac{(x^2)^6(y^4)^6}{x^{-5}(y^3)^{-5}}$

$= \dfrac{x^{(2)(6)}y^{(4)(6)}}{x^{-5}y^{(3)(-5)}}$

$= \dfrac{x^{12}y^{24}}{x^{-5}y^{-15}}$

$= x^{12-(-5)}y^{24-(-15)}$

$= x^{17}y^{39}$

11. $8_\wedge100000. = 8.1 \times 10^6$ The decimal count is positive 6.

12. $1_\wedge46000000000. = 1.46 \times 10^{11}$ The decimal count is positive 11.

13. $0.007_\wedge = 7.0 \times 10^{-3}$ The decimal count is negative 3.

14. $0.0000009_\wedge12 = 9.12 \times 10^{-7}$ The decimal count is negative 7.

15. $3.4 \times 10^{-5} = 0_\wedge00003.4 \times 10^{-5}$

$= 0.000034$

16. $2.71 \times 10^9 = 2.710000000_\wedge \times 10^9$

$= 2{,}710{,}000{,}000$

17. $4.0 \times 10^6 = 4.000000_\wedge \times 10^6$

$= 4{,}000{,}000$

18. $5.97 \times 10^{-8} = 0_\wedge00000005.97 \times 10^{-8}$

$= 0.0000000597$

4.6 SOLUTIONS TO EXERCISES

1. $$\frac{-35x^3yz^2}{-7xy^5z^2} = \frac{-35}{-7}x^{3-1}y^{1-5}z^{2-2}$$
$$= 5x^2y^{-4}z^0$$
$$= \frac{5x^2}{y^4}$$

2. $$\frac{21a^4b}{-3a^6b^5} = \frac{21}{-3}a^{4-6}b^{1-5}$$
$$= -7a^{-2}b^{-4}$$
$$= -\frac{7}{a^2b^4}$$

3. $$\frac{38a^4 - 57a^2}{19a^3} = \frac{38a^4}{19a^3} - \frac{57a^2}{19a^3}$$
$$= 2a - \frac{3}{a}$$

4. $$\frac{-10b^7 + 25b^3}{-5b^9} = \frac{-10b^7}{-5b^9} + \frac{25b^3}{-5b^9}$$
$$= \frac{2}{b^2} - \frac{5}{b^6}$$

5. $$\begin{array}{r} x - 3 \\ x + 8\,\overline{)\,x^2 + 5x - 24} \\ \underline{x^2 + 8x} \\ -3x - 24 \\ \underline{-3x - 24} \\ 0 \end{array}$$

6. $$\begin{array}{r} x - 5 \\ x - 7\,\overline{)\,x^2 - 12x + 35} \\ \underline{x^2 - 7x} \\ -5x + 35 \\ \underline{-5x + 35} \\ 0 \end{array}$$

7. $$\begin{array}{r} 2x - 3 \\ 2x + 1\,\overline{)\,4x^2 - 4x - 3} \\ \underline{4x^2 + 2x} \\ -6x - 3 \\ \underline{-6x - 3} \\ 0 \end{array}$$

8. $$\begin{array}{r} 2x + 3 \\ 4x + 5\,\overline{)\,8x^2 + 22x + 15} \\ \underline{8x^2 + 10x} \\ 12x + 15 \\ \underline{12x + 15} \\ 0 \end{array}$$

9. $$\begin{array}{r} x + 6 \\ x + 5\,\overline{)\,x^2 + 11x + 21} \\ \underline{x^2 + 5x} \\ 6x + 21 \\ \underline{6x + 30} \\ -9 \end{array}$$
$$x + 6 - \frac{9}{x + 5}$$

10. $$\begin{array}{r} x^2 + 1 \\ x - 2\,\overline{)\,x^3 - 2x^2 + x + 6} \\ \underline{x^3 - 2x^2} \\ x + 6 \\ \underline{x - 2} \\ 8 \end{array}$$
$$x^2 + 1 + \frac{8}{x - 2}$$

11. $$\begin{array}{r} x^2 + 5x + 25 \\ x - 5\,\overline{)\,x^3 + 0x^2 + 0x - 125} \\ \underline{x - 5x^2} \\ 5x^2 + 0x \\ \underline{5x^2 - 25x} \\ 25x - 125 \\ \underline{25x - 125} \\ 0 \end{array}$$

12.
$$
\begin{array}{r}
-x^2 - 5x + 1 \\
x - 1\overline{)-x^3 - 4x^2 + 6x - 1} \\
\underline{-x^3 + x^2} \qquad\qquad \\
-5x^2 + 6x \qquad \\
\underline{-5x^2 + 5x} \qquad \\
x - 1 \\
\underline{x - 1} \\
0
\end{array}
$$

13.
$$
\begin{array}{r}
4a^2 - 4a + 1 \\
2a - 1\overline{)8a^3 - 12a^2 + 6a - 5} \\
\underline{8a^3 - 4a^2} \qquad\qquad \\
-8a^2 + 6a \qquad \\
\underline{-8a^2 + 4a} \qquad \\
2a - 5 \\
\underline{2a - 1} \\
-4
\end{array}
$$

$$4a^2 - 4a + 1 - \frac{4}{2a - 1}$$

14.
$$
\begin{array}{r}
-4x + 24 \\
x + 6\overline{)-4x^2 + 0x + 9} \\
\underline{-4x^2 - 24x} \qquad \\
24x + 9 \\
\underline{24x + 144} \\
-135
\end{array}
$$

$$-4x + 24 - \frac{135}{x + 6}$$

15.
$$
\begin{array}{r}
x^2 - 1 \\
x^2 + 0x + 1\overline{)x^4 + 0x^3 + 0x^2 + x + 0} \\
\underline{x^4 + 0x^3 + x^2} \qquad\qquad \\
-x^2 + x + 0 \\
\underline{-x^2 - 0x - 1} \\
x + 1
\end{array}
$$

$$x^2 - 1 + \frac{x + 1}{x^2 + 1}$$

CHAPTER 4 PRACTICE TEST SOLUTIONS

1. $2^6 = 2 \cdot 2 \cdot 2 \cdot 2 \cdot 2 \cdot 2 = 64$

2. $(-4)^4 = (-4)(-4)(-4)(-4) = 256$

3. $-4^4 = -1 \cdot 4 \cdot 4 \cdot 4 \cdot 4 = -256$

4. $3^{-4} = \frac{1}{3^4} = \frac{1}{81}$

5. $\left(\frac{6x^5y^4}{36x^8y^2}\right)^3 = \left(\frac{y^2}{6x^3}\right)^3$
$= \frac{(y^2)^3}{6^3(x^3)^3}$
$= \frac{y^6}{216x^9}$

6. $\frac{8(xy)^5}{(xy)^3} = 8(xy)^{5-3}$
$= 8(xy)^2$
$= 8x^2y^2$

7. $2(x^3y^4)^{-5} = \frac{2}{(x^3y^4)^5}$
$= \frac{2}{(x^3)^5(y^4)^5}$
$= \frac{2}{x^{15}y^{20}}$

8. $\left(\frac{x^3y^4}{x^4y^{-5}}\right)^{-2} = [x^{3-4}y^{4-(-5)}]^{-2}$
$= (x^{-1}y^9)^{-2}$
$= (x^{-1})^{-2}(y^9)^{-2}$
$= x^2y^{-18}$
$= \frac{x^2}{y^{18}}$

9. $\frac{7^3x^{-8}y^{-2}}{7^5x^{-3}y^9} = 7^{3-5}x^{-8-(-3)}y^{-2-9}$
$= 7^{-2}x^{-5}y^{-11}$
$= \frac{1}{7^2x^5y^{11}}$
$= \frac{1}{49x^5y^{11}}$

10. $8_\wedge230{,}000{,}000. = 8.23 \times 10^9$

11. $0.000007_\wedge14 = 7.14 \times 10^{-6}$

12. $2.7 \times 10^{-5} = 0_\wedge00002.7 \times 10^{-5}$
$= 0.000027$

13. $8.3 \times 10^6 = 8.300000_\wedge \times 10^6$
$= 8{,}300{,}000$

14. $(2.1 \times 10^7)(4 \times 10^{-9}) = (2.1)(4) \times 10^{7+(-9)}$
$= 8.4 \times 10^{-2}$
$= 0.084$

15. degree of 1st term: $3 + 4 = 7$
degree of 2nd term: $1 + 2 + 1 = 4$
degree of 3rd term: $4 + 1 + 3 = 8$

degree of polynomial:
8 (the largest of the above degrees)

16. $12x^2yz - 7xy^3 - 8x^2yz + 2xy^3$
$= (12 - 8)x^2yz + (-7 + 2)xy^3$
$= 4x^2yz - 5xy^3$

17. $(4x^3 - 6x^2 + 3x - 9) + (5x^3 - 6x - 3)$
$= (4 + 5)x^3 - 6x^2 + (3 - 6)x + (-9 - 3)$
$= 9x^3 - 6x^2 - 3x - 12$

18.
$$\begin{array}{r} 6x^3 + 2x^2 + x - 3 \\ \underline{-(9x^3 - 3x^2 - 2x + 12)} \end{array}$$

$$\begin{array}{r} 6x^3 + 2x^2 + x - 3 \\ \underline{-9x^3 + 3x^2 + 2x - 12} \\ -3x^3 + 5x^2 + 3x - 15 \end{array}$$

19. $[(6x^2 - 6x + 1) + (x^3 + 5)] - (5x - 1)$
$= (x^3 + 6x^2 - 6x + 6) - (5x - 1)$
$= x^3 + 6x^2 - 6x + 6 - 5x + 1$
$= x^3 + 6x^2 - 11x + 7$

20. $(2x + 3)(x^2 - 4x + 5)$
$= 2x(x^2 - 4x + 5) + 3(x^2 - 4x + 5)$
$= 2x^3 - 8x^2 + 10x + 3x^2 - 12x + 15$
$= 2x^3 - 5x^2 - 2x + 15$

21.
$$\begin{array}{rrrrrr} & & x^3 & + x^2 & - x & - 2 \\ & & & \underline{3x^2} & \underline{- 2x} & \underline{+ 6} \\ & & 6x^3 & + 6x^2 & - 6x & - 12 \\ & -2x^4 & - 2x^3 & + 2x^2 & + 4x & \\ \underline{3x^5} & \underline{+ 3x^4} & \underline{- 3x^3} & \underline{- 6x^2} & & \\ 3x^5 & + x^4 & + x^3 & + 2x^2 & - 2x & - 12 \end{array}$$

22. $(x - 6)(2x + 7) = 2x^2 + 7x - 12x - 42$
$= 2x^2 - 5x - 42$

23. $(2x - 5)(2x + 5) = (2x)^2 - 5^2$
$= 4x^2 - 25$

24. $(3x - 8)^2 = (3x)^2 - 2(3x)(8) + 8^2$
$= 9x^2 - 48x + 64$

25. $(7x + 2)^2 = (7x)^2 + 2(7x)(2) + 2^2$
$= 49x^2 + 28x + 4$

26. $(x^2 - 3b)(x^2 + 3b) = (x^2)^2 - (3b)^2$
$= x^4 - 9b^2$

27. $t = 2$: $-16(2)^2 + 1001 = 937$ feet

$t = 4$: $-16(4)^2 + 1001 = 745$ feet

$t = 6$: $-16(6)^2 + 1001 = 425$ feet

28. $$\frac{12x^2y^3}{3x^4y^5z^2} = \frac{4}{x^2y^2z^2}$$

29. $$\frac{5x^2 + 10xy - 8x}{15xy} = \frac{5x^2}{15xy} + \frac{10xy}{15xy} - \frac{8x}{15xy}$$
$$= \frac{x}{3y} + \frac{2}{3} - \frac{8}{15y}$$

30.
$$\begin{array}{r} x + 6 \\ x + 3\overline{)x^2 + 9x + 18} \\ \underline{x^2 + 3x} \\ 6x + 18 \\ \underline{6x + 18} \\ 0 \end{array}$$

31.
$$\begin{array}{r} 4x^2 - 6x + 9 \\ 2x + 3\overline{)8x^3 + 0x^2 + 0x - 27} \\ \underline{8x^3 + 12x^2} \\ -12x^2 + 0x \\ \underline{-12x^2 - 18x} \\ 18x - 27 \\ \underline{18x + 27} \\ -54 \end{array}$$

$$4x^2 - 6x + 9 - \frac{54}{2x + 3}$$

5.1 SOLUTIONS TO EXERCISES

1. $16 = 2 \cdot 2 \cdot 2 \cdot 2 = 2^4$
   $40 = 2 \cdot 2 \cdot 2 \cdot 5 = 2^3 \cdot 5$
   $72 = 2 \cdot 2 \cdot 2 \cdot 3 \cdot 3 = 2^3 \cdot 3^2$

   GCF $= 2^3 = 8.$

2. The GCF of the numerical coefficients 6, 3, and 15 is 3.
   The GCF of the variable factor a^6, a^3, and a^4 is a^3.
   The GCF of the variable factors b^4, b^9, and b^2 is b^2.

   Thus the GCF of the terms is $3a^3b^2$.

3. $5x - 35 = (5)(x) + (5)(-7)$
   $= 5(x - 7)$

4. $6x - 15y + 21$
   $= (3)(2x) + (3)(-5y) + (3)(7)$
   $= 3(2x - 5y + 7)$

5. $a(b + 4) - 6(b + 4) = (b + 4)(a - 6)$

6. $12(6 - y) + x(6 - y) = (6 - y)(12 + x)$

7. $c(d^2 - 3) - 9(d^2 - 3) = (d^2 - 3)(c - 9)$

8. $-4x^3y^7 - 12x^4y^6 + 6x^5y^2$
   $= (-2x^3y^2)(2y^5) + (-2x^3y^2)(6xy^4)$
   $+ (-2x^3y^2)(-3x^2)$
   $= -2x^3y^2(2y^5 + 6xy^4 - 3x^2)$

9. $xy + 2y + 11x + 22 = (xy + 2y) + (11x + 22)$
   $= y(x + 2) + 11(x + 2)$
   $= (x + 2)(y + 11)$

10. $3ab + 6a - b - 2 = (3ab + 6a) + (-b - 2)$
    $= 3a(b + 2) - 1(b + 2)$
    $= (b + 2)(3a - 1)$

11. $18 - 45x - 14y + 35xy = (18 - 45x) + (-14y + 35xy)$
    $= 9(2 - 5x) - 7y(2 - 5x)$
    $= (2 - 5x)(9 - 7y)$

12. $4x^3 - 3x^2 + 4x - 3$
    $= (4x^3 - 3x^2) + (4x - 3)$
    $= x^2(4x - 3) + 1(4x - 3)$
    $= (4x - 3)(x^2 + 1)$

13. $8ac - 8ad + bc - bd$
    $= (8ac - 8ad) + (bc - bd)$
    $= 8a(c - d) + b(c - d)$
    $= (8a + b)(c - d)$

14. $8xy + 28y - 32x - 112$
    $= 4(2xy + 7y - 8x - 28)$
    $= 4[(2xy + 7y) + (-8x - 28)]$
    $= 4[y(2x + 7) - 4(2x + 7)]$
    $= 4(2x + 7)(y - 4)$

15. $42 - 21b - 14a + 7ab$
    $= 7(6 - 3b - 2a + ab)$
    $= 7[(6 - 3b) + (-2a + ab)]$
    $= 7[3(2 - b) - a(2 - b)]$
    $= 7(2 - b)(3 - a)$

5.2 SOLUTIONS TO EXERCISES

1.

Negative factors of 28	Sum of Factors
−1, −28	−29
−2, −14	−16
−4, −7	**−11**

$x^2 - 11x + 28 = (x - 4)(x - 7)$

2.

Factors of −27	Sum of Factors
1, −27	−26
−1, 27	26
3, −9	**−6**
−3, 9	6

$x^2 - 6x - 27 = (x + 3)(x - 9)$

3.

Negative factors of 8	Sum of Factors
−1, −8	**−9**
−2, −4	−6

$x^2 - 9x + 8 = (x - 1)(x - 8)$

4.

Negative factors of 6	Sum of Factors
−1, −6	−7
−2, −3	−5

Since none of the factors of 6 have a sum of −1, $x^2 - x + 6$ is prime.

5.

Factors of −30	Sum of Factors
1, −30	−29
−1, 30	29
2, −15	−13
−2, 15	13
3, −10	−7
−3, 10	7
5, −6	−1
−5, 6	**1**

$x^2 + x - 30 = (x - 5)(x + 6)$

6.

Factors of −3	Sum of Factors
1, −3	−2
−1, 3	**2**

$x^2 + 2xy - 3y^2 = (x - 1y)(x + 3y)$
$= (x - y)(x + 3y)$

7. $x^3 - 13x^2 + 42x = x(x^2 - 13x + 42)$

Negative factors of 42	Sum of Factors
−1, −42	−43
−2, −21	−23
−3, −14	−17
−6, −7	−13

$x^3 - 13x^2 + 42x = x(x - 6)(x - 7)$

8. $3x^2 + 15xy + 18y^2 = 3(x^2 + 5xy + 6y^2)$

Positive factors of 6	Sum of Factors
1, 6	7
2, 3	**5**

$3x^2 + 15xy + 18y^2 = 3(x + 2y)(x + 3y)$

9.

Factors of −110	Sum of Factors
1, −110	−109
−1, 110	109
2, −55	−53
−2, 55	53
5, −22	−17
−5, 22	17
10, −11	−1
−10, 11	**1**

$x^2 + x - 110 = (x - 10)(x + 11)$

10.

Positive factors of 13	Sum of Factors
1, 13	14

$x^2 + 14x + 13 = (x + 1)(x + 13)$

11.

Factors of −12	Sum of Factors
1, −12	−11
−1, 12	**11**
2, −6	−4
−2, 6	4
3, −4	−1
−3, 4	1

$x^2 + 11xy - 12y^2 = (x - 1y)(x + 12y)$
$= (x - y)(x + 12y)$

12.

Factors of −10	Sum of Factors
1, −10	−9
−1, 10	9
2, −5	−3
−2, 5	3

Since none of the factors of −10 have a sum of −7, $x^2 - 7x - 10$ is prime.

13. $2x^2 - 26x + 80 = 2(x^2 - 13x + 40)$

Negative factors of 40	Sum of Factors
−1, −40	−41
−2, −20	−22
−4, −10	−14
−5, −8	**−13**

$2x^2 - 26x + 80 = 2(x - 5)(x - 8)$

14. $4x^2y - 8xy + 16y = 4y(x^2 - 2x + 4)$

Negative factors of 4	Sum of Factors
-1, -4	-5
-2, -2	-4

Since none of the factors of 4 have a sum of -2, $x^2 - 2x + 4$ will not factor.

15. $5a^2b - 5ab^2 - 30b^3 = 5b(a^2 - 5ab - 6b^2)$

Factors of -6	Sum of Factors
1, -6	**-5**
-1, 6	5
2, -3	-1
-2, 3	1

$$5a^2 - 5ab^2 - 30b^3 = 5b(a + 1b)(a - 6b)$$
$$= 5b(a + b)(a - 6b)$$

5.3 SOLUTIONS TO EXERCISES

1.

Positive factors of $5 \cdot 2 = 10$	Sum of Factors
1, 10	11
2, 5	**7**

$$5x^2 + 7x + 2 = 5x^2 + 2x + 5x + 2$$
$$= x(5x + 2) + 1(5x + 2)$$
$$= (5x + 2)(x + 1)$$

2.

Positive factors of $2 \cdot 9 = 18$	Sum of Factors
1, 18	19
2, 9	11
3, 6	**9**

$$2x^2 + 9x + 9 = 2x^2 + 3x + 6x + 9$$
$$= x(2x + 3) + 3(2x + 3)$$
$$= (2x + 3)(x + 3)$$

3.

Factors of $2 \cdot (-12) = -24$	Sum of Factors
1, -24	-23
-1, 24	23
2, -12	-10
-2, 12	10
3, -8	**-5**
-3, 8	5
4, -6	-2
-4, 6	2

$$2x^2 - 5x - 12 = 2x^2 + 3x - 8x - 12$$
$$= x(2x + 3) - 4(2x + 3)$$
$$= (2x + 3)(x - 4)$$

4.

Factors of $3 \cdot (-2) = -6$	Sum of Factors
1, -6	**-5**
-1, 6	5
2, -3	-1
-2, 3	1

$$3a^2 - 5a - 2 = 3a^2 + a - 6a - 2$$
$$= a(3a + 1) - 2(3a + 1)$$
$$= (3a + 1)(a - 2)$$

5.

Factors of $9 \cdot (-8) = -72$	Sum of Factors
1, -72	-71
-1, 72	71
2, -36	-34
-2, 36	34
3, -24	-21
-3, 24	21
4, -18	-14
-4, 18	14
6, -12	-6
-6, 12	6
8, -9	-1
-8, 9	**1**

$$9y^2 + y - 8 = 9y^2 - 8y + 9y - 8$$
$$= y(9y - 8) + 1(9y - 8)$$
$$= (9y - 8)(y + 1)$$

6.

Positive factors of $4 \cdot 5 = 20$	Sum of Factors
1, 20	21
2, 10	12
4, 5	9

Since none of the factors of 20 have a sum of 1, $4b^2 + b + 5$ is prime.

7. $16a^2 + 8a + 1$
$= (4a)^2 + 2(4a)(1) + (1)^2$
$= (4a + 1)^2$

8. $49a^2 - 28a + 4$
$= (7a)^2 - 2(7a)(2) + (2)^2$
$= (7a - 2)^2$

9. $10x^2 + 58x - 12 = 2(5x^2 + 29x - 6)$

Factors of $5 \cdot (-6) = -30$	Sum of Factors
1, −30	−29
−1, 30	**29**
2, −15	−13
−2, 15	13
3, −10	−7
−3, 10	7
5, −6	−1
−5, 6	1

$10x^2 + 58x - 12 = 2(5x^2 - 1x + 30x - 6)$
$= 2[x(5x - 1) + 6(5x - 1)]$
$= 2(5x - 1)(x + 6)$

10. $10x^2 - 55x + 45 = 5(2x^2 - 11x + 9)$

Negative factors of $2 \cdot 9 = 18$	Sum of Factors
−1, −18	−19
−2, −9	**−11**
−3, −6	−9

$10x^2 - 55x + 45 = 5(2x^2 - 2x - 9x + 9)$
$= 5[2x(x - 1) - 9(x - 1)]$
$= 5(x - 1)(2x - 9)$

11.

Factors of $9 \cdot (-4) = -36$	Sum of Factors
1, −36	−35
−1, 36	35
2, −18	−16
−2, 18	16
3, −12	−9
−3, 12	9
4, −9	−5
−4, 9	5
6, −6	0

Since none of the factors of −36 have a sum of 6, $9y^2 + 6y - 4$ is prime.

12. $9x^2 + 36x + 54 = 9(x^2 + 4x + 6)$

Positive factors of 6	Sum of Factors
1, 6	7
2, 3	5

Since none of the factors of 6 have a sum of 4, $x^2 + 4x + 6$ will not factor.

$9x^2 + 36x + 54 = 9(x^2 + 4x + 6)$

13.

Negative factors of $8 \cdot 15 = 120$	Sum of Factors
−1, −120	−121
−2, −60	−62
−3, −40	−43
−4, −30	−34
−5, −24	−29
−6, −20	**−26**
−8, −15	−23
−10, −12	−22

$8d^2 - 26d + 15 = 8d^2 - 6d - 20d + 15$
$= 2d(4d - 3) - 5(4d - 3)$
$= (4d - 3)(2d - 5)$

14. $6x^3 + 25x^2 + 4x = x(6x^2 + 25x + 4)$

Positive Factors of $6 \cdot 4 = 24$	Sum of Factors
1, 24	25
2, 12	14
3, 8	11
4, 6	10

$$\begin{aligned} &6x^3 + 25x^2 + 4x \\ &= x(6x^2 + x + 24x + 4) \\ &= x[x(6x + 1) + 4(6x + 1)] \\ &= x(6x + 1)(x + 4) \end{aligned}$$

15. $9x^2y^2 - 30xy + 25$
$= (3xy)^2 - 2(3xy)(5) + (5)^2$
$= (3xy - 5)^2$

16. $100 - 180a + 81a^2$
$= 10^2 - 2(10)(9a) + (9a)^2$
$= (10 - 9a)^2$

17. $49x^2 + 84xy + 36y^2$
$= (7x)^2 + 2(7x)(6y) + (6y)^2$
$= (7x + 6y)^2$

5.4 SOLUTIONS TO EXERCISES

1. $9x^2 - 25 = (3x)^2 - (5)^2$
$= (3x + 5)(3x - 5)$

2. $100x^2 + 9$ is prime since it is the sum of two squares.

3. $144a^2 - 49b^2 = (12a)^2 - (7b)^2$
$= (12a + 7b)(12a - 7b)$

4. $y^3 - 64 = (y)^3 - (4)^3$
$= (y - 4)[(y)^2 + (y)(4) + (4)^2]$
$= (y - 4)(y^2 + 4y + 16)$

5. $b^3 + 125 = (b)^3 + (5)^3$
$= (b + 5)[(b)^2 - (b)(5) + (5)^2]$
$= (b + 5)(b^2 - 5b + 25)$

6. $12x^3 + 27 = 3(4x^3 + 9)$

7. $2 - 8a^2 = 2(1 - 4a^2)$
$= 2[(1)^2 - (2a)^2]$
$= 2(1 + 2a)(1 - 2a)$

8. $49x^2 - 121y^2 = (7x)^2 - (11y)^2$
$= (7x + 11y)(7x - 11y)$

9. $24x^3 + 1029$
$= 3(8x^3 + 343)$
$= 3[(2x)^3 + (7)^3]$
$= 3(2x + 7)[(2x)^2 - (2x)(7) + (7)^2]$
$= 3(2x + 7)(4x^2 - 14x + 49)$

10. $27c^3 - 125d^3$
$= (3c)^3 - (5d)^3$
$= (3c - 5d)[(3c)^2 + (3c)(5d) + (5d)^2]$
$= (3c - 5d)(9c^2 + 15cd + 25d^2$

11. $27x^2y^3 + 4x^2y = x^2y(27y^2 + 4)$

12. $4x^2y^2 + 9z^2$ is prime since it is the sum of two squares.

13. $486 - 18a^3$
$= 18(27 - a^3)$
$= 18[(3)^3 - (a)^3]$
$= 18(3 - a)[(3)^2 + (3)(a) + (a)^2]$
$= 18(3 - a)(9 + 3a + a^2)$

14. $16x^4y + 64xy^3 = 16xy(x^3 + 4y^2)$

15. $7 + 56z^3$
$= 7(1 + 8z^3)$
$= 7[(1)^3 + (2z)^3]$
$= 7(1 + 2z)[(1)^2 - (1)(2z) + (2z)^2]$
$= 7(1 + 2z)(1 - 2z + 4z^2)$

5.5 SOLUTIONS TO EXERCISES

1. $x^2 - 4xy + 4y^2 = (x)^2 - 2(x)(2y) + (2y)^2$
$= (x - 2y)^2$

2. Factors of 12 whose sum is -7: -3, -4
$a^2 - 7a + 12 = (a - 3)(a - 4)$

3. Factors of 2 whose sum is 3: 1, 2
$b^2 - 3b + 2 = (b + 1)(b + 2)$

4. $x^2 + 6x + 9 = (x)^2 + 2(x)(3) + (3)^2$
$= (x + 3)^2$

5. $2x^2 + 2x - 112 = 2(x^2 + x - 56)$
Factors of -56 whose sum is 1: -7, 8

$2x^2 + 2x - 112 = 2(x - 7)(x + 8)$

6. $x^3 - 9x^2 + 20x = x(x^2 - 9x + 20)$
Factors of 20 whose sum is -9: -4, -5

$x^3 - 9x^2 + 20x = x(x - 4)(x - 5)$

7. $4x^2 + 49y^2$ is prime since it is the sum of two squares.

8. $10 - 7x + x^2 = x^2 - 7x + 10$
Factors of 10 whose sum is -7: -2, -5

$x^2 - 7x + 10 = (x - 2)(x - 5)$

The answer can also be written as $(2 - x)(5 - x)$.

9. $4a^2 - 49b^2 = (2a)^2 - (7b)^2$
$= (2a + 7b)(2a - 7b)$

10. $d^3 - 64c^3 = (d)^3 - (4c)^3$
$= (d - 4c)[(d)^2 + (d)(4c) + (4c)^2]$
$= (d - 4c)(d^2 + 4cd + 16c^2)$

11. $6x^3 - 150x = 6x(x^2 - 25)$
$= 6x[(x)^2 - (5)^2]$
$= 6x(x + 5)(x - 5)$

12. $3ab - 5a + 6b - 10$
$= (3ab - 5a) + (6b - 10)$
$= a(3b - 5) + 2(3b - 5)$
$= (3b - 5)(a + 2)$

13. Prime, since there are no factors of 8 whose sum is 7.

14. $25x^2 + 40xy + 16y^2$
$= (5x)^2 + 2(5x)(4y) + (4y)^2$
$= (5x + 4y)^2$

15. $x - y + a(x - y) = 1(x - y) + a(x - y)$
$= (x - y)(1 + a)$

16. $x^3 - 16x + 2x^2 - 32$
$= (x^3 - 16x) + (2x^2 - 32)$
$= x(x^2 - 16) + 2(x^2 - 16)$
$= (x^2 - 16)(x + 2)$
$= [(x)^2 - (4)^2](x + 2)$
$= (x + 4)(x - 4)(x + 2)$

5.6 SOLUTIONS TO EXERCISES

1. $(x - 6)(x + 7) = 0$
$x - 6 = 0$ or $x + 7 = 0$
$x = 6$ or $x = -7$

$\{6, -7\}$

2. $3x(2x - 9) = 0$
$3x = 0$ or $2x - 9 = 0$
$x = 0$ or $2x = 9$
$x = \frac{9}{2}$

$\left\{0, \frac{9}{2}\right\}$

3. $(8y + 1)(7y + 2) = 0$
$8y + 1 = 0$ or $7y + 2 = 0$
$8y = -1$ or $7y = -2$

$y = -\frac{1}{8}$ or $y = -\frac{2}{7}$

$\left\{-\frac{1}{8}, -\frac{2}{7}\right\}$

4. $x^2 + 8x + 12 = 0$
$(x + 2)(x + 6) = 0$
$x + 2 = 0$ or $x + 6 = 0$
$x = -2$ or $x = -6$

$\{-2\ -6\}$

5. $b^2 + 4b - 45 = 0$
$(b + 9)(b - 5) = 0$
$b + 9 = 0$ or $b - 5 = 0$
$b = -9$ or $b = 5$

$\{-9, 5\}$

6. $2x^2 + 5x - 7 = 0$
$(2x + 7)(x - 1) = 0$
$2x + 7 = 0$ or $x - 1 = 0$
$2x = -7$ or $x = 1$
$x = -\frac{7}{2}$

$\left\{-\frac{7}{2}, 1\right\}$

7. $z^2 - 3z = -2$
$z^2 - 3z + 2 = 0$
$(z - 2)(z - 1) = 0$
$z - 2 = 0$ or $z - 1 = 0$
$z = 2$ or $z = 1$

$\{2, 1\}$

8. $x^2 + 21 = -10x$
$x^2 + 10x + 21 = 0$
$(x + 7)(x + 3) = 0$
$x + 7 = 0$ or $x + 3 = 0$
$x = -7$ or $x = -3$

$\{-7, -3\}$

9. $25(x + 1) = -6x^2$
$25x + 25 = -6x^2$
$6x^2 + 25x + 25 = 0$
$(2x + 5)(3x + 5) = 0$
$2x + 5 = 0$ or $3x + 5 = 0$
$2x = -5$ or $3x = -5$
$x = -\frac{5}{2}$ or $x = -\frac{5}{3}$

$\left\{-\frac{5}{2}, -\frac{5}{3}\right\}$

10. $a(11 - a) = 0$
$a = 0$ or $11 - a = 0$
$11 = a$

$\{0, 11\}$

11. $25b^2 - 1 = 0$
$(5b + 1)(5b - 1) = 0$
$5b + 1 = 0$ or $5b - 1 = 0$
$5b = -1$ or $5b = 1$
$b = -\frac{1}{5}$ or $b = \frac{1}{5}$

$\left\{-\frac{1}{5}, \frac{1}{5}\right\}$

12. $x^3 + 8x^2 + 7x = 0$
$x(x^2 + 8x + 7) = 0$
$x(x + 7)(x + 1) = 0$
$x = 0$ or $x + 7 = 0$ or $x + 1 = 0$
$x = -7$ or $x = -1$

$\{0, -7, -1\}$

13. $9t + 8 = 12$
$9t = 4$
$t = \frac{4}{9}$

$\left\{\frac{4}{9}\right\}$

14. $12x^2 + 25x + 12 = 0$
$(4x + 3)(3x + 4) = 0$
$4x + 3 = 0$ or $3x + 4 = 0$
$4x = -3$ or $3x = -4$
$x = -\frac{3}{4}$ or $x = -\frac{4}{3}$

$\left\{-\frac{3}{4}, -\frac{4}{3}\right\}$

15. $x^2 - 10x + 10 = -15$
$x^2 - 10x + 25 = 0$
$(x - 5)(x - 5) = 0$
$x - 5 = 0$ or $x - 5 = 0$
$x = 5$ or $x = 5$

$\{5\}$

5.7 SOLUTIONS TO EXERCISES

1. x = first number
$18 - x$ = second number

2. x = age now
$x + 11$ = age 11 years from now

3. x = second side
$3x - 5$ = first side
$x + 9$ = third side

4. x = first even integer
$x + 2$ = second even integer
$(x + 2) + 2 = x + 4$ = third even integer

5. x = width
$2x - 2$ = length

$$\text{Area} = \text{width} \cdot \text{length}$$
$$84 = (x)(2x - 2)$$
$$84 = 2x^2 - 2x$$
$$0 = 2x^2 - 2x - 84$$
$$0 = 2(x^2 - x - 42)$$
$$0 = 2(x - 7)(x + 6)$$
$$x - 7 = 0 \quad \text{or} \quad x + 6 = 0$$
$$x = 7 \quad \text{or} \quad x = -6$$

Discard $x = -6$ since the width must be positive.

$$x = 7$$
$$2x - 2 = 2(7) - 2 = 12$$

The dimensions are 7 cm by 12 cm.

6. Find t when $h = 0$.

$$h = -16t^2 + 576$$
$$0 = -16t^2 + 576$$
$$0 = -16(t^2 - 36)$$
$$0 = -16(t + 6)(t - 6)$$
$$t + 6 = 0 \quad \text{or} \quad t - 6 = 0$$
$$t = -6 \quad \text{or} \quad t = 6$$

Discard $t = -6$ since time is positive.

It will hit the ground in 6 seconds.

7. x = altitude
$4x - 5$ = base

$$\text{Area} = \frac{1}{2} \cdot \text{base} \cdot \text{altitude}$$
$$57 = \frac{1}{2}(4x - 5)(x)$$
$$2(57) = 2\left[\frac{1}{2}(4x - 5)(x)\right]$$
$$114 = (4x - 5)(x)$$
$$114 = 4x^2 - 5x$$
$$0 = 4x^2 - 5x - 114$$
$$0 = (4x + 19)(x - 6)$$
$$4x + 19 = 0 \quad \text{or} \quad x - 6 = 0$$
$$4x = -19 \quad \text{or} \quad x = 6$$
$$x = -\frac{19}{4}$$

Discard $x = -\frac{19}{4}$ since altitude is positive.

$$x = 6$$
$$4x - 5 = 4(6) - 5 = 19$$

The altitude is 6 meters and the base is 19 meters.

8. x = original side length
$x - 4$ = side length decreased by 4

$$\text{Area} = (\text{side})^2$$
$$25 = (x - 4)^2$$
$$25 = x^2 - 8x + 16$$
$$0 = x^2 - 8x - 9$$
$$0 = (x - 9)(x + 1)$$
$$x - 9 = 0 \quad \text{or} \quad x + 1 = 0$$
$$x = 9 \quad \text{or} \quad x = -1$$

Discard $x = -1$ since length is positive.

The original square had sides of length 9 inches.

9. x = hypotenuse
$x - 14$ = short leg
$x - 7$ = long leg

$$(\text{leg})^2 + (\text{leg})^2 = (\text{hypotenuse})^2$$
$$(x - 14)^2 + (x - 7)^2 = (x)^2$$
$$x^2 - 28x + 196 + x^2 - 14x + 49 = x^2$$
$$2x^2 - 42x + 245 = x^2$$
$$x^2 - 42x + 245 = 0$$
$$(x - 7)(x - 35) = 0$$
$$x - 7 = 0 \quad \text{or} \quad x - 35 = 0$$
$$x = 7 \quad \text{or} \quad x = 35$$

If $x = 7$ then $x - 14 = 7 - 14 = -7$ and a leg cannot have a negative length.

$$x = 35$$
$$x - 14 = 35 - 14 = 21$$
$$x - 7 = 35 - 7 = 28$$

The triangle has legs of lengths 21 cm and 28 cm and a hypotenuse of 35 cm.

10. x = the number

$$x + x^2 = 272$$
$$x^2 + x - 272 = 0$$
$$(x + 17)(x - 16) = 0$$
$$x + 17 = 0 \quad \text{or} \quad x - 16 = 0$$
$$x = -17 \quad \text{or} \quad x = 16$$

The number is −17 or 16.

11. x = first number
$17 - x$ = second number

$$x^2 + (17 - x)^2 = 157$$
$$x^2 + 289 - 34x + x^2 = 157$$
$$2x^2 - 34x + 289 = 157$$
$$2x^2 - 34x + 132 = 0$$
$$2(x^2 - 17x + 66) = 0$$
$$2(x - 6)(x - 11) = 0$$
$$x - 6 = 0 \quad \text{or} \quad x - 11 = 0$$
$$x = 6 \quad \text{or} \quad x = 11$$
$$17 - x = 17 - 6 = 11 \qquad 17 - x = 17 - 11 = 6$$

The numbers are 6 and 11.

12. 2 · width + 2 · length = Perimeter

$$2W + 2L = 52$$
$$2(W + L) = 52$$

$$\frac{1}{2}[2(W + L)] = \frac{1}{2}(52)$$

$$W + L = 26$$
$$L = 26 - W$$

We can represent the width by W and the length by $26 - W$.

Area = width · length
$$153 = (W)(26 - W)$$
$$153 = 26W - W^2$$
$$W^2 - 26W + 153 = 0$$
$$(W - 9)(W - 17) = 0$$
$$W - 9 = 0 \quad \text{or} \quad W - 17 = 0$$
$$W = 9 \quad \text{or} \quad W = 17$$
$$26 - W = 26 - 9 = 17 \quad \text{or} \quad 26 - W = 26 - 17 = 9$$

The dimensions are 9 feet by 17 feet.

13. 1st odd integer = x
2nd odd integer = $x + 2$ (larger integer)

$$(x)^2 + (x + 2)^2 = 20(x + 2) - 18$$
$$x^2 + x^2 + 4x + 4 = 20x + 40 - 18$$
$$2x^2 + 4x + 4 = 20x + 22$$
$$2x^2 - 16x - 18 = 0$$
$$2(x^2 - 8x - 9) = 0$$
$$2(x - 9)(x + 1) = 0$$
$$x - 9 = 0 \quad \text{or} \quad x + 1 = 0$$
$$x = 9 \quad \text{or} \quad x = -1$$
$$x + 2 = 9 + 2 = 11 \quad \text{or} \quad x + 2 = -1 + 2 = 1$$

The integers are 9 and 11 or −1 and 1.

14. 1st even integer = x
2nd even integer = $x + 2$

$$(x)(x + 2) = 1088$$
$$x^2 + 2x = 1088$$
$$x^2 + 2x - 1088 = 0$$
$$(x + 34)(x - 32) = 0$$
$$x + 34 = 0 \quad \text{or} \quad x - 32 = 0$$
$$x = -34 \quad \text{or} \quad x = 32$$
$$x + 2 = -34 + 2 = -32 \quad \text{or} \quad x + 2 = 32 + 2 = 34$$

The integers are −34 and −32 or 32 and 34.

15. x = base
$x - 7$ = altitude

$$\text{Area} = \frac{1}{2}\text{ base} \cdot \text{altitude}$$
$$99 = \frac{1}{2}(x)(x - 7)$$
$$2(99) = 2\left[\frac{1}{2}(x)(x - 7)\right]$$
$$198 = x(x - 7)$$
$$198 = x^2 - 7x$$
$$0 = x^2 - 7x - 198$$
$$0 = (x - 18)(x + 11)$$
$$x - 18 = 0 \quad \text{or} \quad x + 11 = 0$$
$$x = 18 \quad \text{or} \quad x = -11$$

Discard $x = -11$ since the base is positive.

$$x = 18$$
$$x - 7 = 18 - 7 = 11$$

The base is 18 inches and the altitude is 11 in.

CHAPTER 5 PRACTICE TEST SOLUTIONS

1. $8x^3 + 29x^2 + 20x$
$= x(8x^2) + x(29x) + x(20)$
$= x(8x^2 + 29x + 20)$

2. $x^2 + x - 9$

Factors of -9	Sum of Factors
1, -9	-8
-1, 9	8
3, -3	0

Since the sum of factors is never 1, $x^2 + x - 9$ is prime.

3. Look for two numbers whose product is -48 and whose sum is -13: -16, 3

$y^2 - 13y - 48 = (y - 16)(y + 3)$

4. $2a^2 + 2ab - 5a - 5b$
$= 2a(a + b) - 5(a + b)$
$= (a + b)(2a - 5)$

5. Look for two numbers whose product is (3)(2) = 6 and whose sum is -7: -1, -6

$3x^2 - 7x + 2 = 3x^2 - x - 6x + 2$
$= x(3x - 1) - 2(3x - 1)$
$= (3x - 1)(x - 2)$

6. $x^2 + 19x + 80$

Positive factors of 80	Sum of Factors
1, 80	81
2, 40	42
4, 20	24
5, 16	21
8, 10	18

Since the sum of factors is never 19, $x^2 + 19x + 80$ is prime.

7. Look for two numbers whose product is 24 and whose sum is 11: 8, 3

$x^2 + 11xy + 24y^2 = (x + 8y)(x + 3y)$

8. $40x^3 + 10x^2 - 16x = 2x(20x^2 + 5x - 8)$

9. $175 - 7x^2 = 7(25 - x^2)$
$= 7(5^2 - x^2)$
$= 7(5 - x)(5 + x)$

10. $27x^3 - 64 = (3x)^3 - 4^3$
$= (3x - 4)[(3x)^2 + (3x)(4) + 4^2]$
$= (3x - 4)(9x^2 + 12x + 16)$

11. Look for two numbers whose product is 4(-7) = -28 and whose sum is 12: 14, -2

$4t^2 + 12t - 7 = 4t^2 + 14t - 2t - 7$
$= 2t(2t + 7) - 1(2t + 7)$
$= (2t + 7)(2t - 1)$

12. $xy^2 - 11y^2 - 9x + 99$
$= y^2(x - 11) - 9(x - 11)$
$= (x - 11)(y^2 - 9)$
$= (x - 11)(y^2 - 3^2)$
$= (x - 11)(y - 3)(y + 3)$

13. $16x - x^5 = x(16 - x^4)$
$= x[4^2 - (x^2)^2]$
$= x(4 - x^2)(4 + x^2)$
$= x(2^2 - x^2)(4 + x^2)$
$= x(2 - x)(2 + x)(4 + x^2)$

14. $x^2 - 2x = 35$
$x^2 - 2x - 35 = 0$
$(x - 7)(x + 5) = 0$
$x - 7 = 0$ or $x + 5 = 0$
$x = 7$ or $x = -5$

{7, -5}

15. $x^2 - 121 = 0$
$(x - 11)(x + 11) = 0$
$x - 11 = 0$ or $x + 11 = 0$
$x = 11$ or $x = -11$

{11, -11}

16. $2x(5x - 1)(4x + 3) = 0$

$2x = 0$ or $5x - 1 = 0$ or $4x + 3 = 0$
$x = 0$ or $5x = 1$ or $4x = -3$

$x = 0$ or $x = \frac{1}{5}$ or $x = -\frac{3}{4}$

$\left\{0, \frac{1}{5}, -\frac{3}{4}\right\}$

17. $8x^2 = -48x$
$8x^2 + 48x = 0$
$8x(x + 6) = 0$
$8x = 0$ or $x + 6 = 0$
$x = 0$ or $x = -6$

$\{0, -6\}$

18. width: x
length: $x + 4$

width · length = Area
$x(x + 4) = 525$
$x^2 + 4x = 525$
$x^2 + 4x - 525 = 0$
$(x + 25)(x - 21) = 0$
$x + 25 = 0$ or $x - 21 = 0$
$x = -25$ or $x = 21$

Since the width is positive, $x = 21$.

width: 21 feet
length: $x + 4 = 21 + 4 = 25$ feet

19. 1st number: x
2nd number: $20 - x$

$x^2 + (20 - x)^2 = 202$
$x^2 + 400 - 40x + x^2 = 202$
$2x^2 - 40x + 198 = 0$
$x^2 - 20x + 99 = 0$
$(x - 9)(x - 11) = 0$
$x - 9 = 0$ or $x - 11 = 0$
$x = 9$ or $x = 11$
$20 - x = 20 - 9 = 11$ or $20 - x = 20 - 11 = 9$

The numbers are 9 and 11.

20. altitude: x
base: $x + 5$

Area $= \frac{1}{2}$ base · height

$33 = \frac{1}{2}(x + 5)(x)$

$2(33) = 2\left[\frac{1}{2}(x + 5)(x)\right]$

$66 = (x + 5)(x)$
$66 = x^2 + 5x$
$0 = x^2 + 5x - 66$
$0 = (x + 11)(x - 6)$
$x + 11 = 0$ or $x - 6 = 0$
$x = -11$ or $x = 6$

Since the altitude is positive, $x = 6$.

altitude: 6 feet
base: $x + 5 = 6 + 5 = 11$ feet

6.1 SOLUTIONS TO EXERCISES

1. $\dfrac{x+6}{x-9} = \dfrac{3+6}{3-9}$

$= \dfrac{9}{-6}$

$= -\dfrac{3}{2}$

2. $\dfrac{2y+3}{-8-y} = \dfrac{2(-1)+3}{-8-(-1)}$

$= \dfrac{-2+3}{-8+1}$

$= \dfrac{1}{-7}$

$= -\dfrac{1}{7}$

3. $\dfrac{2z}{z^2+3z+7} = \dfrac{2(-4)}{(-4)^2+3(-4)+7}$

$= \dfrac{-8}{16-12+7}$

$= \dfrac{-8}{11}$

$= -\dfrac{8}{11}$

4. $\dfrac{-y^3}{y^2+1} = \dfrac{-(-1)^3}{(-1)^2+1}$

$= \dfrac{-(-1)}{1+1}$

$= \dfrac{1}{2}$

5. $x - 8 = 0$
$x = 8$

The expression is undefined when $x = 8$.

6. Since the denominator is 7 it will never equal 0. Hence the expression is defined for all real numbers.

7. $16y - 32 = 0$
$16y = 32$
$y = 2$

The expression is undefined when $y = 2$.

8. $y^2 + 100 = 0$
$y^2 = -100$
No solution.

The denominator never equals 0, hence the expression is defined for all real numbers.

9. $\dfrac{18x^3y^9}{-9x^2y^{12}} = \dfrac{9 \cdot 2 \cdot x^2 \cdot x \cdot y^9}{-1 \cdot 9 \cdot x^2 \cdot y^9 \cdot y^3}$

$= -\dfrac{2x}{y^3}$

10. $\dfrac{(x+2)(x-5)}{4(x-5)} = \dfrac{x+2}{4}$

11. $\dfrac{-7(a-8)}{(a-8)^2} = \dfrac{-7(a-8)}{(a-8)(a-8)}$

$= -\dfrac{7}{a-8}$

or $\dfrac{7}{-(a-8)} = \dfrac{7}{8-a}$

12. $\dfrac{7c-6d}{6d-7c} = \dfrac{-1(6d-7c)}{1(6d-7c)}$

$= \dfrac{-1}{1}$

$= -1$

13. $\dfrac{x+10}{x^2+19x+90} = \dfrac{1(x+10)}{(x+9)(x+10)}$

$= \dfrac{1}{x+9}$

14. $\dfrac{x^2 - 9}{x^2 - 6x + 9} = \dfrac{(x + 3)(x - 3)}{(x - 3)^2}$

$= \dfrac{(x + 3)(x - 3)}{(x - 3)(x - 3)}$

$= \dfrac{x + 3}{x - 3}$

15. $\dfrac{-a^2 - ab}{5a + 5b} = \dfrac{-a(a + b)}{5(a + b)}$

$= -\dfrac{a}{5}$

16. $\dfrac{6d^2 + 24d}{d + 4} = \dfrac{6d(d + 4)}{1(d + 4)}$

$= \dfrac{6d}{1}$

$= 6d$

17. $\dfrac{x^2 + 5x - 24}{x^2 + 15x + 56} = \dfrac{(x + 8)(x - 3)}{(x + 8)(x + 7)}$

$= \dfrac{x - 3}{x + 7}$

18. $\dfrac{14 - 5y - y^2}{y^2 - 9y + 14} = \dfrac{-1(y^2 + 5y - 14)}{y^2 - 9y + 14}$

$= \dfrac{-1(y + 7)(y - 2)}{(y - 7)(y - 2)}$

$= \dfrac{-1(y + 7)}{y - 7}$

$= -\dfrac{y + 7}{y - 7}$

or $\dfrac{y + 7}{-(y - 7)} = \dfrac{y + 7}{7 - y}$

19. $\dfrac{7x^2 - 20x - 3}{6x^2 - 17x - 3} = \dfrac{(7x + 1)(x - 3)}{(6x + 1)(x - 3)}$

$= \dfrac{7x + 1}{6x + 1}$

20. $\dfrac{4 - x}{x^3 - 64} = \dfrac{-1(x - 4)}{(x - 4)(x^2 + 4x + 64)}$

$= \dfrac{-1}{x^2 + 4x + 64}$

$= -\dfrac{1}{x^2 + 4x + 64}$

6.2 SOLUTIONS TO EXERCISES

1. $\dfrac{10x^2}{3y} \cdot \dfrac{9y}{5x^2} = \dfrac{10x^2 \cdot 9y}{3y \cdot 5x^2}$

$= \dfrac{5 \cdot 2x^2 \cdot 3 \cdot 3y}{3y \cdot 5x^2}$

$= \dfrac{2 \cdot 3}{1}$

$= 6$

2. $-\dfrac{7a^3b^2}{49a^3b^4} \cdot b^5 = \dfrac{-7a^3b^2}{49a^3b^4} \cdot \dfrac{b^5}{1}$

$= \dfrac{-7a^3b^2 \cdot b^5}{49a^3b^4 \cdot 1}$

$= \dfrac{-7a^3b^7}{7 \cdot 7a^3b^4}$

$= \dfrac{-1b^3}{7}$

$= -\dfrac{b^3}{7}$

3. $\dfrac{4x^2 - 9y^2}{2x + 3y} \cdot \dfrac{x}{x^2 - xy}$

$= \dfrac{(4x^2 - 9y^2)(x)}{(2x + 3y)(x^2 - xy)}$

$= \dfrac{(2x + 3y)(2x - 3y)(x)}{(2x + 3y)(x)(x - y)}$

$= \dfrac{2x - 3y}{x - y}$

4. $\dfrac{8x^{10}}{3x^4} \div \dfrac{24x^8}{15x} = \dfrac{8x^{10}}{3x^4} \cdot \dfrac{15x}{24x^8}$

$= \dfrac{8x^{10} \cdot 15x}{3x^4 \cdot 24x^8}$

$= \dfrac{8 \cdot 3 \cdot 5 \cdot x^{11}}{3 \cdot 3 \cdot 8x^{12}}$

$= \dfrac{5}{3x}$

5. $\dfrac{(x + 1)^2}{6} \div \dfrac{3x + 3}{18} = \dfrac{(x + 1)^2}{6} \cdot \dfrac{18}{3x + 3}$

$= \dfrac{3 \cdot 6 \cdot (x + 1)^2}{6 \cdot 3 \cdot (x + 1)}$

$= x + 1$

6. $\dfrac{c^2 - 100d^2}{c + 10d} \div \dfrac{2c - 20d}{4c - 8}$

$= \dfrac{c^2 - 100d^2}{c + 10d} \cdot \dfrac{4c - 8}{2c - 20d}$

$= \dfrac{(c + 10d)(c - 10d) \cdot 4(c - 2)}{(c + 10d)(2)(c - 10d)}$

$= 2(c - 2)$
or $2c - 4$

7. $\dfrac{y^2 + 11y}{7} \cdot \dfrac{49}{6y + 66} = \dfrac{y(y + 11) \cdot 7 \cdot 7}{7 \cdot 6(y + 11)}$

$= \dfrac{7y}{6}$

8. $\dfrac{x^2 - x - 2}{x^2 + 4x + 3} \cdot \dfrac{x^2 + 10x + 21}{x^2 + 9x + 14}$

$= \dfrac{(x - 2)(x + 1) \cdot (x + 7)(x + 3)}{(x + 3)(x + 1) \cdot (x + 7)(x + 2)}$

$= \dfrac{x - 2}{x + 2}$

9. $\dfrac{x^2 - 36}{5y} \div \dfrac{6 - x}{10xy} = \dfrac{x^2 - 36}{5y} \cdot \dfrac{10xy}{6 - x}$

$= \dfrac{(x + 6)(x - 6) \cdot 2 \cdot 5xy}{5y \cdot (-1)(x - 6)}$

$= \dfrac{2x(x + 6)}{-1}$

$= -2x(x + 6)$

or $-2x^2 - 12x$

10. $\dfrac{b^2 + 8b + 16}{b^2 - b - 20} \cdot \dfrac{b^2 - 13b + 40}{b^2 + 10b + 24}$

$= \dfrac{(b + 4)(b + 4) \cdot (b - 8)(b - 5)}{(b + 4)(b - 5) \cdot (b + 6)(b + 4)}$

$= \dfrac{b - 8}{b + 6}$

11. $\dfrac{d^2 - 9d + 14}{d^2 + 2d + 1} \div \dfrac{d^2 - d - 42}{d^2 + 7d + 6}$

$= \dfrac{d^2 - 9d + 14}{d^2 + 2d + 1} \cdot \dfrac{d^2 + 7d + 6}{d^2 - d - 42}$

$= \dfrac{(d - 7)(d - 2) \cdot (d + 6)(d + 1)}{(d + 1)(d + 1) \cdot (d - 7)(d + 6)}$

$= \dfrac{d - 2}{d + 1}$

12. $\dfrac{6x^2 + 5x - 4}{2x^2 + 9x - 5} \div \dfrac{12x^2 + 7x - 12}{7x^2 + 36x + 5}$

$= \dfrac{6x^2 + 5x - 4}{2x^2 + 9x - 5} \cdot \dfrac{7x^2 + 36x + 5}{12x^2 + 7x - 12}$

$$= \frac{(2x - 1)(3x + 4) \cdot (7x + 1)(x + 5)}{(2x - 1)(x + 5) \cdot (3x + 4)(4x - 3)}$$

$$= \frac{7x + 1}{4x - 3}$$

13. $$\frac{a^2 - 9b^2}{a^2 - 6ab + 9b^2} \cdot \frac{3b - a}{a + 3b}$$

$$= \frac{(a + 3b)(a - 3b) \cdot (-1)(a - 3b)}{(a - 3b)(a - 3b) \cdot (a + 3b)}$$

$$= -1$$

14. $$\frac{x^2 - xz + xy - yz}{x + z} \div \frac{x + y}{x - z}$$

$$= \frac{x^2 - xz + xy - yz}{x + z} \cdot \frac{x - z}{x + y}$$

$$= \frac{x(x - z) + y(x - z)}{x + z} \cdot \frac{x - z}{x + y}$$

$$= \frac{(x - z)(x + y) \cdot (x - z)}{(x + z) \cdot (x + y)}$$

$$= \frac{(x - z)^2}{x + z}$$

15. $$\frac{x^3 + 27}{x^2 - 3x + 9} \cdot \frac{6}{x^2 - 9}$$

$$= \frac{(x + 3)(x^2 - 3x + 9) \cdot 6}{(x^2 - 3x + 9) \cdot (x + 3)(x - 3)}$$

$$= \frac{6}{x - 3}$$

6.3 SOLUTIONS TO EXERCISES

1. $$\frac{x}{21} + \frac{8}{21} = \frac{x + 8}{21}$$

2. $$\frac{10}{2 + y} + \frac{y + 7}{2 + y} = \frac{10 + y + 7}{2 + y}$$

$$= \frac{y + 17}{2 + y}$$

3. $$\frac{9a}{a - 3} - \frac{27}{a - 3} = \frac{9a - 27}{a - 3}$$

$$= \frac{9(a - 3)}{a - 3}$$

$$= 9$$

4. $$\frac{12}{x + 7} - \frac{5 - x}{x + 7} = \frac{12 - (5 - x)}{x + 7}$$

$$= \frac{12 - 5 + x}{x + 7}$$

$$= \frac{x + 7}{x + 7}$$

$$= 1$$

5. $$\frac{4}{yz} + \frac{10}{yz} = \frac{4 + 10}{yz}$$

$$= \frac{14}{yz}$$

6. $$\frac{8}{a - b} - \frac{7}{a - b} = \frac{8 - 7}{a - b}$$

$$= \frac{1}{a - b}$$

7. $$\frac{x^2 + 3x}{x + 6} + \frac{6x + 18}{x + 6} = \frac{x^2 + 3x + 6x + 18}{x + 6}$$

$$= \frac{x^2 + 9x + 18}{x + 6}$$

$$= \frac{(x + 3)(x + 6)}{x + 6}$$

$$= x + 3$$

8. $$\frac{2x^2 + xy}{x - y} - \frac{2xy + y^2}{x - y}$$

$$= \frac{2x^2 + xy - (2xy + y^2)}{x - y}$$

$$= \frac{2x^2 + xy - 2xy - y^2}{x - y}$$

$$= \frac{2x^2 - xy - y^2}{x - y}$$

$$= \frac{(2x + y)(x - y)}{x - y}$$

$$= 2x + y$$

9. $9 = 3 \cdot 3 = 3^2$
$15 = 3 \cdot 5$

LCD $= 3^2 \cdot 5 = 45$

10. $5 = 5$
$15 = 3 \cdot 5$

LCD of coefficients $= 3 \cdot 5 = 15$

The greatest number of times that the factor x appears in any one denominator is 3.

LCD $= 15x^3$

11. $2x - 10 = 2(x - 5)$
$x^2 - x - 20 = (x - 5)(x + 4)$

LCD $= 2(x - 5)(x + 4)$

12. $8 = 2 \cdot 2 \cdot 2 = 2^3$
$12 = 2 \cdot 2 \cdot 3 = 2^2 \cdot 3$

LCD of coefficients $= 2^3 \cdot 3 = 24$

The greatest number of times that the factor x appears is 1.

The greatest number of times that the factor $x - 7$ appears in any one denominator is 3.

LCD $= 24x(x - 7)^3$

13. $\frac{4}{3x} = \frac{4(4x^2)}{3x(4x^2)} = \frac{16x^2}{12x^3}$

14. $\frac{7}{x - 2} = \frac{7(3)}{(x - 2)(3)} = \frac{21}{3(x - 2)}$

15. $\frac{5a + 2}{4a + 20} = \frac{5a + 2}{4(a + 5)}$

$$= \frac{(5a + 2)(a)}{4(a + 5)(a)}$$

$$= \frac{a(5a + 2)}{4a(a + 5)}$$

or $\frac{5a^2 + 2a}{4a(a + 5)}$

16. $\frac{b}{b^3 + 4b^2 - 5b} = \frac{b}{b(b^2 + 4b - 5)}$

$$= \frac{b}{b(b + 5)(b - 1)}$$

$$= \frac{b(b + 3)}{b(b + 5)(b - 1)(b + 3)}$$

or $\frac{b^2 + 3b}{b(b + 5)(b - 1)(b + 3)}$

6.4 SOLUTIONS TO EXERCISES

1. $\frac{7}{3y} + \frac{9}{4y} = \frac{7(4)}{3y(4)} + \frac{9(3)}{4y(3)}$

$$= \frac{28}{12y} + \frac{27}{12y}$$

$$= \frac{28 + 27}{12y}$$

$$= \frac{55}{12y}$$

2. $\frac{16a}{b} + \frac{3b}{4} = \frac{16a(4)}{b(4)} + \frac{3b(b)}{4(b)}$

$$= \frac{64a}{4b} + \frac{3b^2}{4b}$$

$$= \frac{64a + 3b^2}{4b}$$

3. $$\frac{8}{x+2} + \frac{5}{3x+6} = \frac{8}{x+2} + \frac{5}{3(x+2)}$$
$$= \frac{8(3)}{(x+2)(3)} + \frac{5}{3(x+2)}$$
$$= \frac{24}{3(x+2)} + \frac{5}{3(x+2)}$$
$$= \frac{24+5}{3(x+2)}$$
$$= \frac{29}{3(x+2)}$$

4. $$\frac{12}{x-1} - \frac{17}{5x-5} = \frac{12}{x-1} - \frac{17}{5(x-1)}$$
$$= \frac{12(5)}{(x-1)(5)} - \frac{17}{5(x-1)}$$
$$= \frac{60}{5(x-1)} - \frac{17}{5(x-1)}$$
$$= \frac{60-17}{5(x-1)}$$
$$= \frac{43}{5(x-1)}$$

5. $$\frac{4}{x-2} - \frac{4}{(x-2)^2}$$
$$= \frac{4(x-2)}{(x-2)(x-2)} - \frac{4}{(x-2)^2}$$
$$= \frac{4(x-2)-4}{(x-2)^2}$$
$$= \frac{4x-8-4}{(x-2)^2}$$
$$= \frac{4x-12}{(x-2)^2}$$

6. $$\frac{7}{b} - 4 = \frac{7}{b} - \frac{4}{1}$$
$$= \frac{7}{b} - \frac{4(b)}{1(b)}$$
$$= \frac{7-4b}{b}$$

7. $$\frac{7x+6}{(x+2)(x+3)} - \frac{6}{x+3}$$
$$= \frac{7x+6}{(x+2)(x+3)} - \frac{6(x+2)}{(x+3)(x+2)}$$
$$= \frac{7x+6-6(x+2)}{(x+2)(x+3)}$$
$$= \frac{7x+6-6x-12}{(x+2)(x+3)}$$
$$= \frac{x-6}{(x+2)(x+3)}$$

8. $$\frac{9}{5-x} + \frac{x}{3x-15}$$
$$= \frac{9}{-1(x-5)} + \frac{x}{3(x-5)}$$
$$= \frac{-9}{x-5} + \frac{x}{3(x-5)}$$
$$= \frac{-9(3)}{(x-5)(3)} + \frac{x}{3(x-5)}$$
$$= \frac{-27+x}{3(x-5)}$$
or $$\frac{x-27}{3(x-5)}$$

9. $$\frac{-4}{b-6} + \frac{12}{12-2b}$$
$$= \frac{-4}{b-6} + \frac{12}{-2(b-6)}$$
$$= \frac{-4(-2)}{(b-6)(-2)} + \frac{12}{-2(b-6)}$$

$$= \frac{8 + 12}{-2(b - 6)}$$

$$= \frac{20}{-2(b - 6)}$$

$$= \frac{-10}{b - 6}$$

$$\text{or } \frac{10}{-(b - 6)} = \frac{10}{6 - b}$$

10. $$\frac{-12}{x^2 + 5x + 4} + \frac{4}{x + 1}$$

$$= \frac{-12}{(x + 4)(x + 1)} + \frac{4}{x + 1}$$

$$= \frac{-12}{(x + 4)(x + 1)} + \frac{4(x + 4)}{(x + 1)(x + 4)}$$

$$= \frac{-12 + 4(x + 4)}{(x + 4)(x + 1)}$$

$$= \frac{-12 + 4x + 16}{(x + 4)(x + 1)}$$

$$= \frac{4x + 4}{(x + 4)(x + 1)}$$

$$= \frac{4(x + 1)}{(x + 4)(x + 1)}$$

$$= \frac{4}{x + 4}$$

11. $$\frac{8}{y^2 - 16} + \frac{2}{3(y + 4)}$$

$$= \frac{8}{(y + 4)(y - 4)} + \frac{2}{3(y + 4)}$$

$$= \frac{8(3)}{(y + 4)(y - 4)(3)} + \frac{2(y - 4)}{3(y + 4)(y - 4)}$$

$$= \frac{8(3) + 2(y - 4)}{3(y + 4)(y - 4)}$$

$$= \frac{24 + 2y - 8}{3(y + 4)(y - 4)}$$

$$= \frac{2y + 16}{3(y + 4)(y - 4)}$$

12. $$\frac{162}{a^2 - 81} - \frac{9}{a - 9}$$

$$= \frac{162}{(a + 9)(a - 9)} - \frac{9}{a - 9}$$

$$= \frac{162}{(a + 9)(a - 9)} - \frac{9(a + 9)}{(a - 9)(a + 9)}$$

$$= \frac{162 - 9(a + 9)}{(a + 9)(a - 9)}$$

$$= \frac{162 - 9a - 81}{(a + 9)(a - 9)}$$

$$= \frac{-9a + 81}{(a + 9)(a - 9)}$$

$$= \frac{-9(a - 9)}{(a + 9)(a - 9)}$$

$$= -\frac{9}{a + 9}$$

13. $$\frac{5}{x^2 + 9x + 14} + \frac{x}{x^2 + 10x + 21}$$

$$= \frac{5}{(x + 7)(x + 2)} + \frac{x}{(x + 7)(x + 3)}$$

$$= \frac{5(x + 3)}{(x + 7)(x + 2)(x + 3)} + \frac{x(x + 2)}{(x + 7)(x + 3)(x + 2)}$$

$$= \frac{5(x + 3) + x(x + 2)}{(x + 7)(x + 2)(x + 3)}$$

$$= \frac{5x + 15 + x^2 + 2x}{(x + 7)(x + 2)(x + 3)}$$

$$= \frac{x^2 + 7x + 15}{(x + 7)(x + 2)(x + 3)}$$

14. $$\frac{2y}{y^2 + 11y + 18} - \frac{y + 1}{y^2 + 8y - 9}$$

$$= \frac{2y}{(y + 9)(y + 2)} - \frac{y + 1}{(y + 9)(y - 1)}$$

$$= \frac{2y(y - 1)}{(y + 9)(y + 2)(y - 1)} - \frac{(y + 1)(y + 2)}{(y + 9)(y - 1)(y + 2)}$$

$$= \frac{2y(y - 1) - (y + 1)(y + 2)}{(y + 9)(y + 2)(y - 1)}$$

$$= \frac{2y^2 - 2y - (y^2 + 3y + 2)}{(y + 9)(y + 2)(y - 1)}$$

$$= \frac{y^2 - 5y - 2}{(y + 9)(y + 2)(y - 1)}$$

15. $P = 2W + 2L$

$$P = 2\left(\frac{3}{x}\right) + 2\left(\frac{2}{x + 4}\right)$$

$$P = \frac{6}{x} + \frac{4}{x + 4}$$

$$P = \frac{6(x + 4)}{x(x + 4)} + \frac{4(x)}{(x + 4)(x)}$$

$$P = \frac{6(x + 4) + 4(x)}{x(x + 4)}$$

$$P = \frac{6x + 24 + 4x}{x(x + 4)}$$

$$= \frac{10x + 24}{x(x + 4)} \text{ feet}$$

$A = W \cdot L$

$$A = \left(\frac{3}{x}\right)\left(\frac{2}{x + 4}\right)$$

$$A = \frac{3 \cdot 2}{x \cdot (x + 4)}$$

$$A = \frac{6}{x(x + 4)} \text{ square feet}$$

6.5 SOLUTIONS TO EXERCISES

1. $$\frac{\frac{2}{7}}{-\frac{8}{21}} = \frac{2}{7} \cdot \frac{-21}{8}$$

$$= \frac{2 \cdot (-3)(7)}{7 \cdot 2 \cdot 4}$$

$$= -\frac{3}{4}$$

2. $$\frac{\frac{-4y}{13}}{\frac{10y}{26}} = \frac{-4y}{13} \cdot \frac{26}{10y}$$

$$= \frac{-2 \cdot 2y \cdot 2 \cdot 13}{13 \cdot 2 \cdot 5y}$$

$$= \frac{-4}{5}$$

3. $$\frac{\frac{10x + 1}{9x^2}}{\frac{5x + 1}{18x}} = \frac{10x + 1}{9x^2} \cdot \frac{18x}{5x + 1}$$

$$= \frac{(10x + 1) \cdot 2 \cdot 9x}{9x^2(5x + 1)}$$

$$= \frac{2(10x + 1)}{x(5x + 1)}$$

4. $$\frac{\frac{(a + 3)(a - 3)}{12}}{\frac{(a + 3)(a - 4)}{18}}$$

$$= \frac{(a + 3)(a - 3)}{12} \cdot \frac{18}{(a + 3)(a - 4)}$$

$$= \frac{(a + 3)(a - 3) \cdot 6 \cdot 3}{2 \cdot 6(a + 3)(a - 4)}$$

$$= \frac{3(a - 3)}{2(a - 4)}$$

5. $$\frac{\frac{1}{3} + \frac{2}{5}}{\frac{3}{4} - \frac{1}{2}} = \frac{60\left(\frac{1}{3} + \frac{2}{5}\right)}{60\left(\frac{3}{4} - \frac{1}{2}\right)}$$

$$= \frac{60\left(\frac{1}{3}\right) + 60\left(\frac{2}{5}\right)}{60\left(\frac{3}{4} - \frac{1}{2}\right)}$$

$$= \frac{60\left(\frac{1}{3}\right) + 60\left(\frac{2}{5}\right)}{60\left(\frac{3}{4}\right) - 60\left(\frac{1}{2}\right)}$$

$$= \frac{20 + 24}{45 - 30}$$

$$= \frac{44}{15}$$

6. $$\frac{\frac{b}{3} + 3}{\frac{b}{3} - 3} = \frac{3\left(\frac{b}{3} + 3\right)}{3\left(\frac{b}{3} - 3\right)}$$

$$= \frac{3\left(\frac{b}{3}\right) + 3(3)}{3\left(\frac{b}{3}\right) - 3(3)}$$

$$= \frac{b + 9}{b - 9}$$

7. $$\frac{\frac{1}{6} - \frac{2}{x}}{\frac{5}{12} + \frac{1}{x^2}}$$

$$= \frac{12x^2\left(\frac{1}{6} - \frac{2}{x}\right)}{12x^2\left(\frac{5}{12} + \frac{1}{x^2}\right)}$$

$$= \frac{12x^2\left(\frac{1}{6}\right) - 12x^2\left(\frac{2}{x}\right)}{12x^2\left(\frac{5}{12}\right) + 12x^2\left(\frac{1}{x^2}\right)}$$

$$= \frac{2x^2 - 24x}{5x^2 + 12}$$

8. $$\frac{1 + \frac{1}{x - 5}}{x + \frac{1}{x - 5}}$$

$$= \frac{(x - 5)\left(1 + \frac{1}{x - 5}\right)}{(x - 5)\left(x + \frac{1}{x - 5}\right)}$$

$$= \frac{(x - 5)(1) + (x - 5)\left(\frac{1}{x - 5}\right)}{(x - 5)(x) + (x - 5)\left(\frac{1}{x - 5}\right)}$$

$$= \frac{x - 5 + 1}{x^2 - 5x + 1}$$

$$= \frac{x - 4}{x^2 - 5x + 1}$$

9. $$\frac{\frac{-8}{15x^2}}{\frac{24}{5x^3}} = \frac{-8}{15x^2} \cdot \frac{5x^3}{24}$$

$$= \frac{-8 \cdot 5x^3}{3 \cdot 5x^2 \cdot 8 \cdot 3}$$

$$= -\frac{x}{9}$$

10. $$\frac{\frac{7y-21}{14}}{\frac{2y-6}{6}} = \frac{7y-21}{14} \cdot \frac{6}{2y-6}$$

$$= \frac{7(y-3) \cdot 2 \cdot 3}{7 \cdot 2 \cdot 2(y-3)}$$

$$= \frac{3}{2}$$

11. $$\frac{4}{3+\frac{1}{5}} = \frac{5(4)}{5\left(3+\frac{1}{5}\right)}$$

$$= \frac{5(4)}{5(3)+5\left(\frac{1}{5}\right)}$$

$$= \frac{20}{15+1}$$

$$= \frac{20}{16}$$

$$= \frac{5}{4}$$

12. $$\frac{\frac{6}{7a}+4}{\frac{6}{7a}-4} = \frac{7a\left(\frac{6}{7a}+4\right)}{7a\left(\frac{6}{7a}-4\right)}$$

$$= \frac{7a\left(\frac{6}{7a}\right)+7a(4)}{7a\left(\frac{6}{7a}\right)-7a(4)}$$

$$= \frac{6+28a}{6-28a}$$

$$= \frac{2(3+14a)}{2(3-14a)}$$

$$= \frac{3+14a}{3-14a}$$

13. $$\frac{\frac{ax-ab}{x^2-b^2}}{\frac{x-b}{x+b}} = \frac{ax-ab}{x^2-b^2} \cdot \frac{x+b}{x-b}$$

$$= \frac{a(x-b)(x+b)}{(x+b)(x-b)(x-b)}$$

$$= \frac{a}{x-b}$$

14. $$\frac{\frac{5}{x-3}+7}{\frac{10}{x-3}-7} = \frac{(x-3)\left(\frac{5}{x-3}+7\right)}{(x-3)\left(\frac{10}{x-3}-7\right)}$$

$$= \frac{(x-3)\left(\frac{5}{x-3}\right)+(x-3)(7)}{(x-3)\left(\frac{10}{x-3}\right)-(x-3)(7)}$$

$$= \frac{5+7x-21}{10-7x+21}$$

$$= \frac{7x-16}{-7x+31}$$

15. $$\frac{\frac{-4+y}{6}}{\frac{4-y}{18}} = \frac{-4+y}{6} \cdot \frac{18}{4-y}$$

$$= \frac{-(4-y) \cdot 6 \cdot 3}{6(4-y)}$$

$$= -3$$

6.6 SOLUTIONS TO EXERCISES

1. $$\frac{x}{4} - 2 = 7$$

$$4\left(\frac{x}{4} - 2\right) = 4(7)$$

$$4\left(\frac{x}{4}\right) - 4(2) = 4(7)$$

$$x - 8 = 28$$

$$x = 36$$

$\{36\}$

2. $$\frac{x}{3} + \frac{4x}{6} = \frac{x}{18}$$

$$18\left(\frac{x}{3} + \frac{4x}{6}\right) = 18\left(\frac{x}{18}\right)$$

$$18\left(\frac{x}{3}\right) + 18\left(\frac{4x}{6}\right) = 18\left(\frac{x}{18}\right)$$

$$6x + 12x = x$$
$$18x = x$$
$$17x = 0$$
$$x = 0$$

$\{0\}$

3. $$\frac{5b}{2} = \frac{b - 3}{4}$$

$$4\left(\frac{5b}{2}\right) = 4\left(\frac{b - 3}{4}\right)$$

$$10b = b - 3$$
$$9b = -3$$

$$b = -\frac{1}{3}$$

$\left\{-\frac{1}{3}\right\}$

4. $$\frac{y}{7} = \frac{y + 1}{6}$$

$$42\left(\frac{y}{7}\right) = 42\left(\frac{y + 1}{6}\right)$$

$$6y = 7(y + 1)$$
$$6y = 7y + 7$$
$$-y = 7$$
$$y = -7$$

$\{-7\}$

5. $$\frac{x - 4}{2} + \frac{x - 8}{3} = \frac{1}{6}$$

$$6\left(\frac{x - 4}{2} + \frac{x - 8}{3}\right) = 6\left(\frac{1}{6}\right)$$

$$6\left(\frac{x - 4}{2}\right) + 6\left(\frac{x - 8}{3}\right) = 6\left(\frac{1}{6}\right)$$

$$3(x - 4) + 2(x - 8) = 1$$
$$3x - 12 + 2x - 16 = 1$$
$$5x - 28 = 1$$
$$5x = 29$$

$$x = \frac{29}{5}$$

$\left\{\frac{29}{5}\right\}$

6. $$\frac{10}{3a - 1} = -5$$

$$(3a - 1)\left(\frac{10}{3a - 1}\right) = (3a - 1)(-5)$$

$$10 = -15a + 5$$
$$5 = -15a$$

$$-\frac{1}{3} = a$$

$\left\{-\frac{1}{3}\right\}$

7. $$\frac{z}{z + 6} + \frac{5}{z + 6} = 2$$

$$(z + 6)\left(\frac{z}{z + 6} + \frac{5}{z + 6}\right) = (z + 6)(2)$$

$$(z + 6)\left(\frac{z}{z + 6}\right) + (z + 6)\left(\frac{5}{z + 6}\right) = (z + 6)(2)$$
$$z + 5 = 2z + 12$$
$$5 = z + 12$$
$$-7 = z$$

$\{-7\}$

8. $$\frac{3a}{a - 5} + 4 = \frac{2a}{a - 5}$$

$$(a - 5)\left(\frac{3a}{a - 5} + 4\right) = (a - 5)\left(\frac{2a}{a - 5}\right)$$

$$(a - 5)\left(\frac{3a}{a - 5}\right) + (a - 5)(4) = (a - 5)\left(\frac{2a}{a - 5}\right)$$
$$3a + 4a - 20 = 2a$$
$$7a - 20 = 2a$$
$$-20 = -5a$$
$$4 = a$$

$\{4\}$

9.

$$1 + \frac{2}{x - 1} = \frac{x}{x + 1}$$

$$(x - 1)(x + 1)\left(1 + \frac{2}{x - 1}\right) = (x - 1)(x + 1)\left(\frac{x}{x + 1}\right)$$

$$(x - 1)(x + 1)(1) + (x - 1)(x + 1)\left(\frac{2}{x - 1}\right) = (x - 1)(x + 1)\left(\frac{x}{x + 1}\right)$$

$$(x^2 - 1)1 + 2(x + 1) = x(x - 1)$$
$$x^2 - 1 + 2x + 2 = x^2 - x$$
$$x^2 + 2x + 1 = x^2 - x$$
$$2x + 1 = -x$$
$$1 = -3x$$
$$-\frac{1}{3} = x$$

$\left\{-\frac{1}{3}\right\}$

10.

$$\frac{9}{y^2 + 7y + 10} + \frac{3}{y + 5} = \frac{5}{y + 2}$$

$$\frac{9}{(y + 2)(y + 5)} + \frac{3}{y + 5} = \frac{5}{y + 2}$$

$$(y + 2)(y + 5)\left[\frac{9}{(y + 2)(y + 5)} + \frac{3}{y + 5}\right] = (y + 2)(y + 5)\left(\frac{5}{y + 2}\right)$$

$$(y + 2)(y + 5)\left[\frac{9}{(y + 2)(y + 5)}\right] + (y + 2)(y + 5)\left(\frac{3}{y + 5}\right) = (y + 2)(y + 5)\left(\frac{5}{y + 2}\right)$$

$$9 + 3(y + 2) = 5(y + 5)$$
$$9 + 3y + 6 = 5y + 25$$
$$3y + 15 = 5y + 25$$
$$-10 = 2y$$
$$-5 = y$$

{ }, since $y = -5$ would cause an original denominator to equal 0.

11.

$$\frac{1 - a}{1 + a} + \frac{4}{a^2 - 1} = \frac{a + 7}{a - 1}$$

$$\frac{1 - a}{1 + a} + \frac{4}{(a + 1)(a - 1)} = \frac{a + 7}{a - 1}$$

$$(a + 1)(a - 1)\left[\frac{1 - a}{1 + a} + \frac{4}{(a + 1)(a - 1)}\right] = (a + 1)(a - 1)\left(\frac{a + 7}{a - 1}\right)$$

$$(a + 1)(a - 1)\left(\frac{1 - a}{1 + a}\right) + (a + 1)(a - 1)\left[\frac{4}{(a + 1)(a - 1)}\right] = (a + 1)(a - 1)\left(\frac{a + 7}{a - 1}\right)$$

$$(a - 1)(1 - a) + 4 = (a + 1)(a + 7)$$
$$-a^2 + 2a - 1 + 4 = a^2 + 8a + 7$$
$$-a^2 + 2a + 3 = a^2 + 8a + 7$$
$$0 = 2a^2 + 6a + 4$$
$$0 = 2(a^2 + 3a + 2)$$
$$0 = 2(a + 2)(a + 1)$$
$$a + 2 = 0 \quad \text{or} \quad a + 1 = 0$$
$$a = -2 \quad \text{or} \quad a = -1$$

$a = -2$ is the only solution since $a = -1$ causes an original denominator to equal 0.

$\{-2\}$

12.

$$\frac{8}{x^2 + 4x} + \frac{2}{x + 4} = 2$$

$$\frac{8}{x(x + 4)} + \frac{2}{x + 4} = 2$$

$$x(x + 4)\left[\frac{8}{x(x + 4)} + \frac{2}{x + 4}\right] = x(x + 4)(2)$$

$$x(x + 4)\left[\frac{8}{x(x + 4)}\right] + x(x + 4)\left(\frac{2}{x + 4}\right) = 2x(x + 4)$$

$$8 + 2x = 2x^2 + 8x$$
$$0 = 2x^2 + 6x - 8$$
$$0 = 2(x^2 + 3x - 4)$$
$$0 = 2(x + 4)(x - 1)$$
$$x + 4 = 0 \quad \text{or} \quad x - 1 = 0$$
$$x = -4 \quad \text{or} \quad x = 1$$

The only solution is $x = 1$ since $x = -4$ causes an original denominator to equal 0.

$\{1\}$

13.

$$P = 2w + 2l$$
$$P - 2w = 2l$$
$$\frac{P - 2w}{2} = \frac{2l}{2}$$
$$\frac{P - 2w}{2} = l$$

14.

$$\frac{7}{x} = \frac{4y}{x - 5}$$

$$x(x - 5)\left(\frac{7}{x}\right) = x(x - 5)\left(\frac{4y}{x - 5}\right)$$

$$7(x - 5) = 4xy$$
$$7x - 35 = 4xy$$

$$-35 = 4xy - 7x$$
$$-35 = x(4y - 7)$$

$$\frac{-35}{4y - 7} = \frac{x(4y - 7)}{4y - 7}$$

$$-\frac{35}{4y - 7} = x$$

$$\text{or } \frac{35}{7 - 4y} = x$$

15. $$\frac{1}{5} - \frac{2}{x} = \frac{4}{y}$$

$$5xy\left(\frac{1}{5} - \frac{2}{x}\right) = 5xy\left(\frac{4}{y}\right)$$

$$5xy\left(\frac{1}{5}\right) - 5xy\left(\frac{2}{x}\right) = 5xy\left(\frac{4}{y}\right)$$

$$xy - 10y = 20x$$
$$-10y = 20x - xy$$
$$-10y = x(20 - y)$$

$$\frac{-10y}{20 - y} = \frac{x(20 - y)}{20 - y}$$

$$-\frac{10y}{20 - y} = x$$

$$\text{or } \frac{10y}{y - 20} = x$$

6.7 SOLUTIONS TO EXERCISES

1. $\frac{4}{16} = \frac{1}{4}$

2. 5 gallons = 5(4) quarts = 20 quarts

   $\frac{7}{20}$

3. 6 dollars = 6(10) dimes = 60 dimes

   $\frac{3}{60} = \frac{1}{20}$

4. 2 hours = 2(60) minutes = 120 minutes

   $\frac{200}{120} = \frac{5}{3}$

5. $$\frac{5}{7} = \frac{x}{35}$$

$$5 \cdot 35 = 7x$$
$$175 = 7x$$
$$25 = x$$

{25}

6. $$\frac{y}{4} = \frac{36}{48}$$

$$y \cdot 48 = 4 \cdot 36$$
$$48y = 144$$
$$y = 3$$

{3}

7. $$\frac{3x}{8} = \frac{9}{4}$$

$$3x \cdot 4 = 8 \cdot 9$$
$$12x = 72$$
$$x = 6$$

{6}

8. $$\frac{y}{y - 11} = \frac{2}{3}$$

$$y \cdot 3 = (y - 11) \cdot 2$$
$$3y = 2y - 22$$
$$y = -22$$

{-22}

9. $$\frac{2x}{x + 1} = \frac{20}{12}$$

$$2x \cdot 12 = (x + 1) \cdot 20$$
$$24x = 20x + 20$$
$$4x = 20$$
$$x = 5$$

{5}

10. $\frac{x+1}{x-2} = \frac{7}{5}$

$$\begin{aligned}(x+1)\cdot 5 &= (x-2)\cdot 7\\ 5x+5 &= 7x-14\\ 5 &= 2x-14\\ 19 &= 2x\\ \frac{19}{2} &= x\end{aligned}$$

$\left\{\frac{19}{2}\right\}$

11. $\frac{1 \text{ inch}}{150 \text{ miles}} = \frac{4\frac{3}{5} \text{ inches}}{x \text{ miles}}$

$$\frac{1}{150} = \frac{\frac{23}{5}}{x}$$

$$\begin{aligned}1\cdot x &= 150\cdot\frac{23}{5}\\ x &= 690\end{aligned}$$

It represents 690 miles.

12. $\frac{28 \text{ miles}}{1 \text{ gal}} = \frac{x \text{ miles}}{10.5 \text{ gal}}$

$$\frac{28}{1} = \frac{x}{10.5}$$

$$\begin{aligned}28\cdot 10.5 &= 1\cdot x\\ 294 &= x\end{aligned}$$

She can drive 294 miles.

13. $\frac{9 \text{ sq. ft.}}{1 \text{ student}} = \frac{x \text{ sq. ft.}}{33 \text{ students}}$

$$\frac{9}{1} = \frac{x}{33}$$

$$\begin{aligned}9\cdot 33 &= 1\cdot x\\ 297 &= x\end{aligned}$$

The minimum floor space is 297 sq. feet.

14. $\frac{416 \text{ miles}}{8 \text{ hours}} = \frac{x \text{ miles}}{3 \text{ hours}}$

$$\frac{416}{8} = \frac{x}{3}$$

$$\begin{aligned}416\cdot 3 &= 8\cdot x\\ 1248 &= 8x\\ 156 &= x\end{aligned}$$

She can travel 156 miles.

15. $\frac{\$208}{32 \text{ hours}} = \frac{\$x}{17 \text{ hours}}$

$$\frac{208}{32} = \frac{x}{17}$$

$$\begin{aligned}208\cdot 17 &= 32\cdot x\\ 3536 &= 32x\\ 110.5 &= x\end{aligned}$$

He will earn \$110.50.

6.8 SOLUTIONS TO EXERCISES

1. x = the unknown number

$$5\left(\frac{1}{x}\right) = 10\left(\frac{1}{8}\right)$$

$$8x\left[5\left(\frac{1}{x}\right)\right] = 8x\left[10\left(\frac{1}{8}\right)\right]$$

$$\begin{aligned}40 &= 10x\\ 4 &= x\end{aligned}$$

The number is 4.

2. $\frac{6}{x+3} = \frac{4}{x-3}$

$$\begin{aligned}6(x-3) &= (x+3)\cdot 4\\ 6x-18 &= 4x+12\\ 2x-18 &= 12\\ 2x &= 30\\ x &= 15\end{aligned}$$

$\{15\}$

3. x = the unknown number

$$x + 20\left(\frac{1}{x}\right) = 12$$

$$x\left[x + 20\left(\frac{1}{x}\right)\right] = x(12)$$

$$x(x) + x\left[20\left(\frac{1}{x}\right)\right] = 12x$$

$$x^2 + 20 = 12x$$
$$x^2 - 12x + 20 = 0$$
$$(x - 10)(x - 2) = 0$$
$$x - 10 = 0 \quad \text{or} \quad x - 2 = 0$$
$$x = 10 \quad \text{or} \quad x = 2$$

The number is 10 or 2.

4.	Hours to Complete Total Job	Part of Job Completed in 1 hr
bricklayer	5	1/5
apprentice	8	1/8
Together	x	$1/x$

Part of job bricklayer completed in 1 hour	added to	part of job apprentice completed in 1 hour	is equal to	part of job they completed together in 1 hour
$\frac{1}{5}$	+	$\frac{1}{8}$	=	$\frac{1}{x}$

$$\frac{1}{5} + \frac{1}{8} = \frac{1}{x}$$

$$40x\left(\frac{1}{5} + \frac{1}{8}\right) = 40x\left(\frac{1}{x}\right)$$

$$40x\left(\frac{1}{5}\right) + 40x\left(\frac{1}{8}\right) = 40x\left(\frac{1}{x}\right)$$
$$8x + 5x = 40$$
$$13x = 40$$
$$x = \frac{40}{13}$$

Together it will take them $\frac{40}{13}$ or $3\frac{1}{13}$ hours.

5. x = the speed of the boat in still water

	Distance	Rate	Time = D/R
downstream	6	$x + 3$	$6/(x + 3)$
upstream	4	$x - 3$	$4/(x - 3)$

Time downstream equals time upstream

$$\frac{6}{x+3} = \frac{4}{x-3}$$

$$\frac{6}{x+3} = \frac{4}{x-3}$$

$$\begin{aligned} 6(x - 3) &= 4(x + 3) \\ 6x - 18 &= 4x + 12 \\ 2x - 18 &= 12 \\ 2x &= 30 \\ x &= 15 \end{aligned}$$

The boat's rate in still water is 15 mph.

6.	Hours to Complete Total Job	Part of Job Completed in 1 hr
Pipeline A	30	1/30
Pipeline B	x	$1/x$
Together	20	1/20

Part of job Pipeline A completed in 1 hour added to part of job Pipeline B completed in 1 hour is equal to part of job they completed together in 1 hour

$$\frac{1}{30} + \frac{1}{x} = \frac{1}{20}$$

$$\frac{1}{30} + \frac{1}{x} = \frac{1}{20}$$

$$60x\left(\frac{1}{30} + \frac{1}{x}\right) = 60x\left(\frac{1}{20}\right)$$

$$60x\left(\frac{1}{30}\right) + 60x\left(\frac{1}{x}\right) = 60x\left(\frac{1}{20}\right)$$

$$\begin{aligned} 2x + 60 &= 3x \\ 60 &= x \end{aligned}$$

Pipeline B alone will take 60 hours.

7. $\frac{22}{10} = \frac{26}{x}$

$$22x = 10 \cdot 26$$
$$22x = 260$$
$$x = \frac{130}{11} \approx 11.8$$

$$\frac{22}{10} = \frac{22}{y}$$

$$22y = 10 \cdot 22$$
$$22y = 220$$
$$y = 10$$

$x = 11.8''$ and $y = 10''$

8.

	Distance	Time	Rate = D/T
Part by jet	1440	x	$1440/x$
Part by car	360	$x + 2$	$360/(x + 2)$

Rate of Jet is 6 times rate of car

$$\frac{1440}{x} = 6 \cdot \frac{360}{x+2}$$

$$\frac{1440}{x} = 6 \cdot \frac{360}{x+2}$$

$$x(x+2)\left(\frac{1440}{x}\right) = x(x+2)(6)\left(\frac{360}{x+2}\right)$$

$$1440(x + 2) = 2160x$$
$$1440x + 2880 = 2160x$$
$$2880 = 720x$$
$$4 = x$$
$$x + 2 = 4 + 2 = 6$$

She travels 4 hours by jet and 6 hours by car.

9. x = rate of the wind

	Distance	Rate	Time = D/R
with wind	42	$15 + x$	$42/(15 + x)$
into wind	18	$15 - x$	$18/(15 - x)$

Time with wind equals Time into wind

$$\frac{42}{15+x} = \frac{18}{15-x}$$

$$\frac{42}{15+x} = \frac{18}{15-x}$$

$$42(15 - x) = (15 + x) \cdot 18$$
$$630 - 42x = 270 + 18x$$
$$630 = 270 + 60x$$
$$360 = 60x$$
$$6 = x$$

The wind's rate is 6 mph.

10. x = the unknown number

$$\frac{x-1}{x+2} = \frac{2}{3}$$

$$(x - 1) \cdot 3 = (x + 2) \cdot 2$$
$$3x - 3 = 2x + 4$$
$$x - 3 = 4$$
$$x = 7$$

The number is 7.

11. x = the unknown number

$$2 - \frac{1}{x} = 3\left(\frac{1}{x^2}\right)$$

$$x^2\left(2 - \frac{1}{x}\right) = x^2\left[3\left(\frac{1}{x^2}\right)\right]$$

$$x^2(2) - x^2\left(\frac{1}{x}\right) = x^2\left[3\left(\frac{1}{x^2}\right)\right]$$

$$2x^2 - x = 3$$
$$2x^2 - x - 3 = 0$$
$$(2x - 3)(x + 1) = 0$$
$$2x - 3 = 0 \quad \text{or} \quad x + 1 = 0$$
$$2x = 3 \quad \text{or} \quad x = -1$$

$$x = \frac{3}{2}$$

The number is $\frac{3}{2}$ or -1.

12. $\frac{2}{5} = \frac{8}{x}$

$$2 \cdot x = 5 \cdot 8$$
$$2x = 40$$
$$x = 20$$

$$\frac{2}{5} = \frac{10}{y}$$

$$2 \cdot y = 5 \cdot 10$$
$$2y = 50$$
$$y = 25$$

$x = 20''$ and $y = 25''$

13.	Hours to Complete Total Job	Part of Job Completed in 1 hr
Mrs. Rath	6	1/6
Assistant	2(6) = 12	1/12
Together	x	$1/x$

Part of job Mrs.Rath completed in 1 hour	added to	part of job assistant completed in 1 hour	is equal to	part of job they completed together in 1 hour
$\frac{1}{6}$	+	$\frac{1}{12}$	=	$\frac{1}{x}$

$$\frac{1}{6} + \frac{1}{12} = \frac{1}{x}$$

$$12x\left(\frac{1}{6} + \frac{1}{12}\right) = 12x\left(\frac{1}{x}\right)$$

$$12x\left(\frac{1}{6}\right) + 12x\left(\frac{1}{12}\right) = 12x\left(\frac{1}{x}\right)$$

$$2x + x = 12$$
$$3x = 12$$
$$x = 4$$

Together it would take them 4 hours.

14.

	Rate ·	Time =	Distance
before noon	60	x	$60x$
after noon	55	$8 - x$	$55(8 - x)$

Distance driver before noon	added to	Distance driven after noon	equals	Total distance driven
$60x$	$+$	$55(8 - x)$	$=$	455

$$
\begin{aligned}
60x + 55(8 - x) &= 455 \\
60x + 440 - 55x &= 455 \\
5x + 440 &= 455 \\
5x &= 15 \\
x &= 3 \\
8 - x = 8 - 3 &= 5
\end{aligned}
$$

He drove 3 hours before noon and 5 hours after noon.

15.	Hours to Complete Total Job	Part of Job Completed in 1 hr
Kara	$x + 7$	$1/(x + 7)$
Wilma	x	$1/x$
Together	$x - 9$	$1/(x - 9)$

Part of job Kara completed in 1 hour	added to	part of job Wilma completed in 1 hour	is equal to	part of job they completed together in 1 hour
$\frac{1}{x + 7}$	$+$	$\frac{1}{x}$	$=$	$\frac{1}{x - 9}$

$$\frac{1}{x + 7} + \frac{1}{x} = \frac{1}{x - 9}$$

$$x(x + 7)(x - 9)\left(\frac{1}{x + 7} + \frac{1}{x}\right) = x(x + 7)(x - 9)\left(\frac{1}{x - 9}\right)$$

$$x(x + 7)(x - 9)\left(\frac{1}{x + 7}\right) + x(x + 7)(x - 9)\left(\frac{1}{x}\right) = x(x + 7)(x - 9)\left(\frac{1}{x - 9}\right)$$

$$
\begin{aligned}
x(x - 9) + (x + 7)(x - 9) &= x(x + 7) \\
x^2 - 9x + x^2 - 2x - 63 &= x^2 + 7x \\
2x^2 - 11x - 63 &= x^2 + 7x
\end{aligned}
$$

$$x^2 - 18x - 63 = 0$$
$$(x - 21)(x + 3) = 0$$
$$x - 21 = 0 \quad \text{or} \quad x + 3 = 0$$
$$x = 21 \quad \text{or} \quad x = -3$$

Since time cannot be negative, discard $x = -3$.

$$x = 21$$
$$x - 9 = 21 - 9 = 12$$

It will take them 12 hours working together.

CHAPTER 6 PRACTICE TEST SOLUTIONS

1. $x^2 - 3x + 2 = 0$
$(x - 2)(x - 1) = 0$
$x - 2 = 0$ or $x - 1 = 0$
$x = 2$ or $x = 1$

The denominator is 0 when $x = 2$ or $x = 1$ causing $\dfrac{x - 4}{x^2 - 3x + 2}$ to be undefined.

2. a.) $C = \dfrac{100(300) + 4000}{300}$

$= \dfrac{34000}{300}$

$= 113.33$

\$113.33

b.) $C = \dfrac{100(2000) + 4000}{2000}$

$= \dfrac{204000}{2000}$

$= 102$

\$102

3. $\dfrac{2x - 6}{7x - 21} = \dfrac{2(x - 3)}{7(x - 3)}$

$= \dfrac{2}{7}$

4. $\dfrac{x + 8}{x^2 - 64} = \dfrac{x + 8}{(x + 8)(x - 8)}$

$= \dfrac{1}{x - 8}$

5. $\dfrac{x + 5}{x^2 + 10x + 25} = \dfrac{x + 5}{(x + 5)^2}$

$= \dfrac{1}{x + 5}$

6. $\dfrac{x - 2}{x^3 - 8} = \dfrac{x - 2}{(x - 2)(x^2 + 2x + 4)}$

$= \dfrac{1}{x^2 + 2x + 4}$

7. $\dfrac{3m^3 + 15m^2 + 18m}{m^2 + 7m + 10} = \dfrac{3m(m + 2)(m + 3)}{(m + 5)(m + 2)}$

$= \dfrac{3m(m + 3)}{m + 5}$

8. $\dfrac{ay + 4a - 2y - 8}{ay + 4a + 6y + 24} = \dfrac{(a - 2)(y + 4)}{(a + 6)(y + 4)}$

$= \dfrac{a - 2}{a + 6}$

9. $\dfrac{y + x}{x^2 - y^2} = \dfrac{x + y}{(x + y)(x - y)}$

$= \dfrac{1}{x - y}$

10. $\dfrac{x^2 + x - 30}{x^2 + 7x + 10} \div \dfrac{x^2 + 3x - 4}{x^2 + x - 2}$

$= \dfrac{x^2 + x - 30}{x^2 + 7x + 10} \cdot \dfrac{x^2 + x - 2}{x^2 + 3x - 4}$

$= \dfrac{(x + 6)(x - 5)}{(x + 5)(x + 2)} \cdot \dfrac{(x + 2)(x - 1)}{(x + 4)(x - 1)}$

$= \dfrac{(x + 6)(x - 5)}{(x + 5)(x + 4)}$

11. $\dfrac{2}{x - 3} \cdot (8x - 24) = \dfrac{2}{x - 3} \cdot \dfrac{8(x - 3)}{1}$

$= 16$

12. $$\frac{y^2 + 5y + 6}{2y - 4} \cdot \frac{y - 2}{5y + 15}$$

$$= \frac{(y + 2)(y + 3)}{2(y - 2)} \cdot \frac{y - 2}{5(y + 3)}$$

$$= \frac{y + 2}{10}$$

13. $$\frac{4}{7x - 3} - \frac{5}{7x - 3} = \frac{4 - 5}{7x - 3}$$

$$= \frac{-1}{7x - 3}$$

$$= \frac{1}{-(7x - 3)}$$

$$= \frac{1}{-7x + 3}$$

14. $$\frac{6a}{a^2 - 5a + 6} - \frac{1}{a - 2}$$

$$= \frac{6a}{(a - 2)(a - 3)} - \frac{1}{a - 2}$$

$$= \frac{6a}{(a - 2)(a - 3)} - \frac{1(a - 3)}{(a - 2)(a - 3)}$$

$$= \frac{6a - (a - 3)}{(a - 2)(a - 3)}$$

$$= \frac{6a - a + 3}{(a - 2)(a - 3)}$$

$$= \frac{5a + 3}{(a - 2)(a - 3)}$$

15. $$\frac{8}{x^2 - 9} + \frac{5}{x + 3}$$

$$= \frac{8}{(x + 3)(x - 3)} + \frac{5}{x + 3}$$

$$= \frac{8}{(x + 3)(x - 3)} + \frac{5(x - 3)}{(x + 3)(x - 3)}$$

$$= \frac{8 + 5(x - 3)}{(x + 3)(x - 3)}$$

$$= \frac{8 + 5x - 15}{(x + 3)(x - 3)}$$

$$= \frac{5x - 7}{(x + 3)(x - 3)}$$

16. $$\frac{x^2 - 16}{x^2 - 4x} \div \frac{xy + 7x + 4y + 28}{3x + 21}$$

$$= \frac{x^2 - 16}{x^2 - 4x} \cdot \frac{3x + 21}{xy + 7x + 4y + 28}$$

$$= \frac{(x + 4)(x - 4)}{x(x - 4)} \cdot \frac{3(x + 7)}{(x + 4)(y + 7)}$$

$$= \frac{3(x + 7)}{x(y + 7)}$$

17. $$\frac{x + 3}{x^2 + 11x + 24} + \frac{6}{x^2 - x - 12}$$

$$= \frac{x + 3}{(x + 8)(x + 3)} + \frac{6}{(x - 4)(x + 3)}$$

$$= \frac{(x + 3)(x - 4)}{(x + 8)(x + 3)(x - 4)} + \frac{6(x + 8)}{(x + 8)(x + 3)(x - 4)}$$

$$= \frac{(x + 3)(x - 4) + 6(x + 8)}{(x + 8)(x + 3)(x - 4)}$$

$$= \frac{x^2 - x - 12 + 6x + 48}{(x + 8)(x + 3)(x - 4)}$$

$$= \frac{x^2 + 5x + 36}{(x + 8)(x + 3)(x - 4)}$$

18. $$\frac{2y}{y^2 + 3y - 4} - \frac{9}{y^2 + 4y - 5}$$

$$= \frac{2y}{(y + 4)(y - 1)} - \frac{9}{(y + 5)(y - 1)}$$

$$= \frac{2y(y + 5)}{(y + 4)(y - 1)(y + 5)} - \frac{9(y + 4)}{(y + 4)(y - 1)(y + 5)}$$

$$= \frac{2y(y + 5) - 9(y + 4)}{(y + 4)(y - 1)(y + 5)}$$

$$= \frac{2y^2 + 10y - 9y - 36}{(y + 4)(y - 1)(y + 5)}$$

$$= \frac{2y^2 + y - 36}{(y + 4)(y - 1)(y + 5)}$$

19.
$$\frac{3}{y} - \frac{4}{7} = \frac{-1}{6}$$

μ?â $$42y\left(\frac{3}{y} - \frac{4}{7}\right) = 42y\left(\frac{-1}{6}\right)$$

$$\frac{42y}{1} \cdot \frac{3}{y} - \frac{42y}{1} \cdot \frac{4}{7} = \frac{42y}{1} \cdot \frac{-1}{6}$$

$$126 - 24y = -7y$$
$$126 = 17y$$
$$\frac{126}{17} = y$$

20.
$$\frac{4}{y + 3} = \frac{5}{y + 4}$$

$$\frac{(y + 3)(y + 4)}{1} \cdot \frac{4}{y + 3} = \frac{(y + 3)(y + 4)}{1} \cdot \frac{5}{y + 4}$$

$$4(y + 4) = 5(y + 3)$$
$$4y + 16 = 5y + 15$$
$$16 = y + 15$$
$$1 = y$$

21.
$$\frac{a}{a - 6} = \frac{2}{a - 6} - \frac{1}{3}$$

$$\frac{3(a - 6)}{1} \cdot \frac{a}{a - 6} = \frac{3(a - 6)}{1}\left(\frac{2}{a - 6} - \frac{1}{3}\right)$$

$$3a = \frac{3(a - 6)}{1} \cdot \frac{2}{a - 6} - \frac{3(a - 6)}{1} \cdot \frac{1}{3}$$

$$3a = 6 - (a - 6)$$
$$3a = 6 - a + 6$$
$$3a = 12 - a$$
$$4a = 12$$
$$a = 3$$

22.
$$\frac{9}{x^2 - 16} = \frac{5}{x + 4} + \frac{2}{x - 4}$$
$$\frac{9}{(x + 4)(x - 4)} = \frac{5}{x + 4} + \frac{2}{x - 4}$$
$$\frac{(x + 4)(x - 4)}{1}\left[\frac{9}{(x + 4)(x - 4)}\right] = \frac{(x + 4)(x - 4)}{1}\left(\frac{5}{x + 4} + \frac{2}{x - 4}\right)$$
$$9 = \frac{(x + 4)(x - 4)}{1} \cdot \frac{5}{x + 4} + \frac{(x + 4)(x - 4)}{1} \cdot \frac{2}{x - 4}$$
$$9 = 5(x - 4) + 2(x + 4)$$
$$9 = 5x - 20 + 2x + 8$$
$$9 = 7x - 12$$
$$21 = 7x$$
$$3 = x$$

23.
$$\frac{\frac{8x^3}{y^2z}}{\frac{16x^2}{z^3}} = \frac{8x^3}{y^2z} \cdot \frac{z^3}{16x^2}$$
$$= \frac{xz^2}{2y^2}$$

24.
$$\frac{\frac{a}{b} + \frac{b}{a}}{\frac{a}{b} + \frac{a}{b}} = \frac{\frac{a}{b} + \frac{b}{a}}{\frac{2a}{b}} \cdot \frac{ab}{ab}$$
$$= \frac{\frac{a}{b} \cdot \frac{ab}{1} + \frac{b}{a} \cdot \frac{ab}{1}}{\frac{2a}{b} \cdot \frac{ab}{1}}$$
$$= \frac{a^2 + b^2}{2a^2}$$

25.
$$\frac{4 - \frac{2}{y^2}}{\frac{3}{y} + \frac{5}{y^2}} = \frac{4 - \frac{2}{y^2}}{\frac{3}{y} + \frac{5}{y^2}} \cdot \frac{y^2}{y^2}$$
$$= \frac{4y^2 - \frac{2}{y^2} \cdot \frac{y^2}{1}}{\frac{3}{y} \cdot \frac{y^2}{1} + \frac{5}{y^2} \cdot \frac{y^2}{1}}$$
$$= \frac{4y^2 - 2}{3y + 5}$$

26. $\frac{\text{defective}}{\text{total}}$: $\frac{6}{110} = \frac{x}{1210}$
$$6(1210) = 110x$$
$$7260 = 110x$$
$$66 = x$$
66 defective bulbs

27. number: x
$$x + 3\left(\frac{1}{x}\right) = 4$$
$$x\left(x + \frac{3}{x}\right) = x(4)$$
$$x(x) + \frac{x}{1} \cdot \frac{3}{x} = 4x$$

$$x^2 + 3 = 4x$$
$$x^2 - 4x + 3 = 0$$
$$(x - 3)(x - 1) = 0$$
$$x - 3 = 0 \quad \text{or} \quad x - 1 = 0$$
$$x = 3 \quad \text{or} \quad x = 1$$

The number is 3 or 1.

28. speed of boat in still water: x

	distance	rate	time = d/r
upstream	16	$x - 3$	$16/(x - 3)$
downstream	19	$x + 3$	$19/(x + 3)$

$$\frac{16}{x-3} = \frac{19}{x+3}$$

$$\frac{(x-3)(x+3)}{1} \cdot \frac{16}{x-3} = \frac{(x-3)(x+3)}{1} \cdot \frac{19}{x+3}$$

$$16(x + 3) = 19(x - 3)$$

$$16x + 48 = 19x - 57$$
$$16x + 105 = 19x$$
$$105 = 3x$$
$$35 = x$$

35 mph

29.	time for whole job	part done in 1 hr
1st pipe	16	1/16
2nd pipe	20	1/20
together	x	$1/x$

$$\frac{1}{16} + \frac{1}{20} = \frac{1}{x}$$

$$80x\left(\frac{1}{16} + \frac{1}{20}\right) = 80x\left(\frac{1}{x}\right)$$

$$\frac{80x}{1} \cdot \frac{1}{16} + \frac{80x}{1} \cdot \frac{1}{20} = 80$$

$$5x + 4x = 80$$
$$9x = 80$$

$$x = \frac{80}{9}$$

$8\frac{8}{9}$ hours

30. Determine price per oz. for each:

$\frac{1.80}{8} = 0.225$; $\frac{2.43}{12} = 0.2025$;

$\frac{2.97}{15} = 0.198$

15 oz. for \$2.97 is the best buy.

7.1 SOLUTIONS TO EXERCISES

1. $7x + y = 10$
   $y = -7x + 10$

   slope: -7; y-intercept: 10

2. $-12x - y = 7$
   $-y = 12x + 7$
   $y = -12x - 7$

   slope: -12; y-intercept: -7

3. $2x - 5y = -10$
   $-5y = -2x - 10$
   $y = \frac{2}{5}x + 2$

   slope: $\frac{2}{5}$; y-intercept: 2

4. $6x + 7y = 42$
   $7y = -6x + 42$
   $y = -\frac{6}{7}x + 6$

   slope: $-\frac{6}{7}$; y-intercept: 6

5. $x - 3 = 0$
   $x = 3$

   This is a vertical line.
   slope: undefined; no y-intercept

6. $y = 15$

   This is a horizontal line.
   slope: 0; y-intercept: 15

7. $2x + y = 8$
   $y = -2x + 8$
   $m_1 = -2$

   $3x + 6y = 12$
   $6y = -3x + 12$
   $y = -\frac{1}{2}x + 2$

   $m_2 = -\frac{1}{2}$

   $m_1 \neq m_2$, so the lines are not parallel.
   $m_1 m_2 \neq -1$, so the lines are not perpendicular.

   Neither

8. $5x + 2y = 20$
   $2y = -5x + 20$
   $y = -\frac{5}{2}x + 10$

   $m_1 = -\frac{5}{2}$

   $2y - 16 = -5x$
   $2y = -5x + 16$
   $y = -\frac{5}{2}x + 8$

   $m_2 = -\frac{5}{2}$

   $m_1 = m_2$, and the y-intercepts are different, so the lines are parallel.

9. $13x - 26y = 52$
   $-26y = -13x + 52$
   $y = \frac{1}{2}x - 2$

   $m_1 = \frac{1}{2}$

   $8x + 4y = 7$
   $4y = -8x + 7$
   $y = -2x + \frac{7}{4}$

   $m_2 = -2$

   $m_1 m_2 = \left(\frac{1}{2}\right)(-2) = -1$, so the lines are perpendicular.

10.
$$y = 10x + 9$$
$$m_1 = 10$$

$$2y + 20x = -3$$
$$2y = -20x - 3$$
$$y = -10x - \frac{3}{2}$$
$$m_2 = -10$$

$m_1 \neq m_2$, so the lines are not parallel.
$m_1 m_2 \neq -1$, so the lines are not perpendicular.

Neither

11.
$$y = mx + b$$
$$y = -2x + 5$$

12.
$$y = mx + b$$
$$y = -\frac{3}{4}x + 2$$

13.
$$y = mx + b$$
$$y = \frac{5}{11}x + \left(-\frac{1}{6}\right)$$
$$y = \frac{5}{11}x - \frac{1}{6}$$

14.
$$y = mx + b$$
$$y = 8x + 0$$
$$y = 8x$$

15.
$$y = \frac{3}{4}x - 1$$

Plot (0, −1) and use $m = \frac{3}{4}$ to locate another point (4, 2).

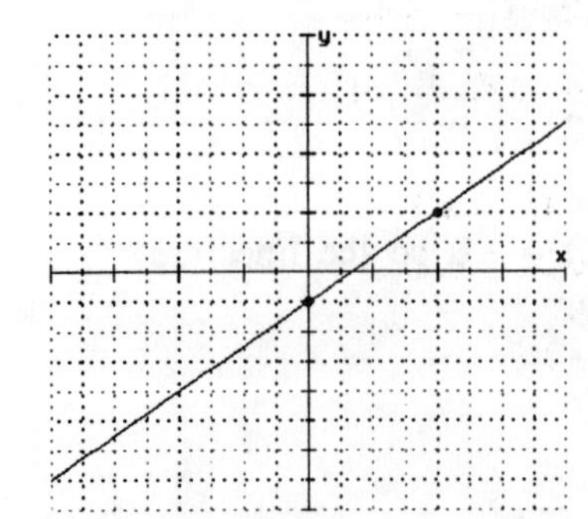

16.
$$y = 4x$$
$$y = 4x + 0$$

Plot (0, 0) and use $m = \frac{4}{1}$ to locate another point (1, 4).

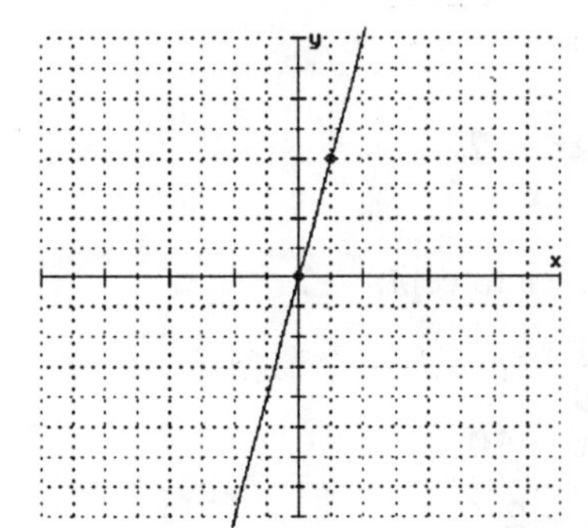

17.
$$2x + 3y = 12$$
$$3y = -2x + 12$$
$$y = -\frac{2}{3}x + 4$$

Plot (0, 4) and use $m = -\frac{2}{3} = \frac{-2}{3}$ to locate another point (3, 2).

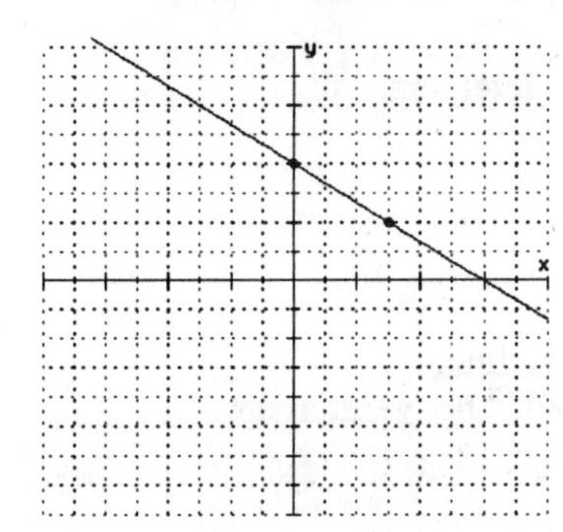

18.
$$4x - 3y = 6$$
$$-3y = -4x + 6$$
$$y = \frac{4}{3}x - 2$$

Plot (0, −2) and use $m = \frac{4}{3}$ to locate another point (3, 2).

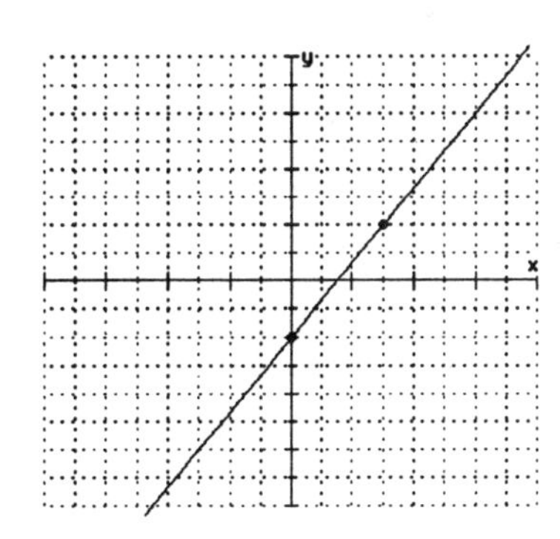

7.2 SOLUTIONS TO EXERCISES

1. $$\begin{aligned} y - y_1 &= m(x - x_1) \\ y - 5 &= -6(x - 3) \\ y - 5 &= -6x + 18 \\ y &= -6x + 23 \\ 6x + y &= 23 \end{aligned}$$

2. $$\begin{aligned} y - y_1 &= m(x - x_1) \\ y - (-1) &= \frac{2}{3}[x - (-4)] \\ y + 1 &= \frac{2}{3}(x + 4) \\ 3(y + 1) &= 3 \cdot \frac{2}{3}(x + 4) \\ 3y + 3 &= 2(x + 4) \\ 3y + 3 &= 2x + 8 \\ 3 &= 2x + 8 - 3y \\ -5 &= 2x - 3y \end{aligned}$$

3. $$\begin{aligned} y - y_1 &= m(x - x_1) \\ y - (-7) &= -\frac{5}{2}[x - (-2)] \\ y + 7 &= -\frac{5}{2}(x + 2) \\ 2(y + 7) &= 2\left(-\frac{5}{2}\right)(x + 2) \\ 2y + 14 &= -5(x + 2) \\ 2y + 14 &= -5x - 10 \\ 5x + 2y + 14 &= -10 \\ 5x + 2y &= -24 \end{aligned}$$

4. $$m = \frac{y_2 - y_1}{x_2 - x_1} = \frac{6 - 1}{3 - 2} = \frac{5}{1} = 5$$
$$\begin{aligned} y - y_1 &= m(x - x_1) \\ y - 1 &= 5(x - 2) \\ y - 1 &= 5x - 10 \\ -1 &= 5x - 10 - y \\ 9 &= 5x - y \end{aligned}$$

5. $$m = \frac{y_2 - y_1}{x_2 - x_1} = \frac{4 - (-8)}{-5 - (-7)} = \frac{12}{2} = 6$$
$$\begin{aligned} y - y_1 &= m(x - x_1) \\ y - (-8) &= 6[x - (-7)] \\ y + 8 &= 6(x + 7) \\ y + 8 &= 6x + 42 \\ 8 &= 6x + 42 - y \\ -34 &= 6x - y \end{aligned}$$

6. $$m = \frac{y_2 - y_1}{x_2 - x_1} = \frac{\frac{3}{4} - 0}{\frac{2}{3} - 0} = \frac{\frac{3}{4}}{\frac{2}{3}} = \frac{3}{4} \cdot \frac{3}{2} = \frac{9}{8}$$
$$\begin{aligned} y - y_1 &= m(x - x_1) \\ y - 0 &= \frac{9}{8}(x - 0) \\ y &= \frac{9}{8}x \\ 0 &= \frac{9}{8}x - y \end{aligned}$$

7. vertical line: $x = a$
   $x = 5$

8. horizontal line: $y = b$
   $y = 16$

9. $y = 10$ is a horizontal line. A parallel line will also be horizontal. Horizontal line through (7, 13): $y = 13$

10. $y = -8$ is horizontal, hence a line perpendicular will be vertical.

    Vertical line through (4, 6): $x = 4$

11. $$\begin{aligned} y - y_1 &= m(x - x_1) \\ y - (-7.6) &= 0(x - 8.1) \\ y + 7.6 &= 0 \\ y &= -7.6 \end{aligned}$$

12. A line with undefined slope is vertical.

Vertical line through $\left(\frac{1}{7}, \frac{1}{6}\right)$: $x = \frac{1}{7}$

$$7 \cdot x = 7 \cdot \frac{1}{7}$$

$$7x = 1$$

13.
$$y = mx + b$$
$$y = -\frac{3}{7}x - 2$$

$$7y = 7\left(-\frac{3}{7}x - 2\right)$$
$$7 \quad 7y = -3x - 14$$
$$3x + 7y = -14$$

14. A line parallel to the x-axis is a horizontal line.

Horizontal line through (6, 1): $y = 1$

15. a. (3, 2200), (6, 1675)

$$m = \frac{2200 - 1675}{3 - 6} = \frac{525}{-3} = -175$$

$$y - y_1 = m(x - x_1)$$
$$y - 2200 = -175(x - 3)$$
$$y - 2200 = -175x + 525$$
$$y = -175x + 2725$$

b. $x = 2002 - 1992 = 10$

$$y = -175x + 2725$$
$$y = -175(10) + 2725$$
$$= -1750 + 2725$$
$$= 975$$

Its value will be $975.

7.3 SOLUTIONS TO EXERCISES

1. $y = x^2 - 3$

x	$y = x^2 - 3$
-3	$(-3)^2 - 3 = 6$
-2	$(-2)^2 - 3 = 1$
-1	$(-1)^2 - 3 = -2$
0	$0^2 - 3 = -3$
1	$1^2 - 3 = -2$
2	$2^2 - 3 = 1$
3	$3^2 - 3 = 6$

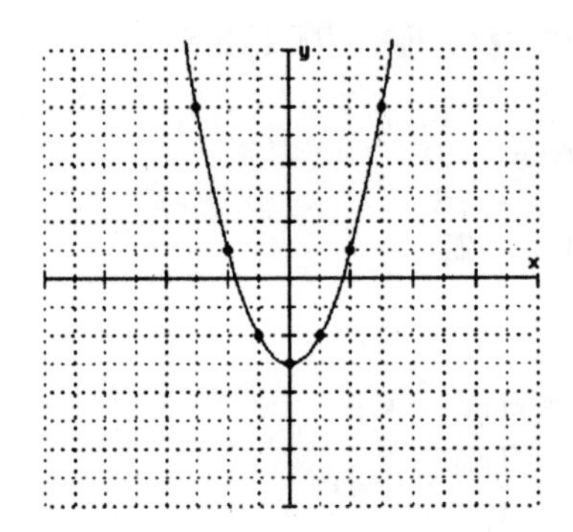

2. $y = 3x^2 + 1$

x	$y = 3x^2 + 1$
-3	$3(-3)^2 + 1 = 28$
-2	$3(-2)^2 + 1 = 13$
-1	$3(-1)^2 + 1 = 4$
0	$3(0)^2 + 1 = 1$
1	$3(1)^2 + 1 = 4$
2	$3(2)^2 + 1 = 13$
3	$3(3)^2 + 1 = 28$

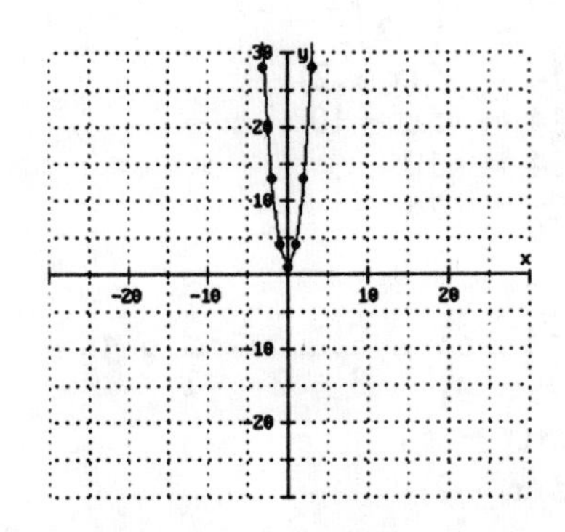

3. $y = |x| + 4$

x	$y = \|x\| + 4$
-3	$\|-3\| + 4 = 7$
-2	$\|-2\| + 4 = 6$
-1	$\|-1\| + 4 = 5$
0	$\|0\| + 4 = 4$
1	$\|1\| + 4 = 5$
2	$\|2\| + 4 = 6$
3	$\|3\| + 4 = 7$

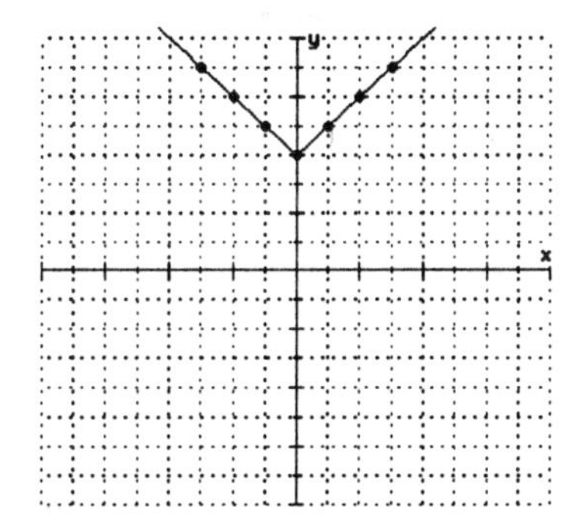

4. $y = |x - 3|$

x	$y = \|x - 3\|$
-2	$\|-2 - 3\| = 5$
-1	$\|-1 - 3\| = 4$
0	$\|0 - 3\| = 3$
1	$\|1 - 3\| = 2$
2	$\|2 - 3\| = 1$
3	$\|3 - 3\| = 0$
4	$\|4 - 3\| = 1$

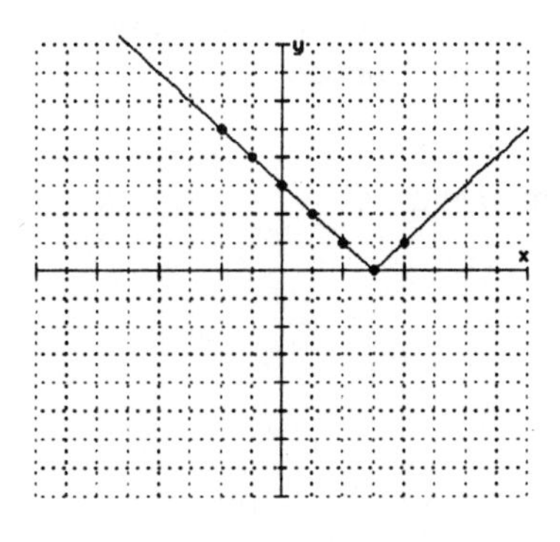

5. $y = -(x - 1)^2$

x	$y = -(x - 1)^2$
-3	$-(-3 - 1)^2 = -16$
-2	$-(-2 - 1)^2 = -9$
-1	$-(-1 - 1)^2 = -4$
0	$-(0 - 1)^2 = -1$
1	$-(1 - 1)^2 = 0$
2	$-(2 - 1)^2 = -1$
3	$-(3 - 1)^2 = -4$

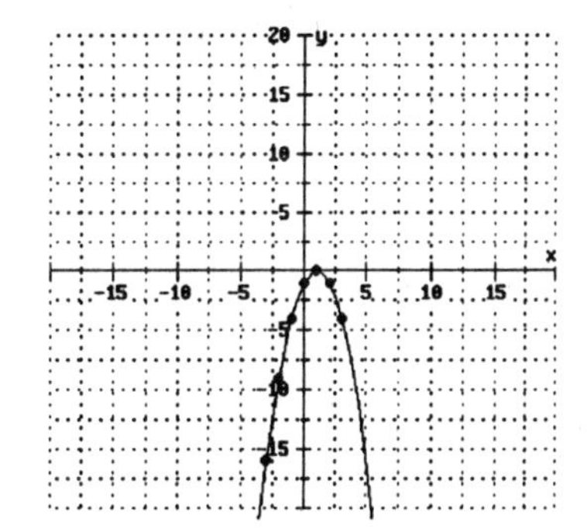

6. $y = 2 - |x|$

x	$y = 2 - \|x\|$
-3	$2 - \|-3\| = -1$
-2	$2 - \|-2\| = 0$
-1	$2 - \|-1\| = 1$
0	$2 - \|0\| = 2$
1	$2 - \|1\| = 1$
2	$2 - \|2\| = 0$
3	$2 - \|3\| = -1$

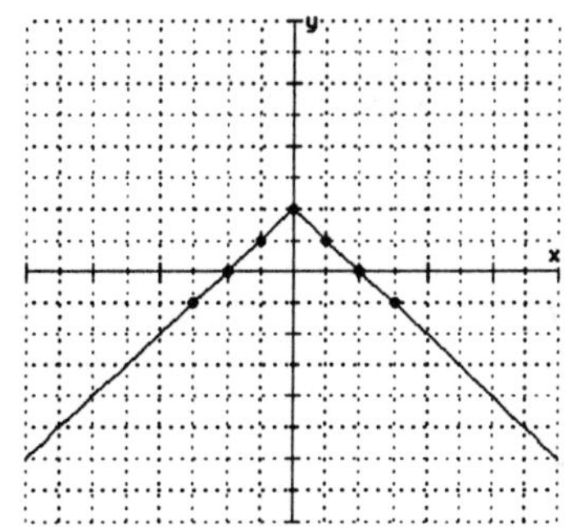

7. Locate 0 on the x-axis, then move vertically until the graph is reached. From the point on the graph move horizontally until the y-axis is reached. The y-value is -8.

8. Locate 0 on the y-axis, then move horizontally until the graph is reached. There are two points and they are both on the x-axis. The x-values are -2 and 2.

9. Locate 1 on the x-axis, then move vertically until the graph is reached. From the point on the graph move horizontally until the y-axis is reached. The y-value is -6.

10. This equation is of the form $Ax + By - C$, so it is linear.

$$x + 2y = 6$$
$$2y = -x + 6$$
$$y = -\frac{1}{2}x + 3$$

$m = -\frac{1}{2}$; y-intercept: 3

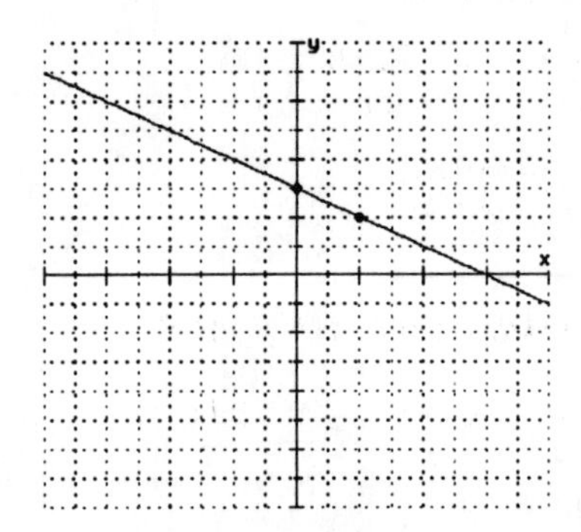

11. This equation is not of the form $Ax + By = C$, so it is nonlinear.

x	$y = \lvert x \rvert - 7$
-3	$\lvert -3 \rvert - 7 = -4$
-2	$\lvert -2 \rvert - 7 = -5$
-1	$\lvert -1 \rvert - 7 = -6$
0	$\lvert 0 \rvert - 7 = -7$
1	$\lvert 1 \rvert - 7 = -6$
2	$\lvert 2 \rvert - 7 = -5$
3	$\lvert 3 \rvert - 7 = -4$

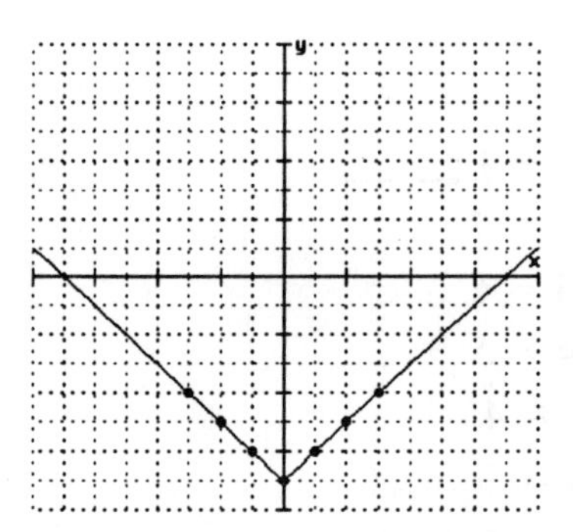

12. This equation is not of the form $Ax + By = C$, so it is nonlinear.

x	$y = (x + 1)^2$
-3	$(-3 + 1)^2 = 4$
-2	$(-2 + 1)^2 = 1$
-1	$(-1 + 1)^2 = 0$
0	$(0 + 1)^2 = 1$
1	$(1 + 1)^2 = 4$
2	$(2 + 1)^2 = 9$

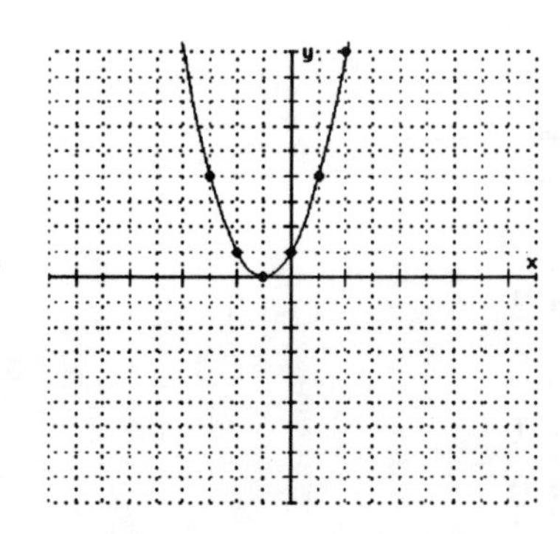

13. This equation can be written in the form $Ax + By = C$, so it is linear.

$y = -5x + 4$
$m = -5$; y-intercept: 4

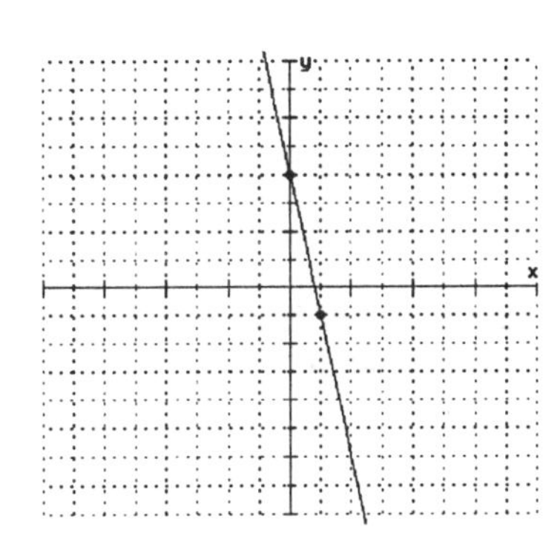

14. This equation is not of the form $Ax + By = C$, so it is nonlinear.

x	$y = x^2 + 6x$
-5	$(-5)^2 + 6(-5) = -5$
-4	$(-4)^2 + 6(-4) = -8$
-3	$(-3)^2 + 6(-3) = -9$
-2	$(-2)^2 + 6(-2) = -8$
-1	$(-1)^2 + 6(-1) = -5$
0	$0^2 + 6(0) = 0$
1	$1^2 + 6(1) = 7$

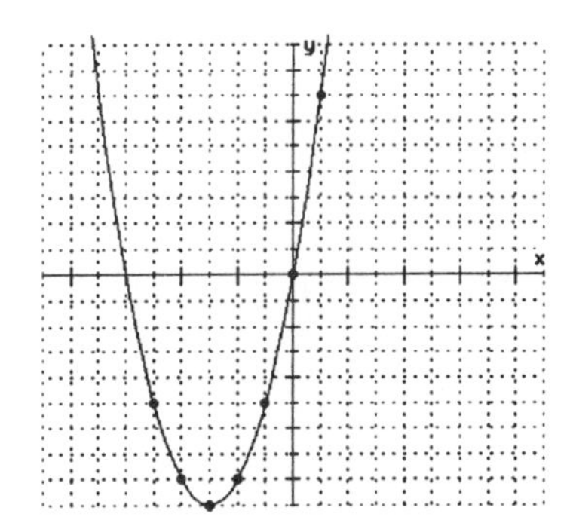

15. This equation is not of the form $Ax + By = C$, so it is nonlinear.

x	$y = -x^2 - 2x$
-4	$-(-4)^2 - 2(-4) = -8$
-3	$-(-3)^2 - 2(-3) = -3$
-2	$-(-2)^2 - 2(-2) = 0$
-1	$-(-1)^2 - 2(-1) = 1$
0	$-0^2 - 2(0) = 0$
1	$-1^2 - 2(1) = -3$
2	$-2^2 - 2(2) = -8$

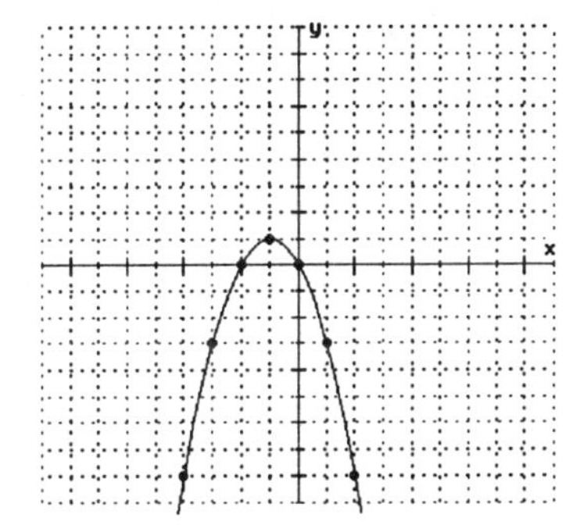

7.4 SOLUTIONS TO EXERCISES

1. Domain: $\{-3, 1, 0, 4\}$ (x-values)
   Range: $\{-8, 2, 6, 7\}$ (y-values)

2. Domain: $\{0, 2, 3\}$ (x-values)
   Range: $\{-1, 1\}$ (y-values)

3. Function (no x-value is used more than once)

4. Not a function; there are two y-values assigned to the x-value -9.

5. $y = 3x - 5$

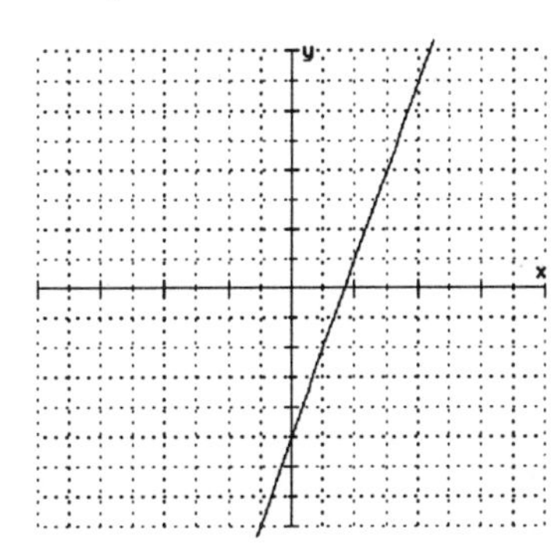

This graph passes the vertical line test so it is a function.

6. $x = -2y^2$

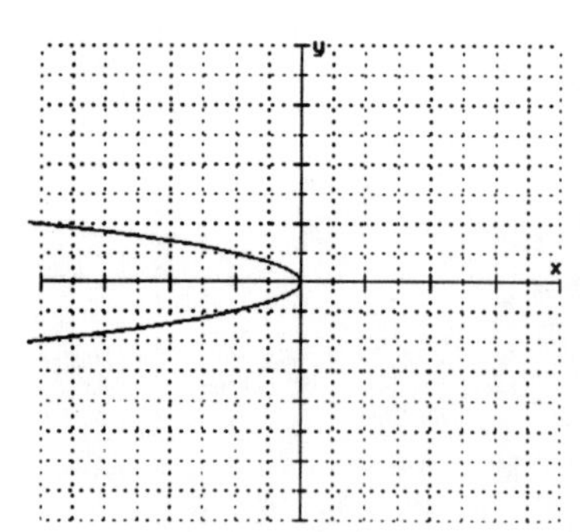

Note that vertical lines would intersect the graph in more than one point so it is not a function.

7. $x = 4$

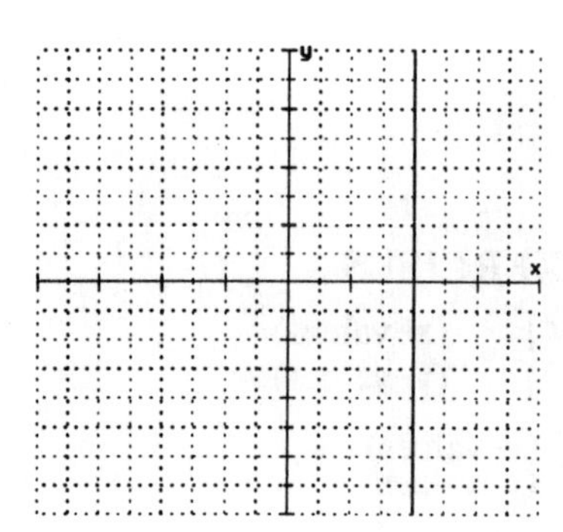

This graph fails the vertical line test so it is not a function.

8. $y < 3x - 2$

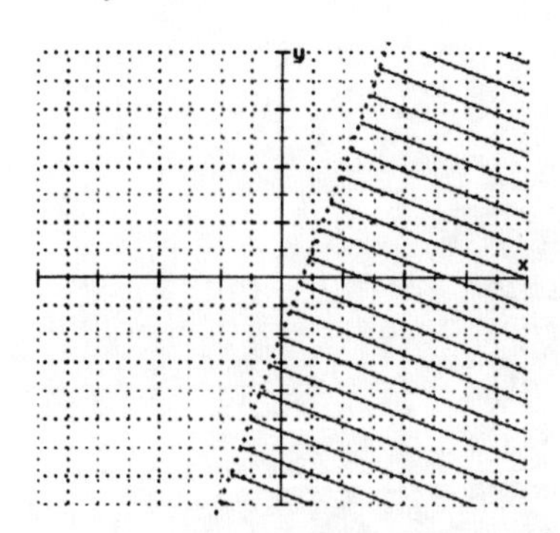

This graph fails the vertical line test so it is not a function.

9. Function

10. Not a function; vertical lines intersect the graph in more than one point.

11. Not a function; vertical lines intersect the graph in more than one point.

12. $g(x) = 4 - 3x$
$g(-1) = 4 - 3(-1) = 7$
$g(0) = 4 - 3(0) = 4$
$g(4) = 4 - 3(4) = -8$

13. $g(x) = 2|x| + 1$
$g(-1) = 2|-1| + 1 = 3$
$g(0) = 2|0| + 1 = 1$
$g(4) = 2|4| + 1 = 9$

14. $g(x) = -x^3 + 5$
$g(-1) = -(-1)^3 + 5 = -(-1) + 5 = 6$
$g(0) = -0^3 + 5 = 0 + 5 = 5$
$g(4) = -4^3 + 5 = -64 + 5 = -59$

15. $g(x) = -10$
$g(-1) = -10$
$g(0) = -10$
$g(4) = -10$

CHAPTER 7 PRACTICE TEST SOLUTIONS

1. $6x - 5y = 8$

$$-5y = -6x + 8$$
$$y = \frac{6}{5}x - \frac{8}{5}$$
$$m = \frac{6}{5}; \quad b = -\frac{8}{5}$$

2. $y = 3x - 7 \quad ; \quad -9x = 3y$

$$m_1 = 3 \qquad -3x = y$$
$$m_2 = -3$$

Neither, since $m_1 \neq m_2$ and $m_1 \neq -\frac{1}{m_2}$.

3. $y - y_1 = m(x - x_1)$

$$y - (-6) = -\frac{1}{3}(x - 6)$$
$$y + 6 = -\frac{1}{3}(x - 6)$$
$$3(y + 6) = 3\left[-\frac{1}{3}(x - 6)\right]$$
$$3y + 18 = -(x - 6)$$
$$3y + 18 = -x + 6$$
$$3y = -x + -12$$
$$x + 3y = -12$$

4. $m = \frac{y_2 - y_1}{x_2 - x_1}$

$$= \frac{3 - 0}{-1 - 0}$$
$$= -3$$
$$y - y_1 = m(x - x_1)$$
$$y - 0 = -3(x - 0)$$
$$y = -3x$$
$$3x + y = 0$$

5. $m = \frac{y_2 - y_1}{x_2 - x_1}$

$$= \frac{-5 - 4}{8 - 2}$$
$$= -\frac{9}{6} = -\frac{3}{2}$$
$$y - y_1 = m(x - x_1)$$
$$y - 4 = -\frac{3}{2}(x - 2)$$
$$2(y - 4) = 2\left(-\frac{3}{2}\right)(x - 2)$$
$$2y - 8 = -3(x - 2)$$
$$2y - 8 = -3x + 6$$
$$2y = -3x + 14$$
$$3x + 2y = 14$$

6. Vertical line through $(-12, 5)$: $x = -12$

7. $y = mx + b$

$$y = \frac{1}{9}x + 6$$
$$9y = 9\left(\frac{1}{9}x + 6\right)$$
$$9y = x + 54$$
$$-x + 9y = 54$$
$$x - 9y = -54$$

8. $x - 3y = 6$

$$-3y = -x + 6$$
$$y = \frac{1}{3}x - 2$$

x	y
-3	(1\3)(-3) - 2 = -3
0	(1/3)(0) - 2 = -2
3	(1/3)(3) - 2 = -1

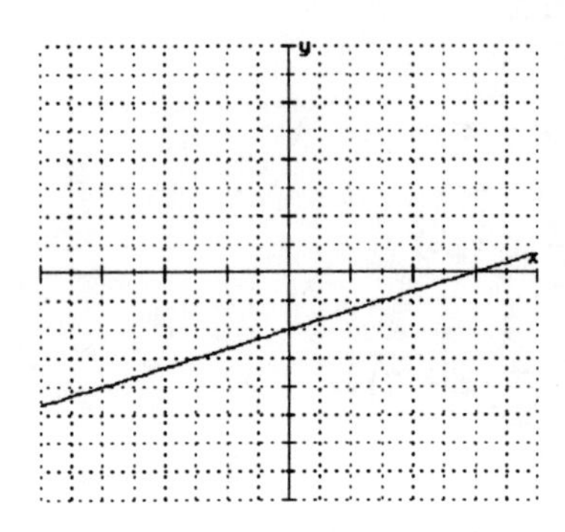

9. $y = -5x$

x	y
-1	-5(-1) = 5
0	-5(0) = 0
1	-5(1) = -5

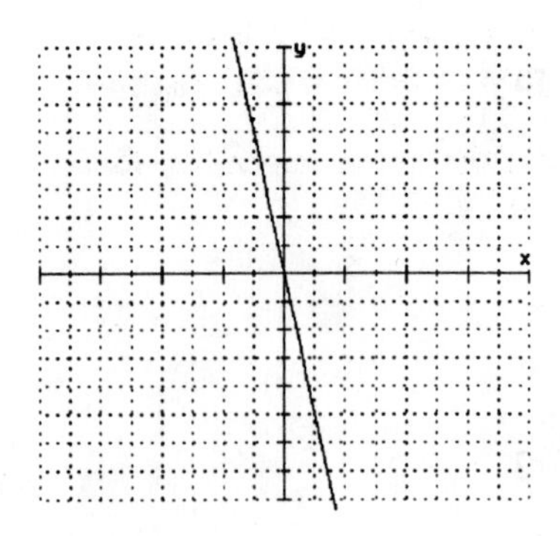

10. $y = |x - 1|$

x	y
-1	$\lvert -1 - 1 \rvert = \lvert -2 \rvert = 2$
0	$\lvert 0 - 1 \rvert = \lvert -1 \rvert = 1$
1	$\lvert 1 - 1 \rvert = \lvert 0 \rvert = 0$
2	$\lvert 2 - 1 \rvert = \lvert 1 \rvert = 1$
3	$\lvert 3 - 1 \rvert = \lvert 2 \rvert = 2$

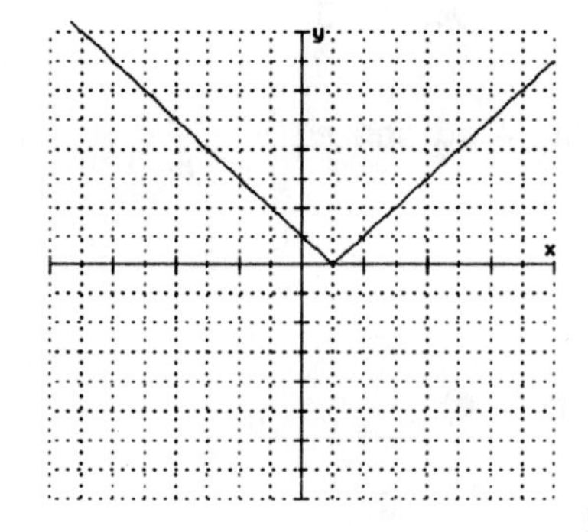

11. $y = x^2 + 2$

x	y
-2	$(-2)^2 + 2 = 6$
-1	$(-1)^2 + 2 = 3$
0	$0^2 + 2 = 2$
1	$1^2 + 2 = 3$
2	$2^2 + 2 = 6$

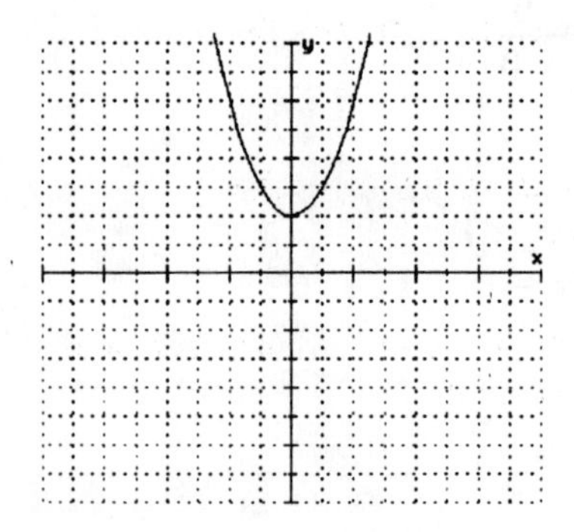

12. $8x - 3y = 7$
$$-3y = -8x + 7$$
$$y = \frac{8}{3}x - \frac{7}{3}$$

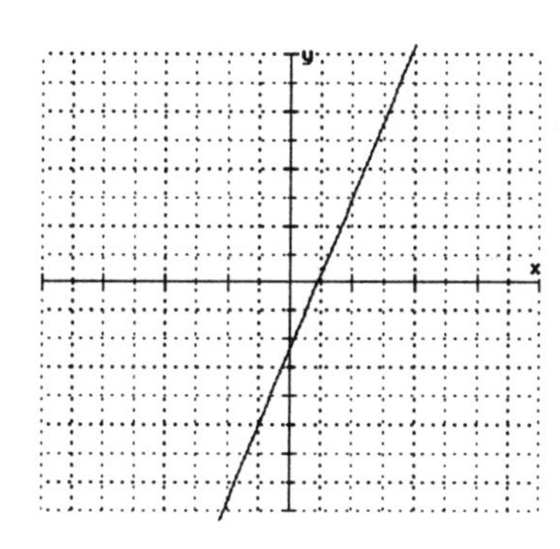

It is a function since it passes the vertical line test.

13. $y = \dfrac{1}{x - 2}$

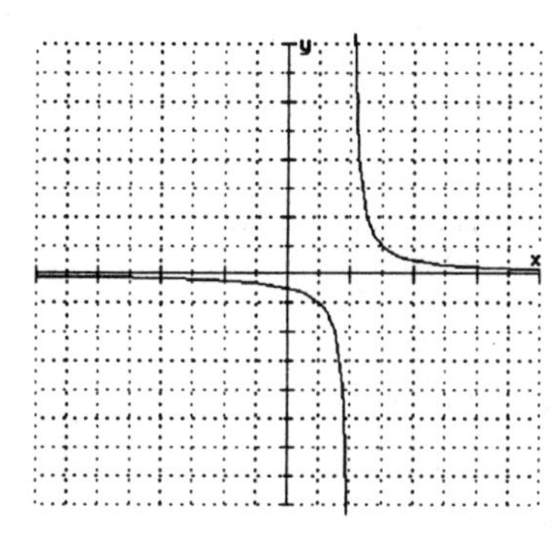

It is a function since it passes the vertical line test.

14. It is a function since it passes the vertical line test.

15. It is not a function. Vertical lines intersect at more than one point.

16. $f(x) = 3x + 5$

(a) $f(-1) = 3(-1) + 5 = 2$

(b) $f(0.3) = 3(0.3) + 5 = 5.9$

(c) $f(0) = 3(0) + 5 = 5$

17. $h(x) = -x^3 + 2x$

(a) $h(0) = -0^3 + 2(0) = 0$

(b) $h(-1) = -(-1)^3 + 2(-1)$
$= -(-1) - 2$
$= 1 - 2$
$= -1$

(c) $h(3) = -3^3 + 2(3)$
$= -27 + 6$
$= -21$

18. $g(x) = -7$

(a) $g(0) = -7$

(b) $g(-12) = -7$

(c) $g(a) = -7$

19. $x - 4 \neq 0$
$x \neq 4$

All real numbers except $x = 4$.

20. Domain: all real numbers (The parabola opens up infinitely to the left and right.)

Range: $y \geq -4$ (The lowest point on the graph is (-3, -4).)

8.1 SOLUTIONS TO EXERCISES

1. a.
$$\begin{aligned} x + y &= 6 \\ 4 + 2 &= 6 \\ 6 &= 6 \end{aligned}$$
True
$$\begin{aligned} 5x + 3y &= 26 \\ 5(4) + 3(2) &= 26 \\ 20 + 6 &= 26 \\ 26 &= 26 \end{aligned}$$
True

(4, 2) is a solution.

b.
$$\begin{aligned} x + y &= 6 \\ -3 + 9 &= 6 \\ 6 &= 6 \end{aligned}$$
True
$$\begin{aligned} 5x + 3y &= 26 \\ 5(-3) + 3(9) &= 26 \\ -15 + 27 &= 26 \\ 12 &= 26 \end{aligned}$$
False

(−3, 9) is not a solution.

2. a.
$$\begin{aligned} x - 9y &= -1 \\ 9 - 9(1) &= -1 \\ 9 - 9 &= -1 \\ 0 &= -1 \end{aligned}$$
False

Since (9, 1) doesn't satisfy the first equation it is not a solution to the system and there is no need to check the second equation.

b.
$$\begin{aligned} x - 9y &= -1 \\ 1 - 9(0) &= -1 \\ 1 &= -1 \end{aligned}$$
False

(1, 0) is not a solution.

3. a.
$$\begin{aligned} 3y &= 6x \\ 3(2) &= 6(1) \\ 6 &= 6 \end{aligned}$$
True
$$\begin{aligned} 3x - y &= 1 \\ 3(1) - 2 &= 1 \\ 3 - 2 &= 1 \\ 1 &= 1 \end{aligned}$$
True

(1, 2) is a solution.

b.
$$\begin{aligned} 3y &= 6x \\ 3(4) &= 6(2) \\ 12 &= 12 \end{aligned}$$
True
$$\begin{aligned} 3x - y &= 1 \\ 3(2) - 4 &= 1 \\ 6 - 4 &= 1 \\ 2 &= 1 \end{aligned}$$
False

(2, 4) is not a solution.

4. a.
$$\begin{aligned} 5y &= 10x \\ 5(-4) &= 10(-2) \\ -20 &= -20 \end{aligned}$$
True
$$\begin{aligned} y - 2x &= 0 \\ -4 - 2(-2) &= 0 \\ -4 + 4 &= 0 \\ 0 &= 0 \end{aligned}$$
True

(−2, −4) is a solution.

b.
$$\begin{aligned} 5y &= 10x \\ 5(6) &= 10(3) \\ 30 &= 30 \end{aligned}$$
True
$$\begin{aligned} y - 2x &= 0 \\ 6 - 2(3) &= 0 \\ 6 - 6 &= 0 \\ 0 &= 0 \end{aligned}$$
True

(3, 6) is a solution.

5. $y = x - 1$

x	y
0	-1
1	0
2	1

$y = -3x + 19$

x	y
5	4
6	1
7	-2

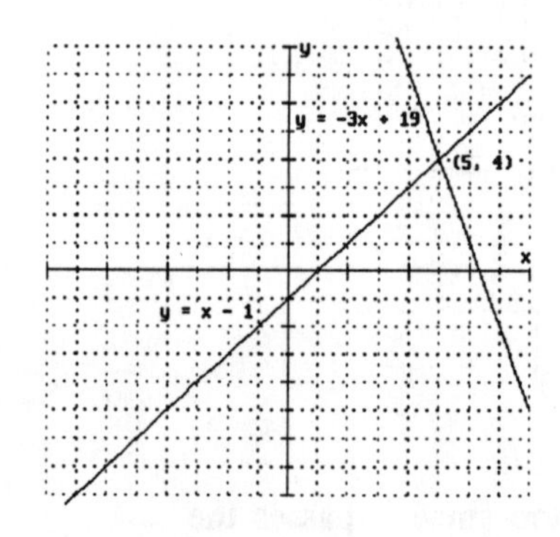

The solution is (5, 4), the system is consistent, and the equations are independent.

6. $y = 3x + 1$ $\qquad$ $x + y = -3$

x	y
0	1
1	4
2	7

x	y
0	-3
-3	0
1	-4

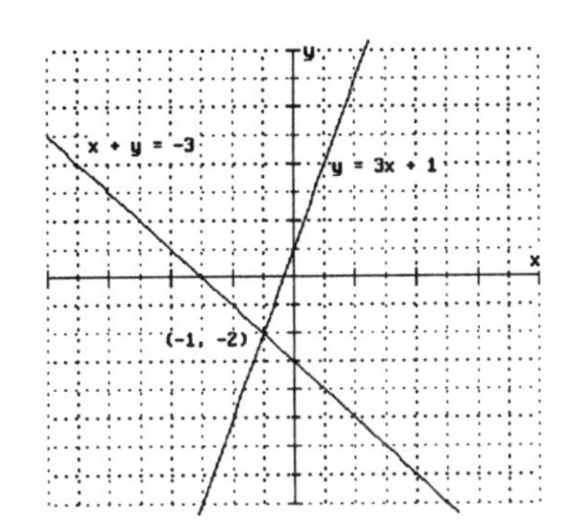

The solution is $(-1, -2)$, the system is consistent, and the equations are independent.

8. $x - y = 4$ $\qquad$ $y - x = 5$

x	y
0	-4
4	0
1	-3

x	y
0	5
-5	0
1	6

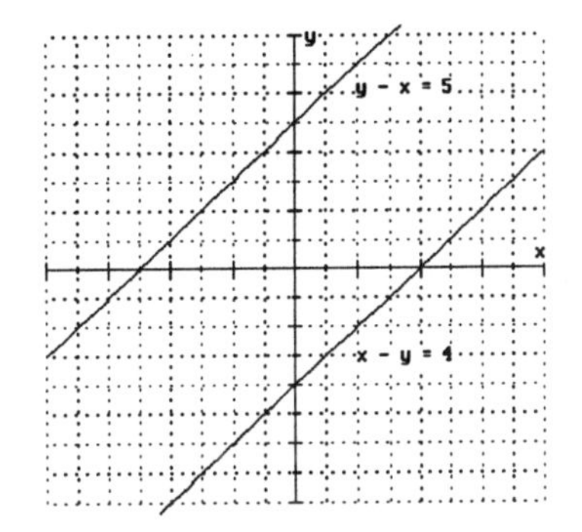

The lines are parallel. There is no solution, the system is inconsistent, and the equations are independent.

7. $x - 2y = 4$ $\qquad$ $y = -1$

x	y
0	-2
4	0
2	-1

x	y
-1	-1
0	-1
1	-1

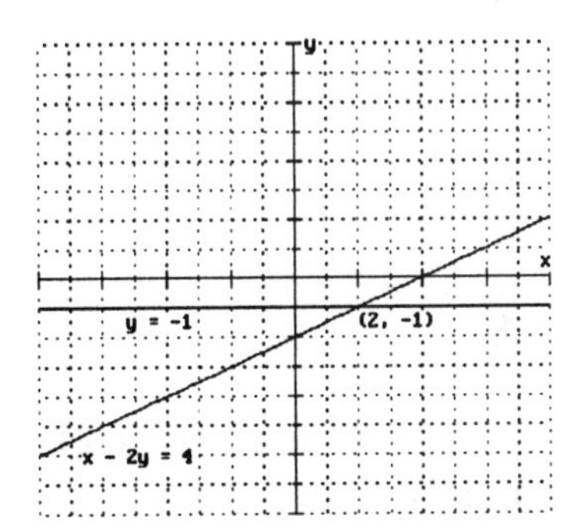

The solution is $(2, -1)$, the system is consistent, and the equations are independent.

9. $x = y$ $\qquad$ $x = 3$

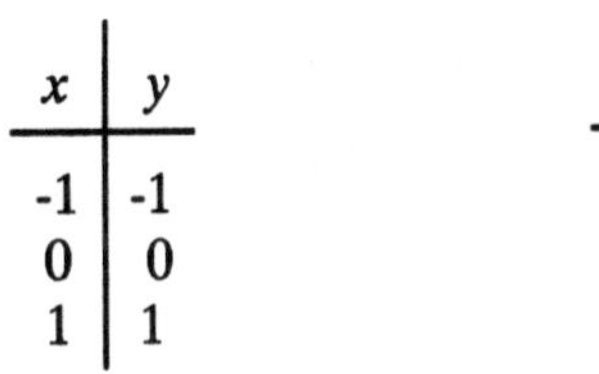

x	y
-1	-1
0	0
1	1

x	y
3	0
3	1
3	2

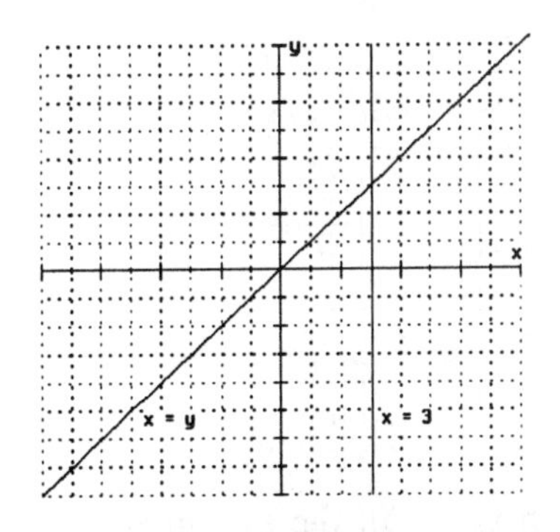

The solution is $(3, 3)$ the system is consistent, and the equations are independent.

10. $y = \frac{1}{2}x - 3$

x	y
0	-3
2	-2
4	-1

$x = 2y + 6$

x	y
0	-3
6	0
8	1

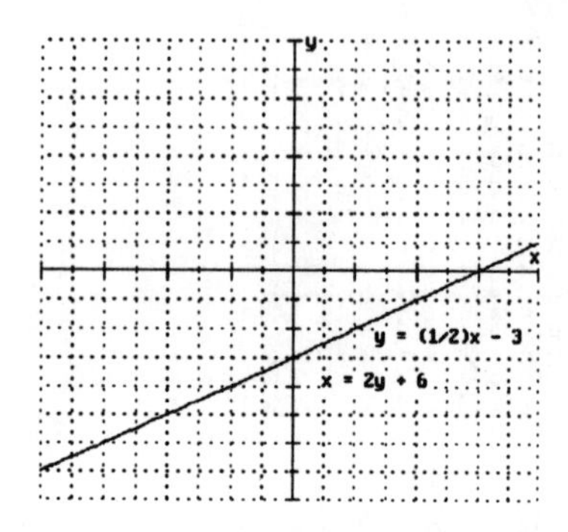

These are the same line. There are an infinite number of solutions, and the equations are dependent.

11. $\frac{2}{3}x + y = 5$

x	y
-3	7
0	5
3	3

$3x - 4y = -20$

x	y
0	5
1	23/4
-1	17/4

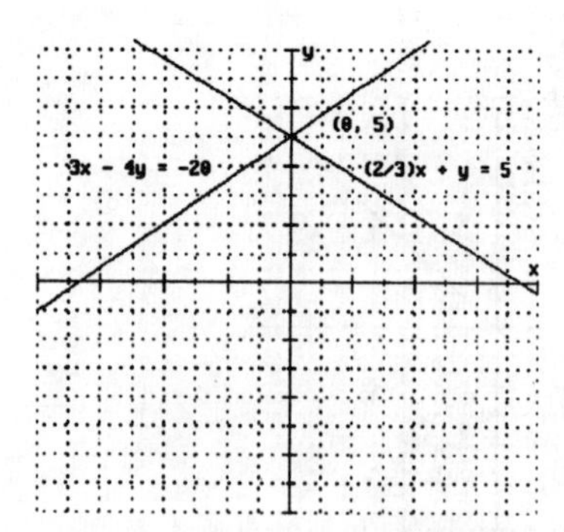

The solution is (0, 5), the system is consistent, and the equations are independent.

12. $4x - 3y = 6$

x	y
0	-2
3/2	0
3	2

$-8x + 6y = 12$

x	y
0	2
-3/2	0
-3	-2

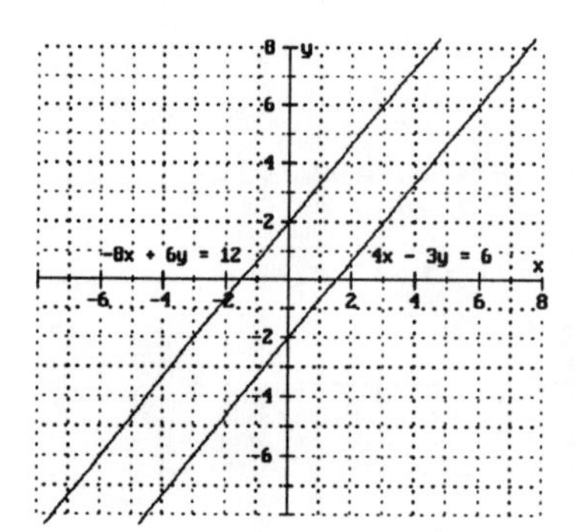

The lines are parallel. There is no solution, the system is inconsistent, and the equations are independent.

13. a. $x + y = 10$

$y = -x + 10$

$m_1 = -1$

$x - y = 6$

$-y = -x + 6$

$y = x - 6$

$m_2 = 1$

Since $m_1 \neq m_2$, the lines intersect at a single point.

b. One solution

14. a. $4x - y = 2$

$-y = -4x + 2$

$y = 4x - 2$

$m_1 = 4$

y-intercept $= -2$

$2x - \frac{1}{2}y = 4$

$-\frac{1}{2}y = -2x + 4$

$y = 4x - 8$

$m_2 = 4$

y-intercept $= -8$

Since $m_1 = m_2$, but the y-intercepts are different, these are parallel lines.

b. No solution

15. a.
$$3x + 2y = 1$$
$$2y = -3x + 1$$
$$y = -\frac{3}{2}x + \frac{1}{2}$$
$$m_1 = -\frac{3}{2}$$
$$y\text{-intercept} = \frac{1}{2}$$
$$4y = 2 - 6x$$
$$y = \frac{1}{2} - \frac{3}{2}x$$
$$m_2 = -\frac{3}{2}$$
$$y\text{-intercept} = \frac{1}{2}$$

Since $m_1 = m_2$ and the y-intercepts are the same, these are identical lines.

b. Infinite number of solutions

16. a. $x = 3$ undefined slope; $y = -8$, $m = 0$

Since $m_1 \neq m_2$ these lines intersect at a single point.

b. One solution

8.2 SOLUTIONS TO EXERCISES

1. Let $y = -2x + 2$ in the 1st equation.
$$x + y = -5$$
$$x + (-2x + 2) = -5$$
$$x - 2x + 2 = -5$$
$$-x + 2 = -5$$
$$-x = -7$$
$$x = 7$$
$$y = -2(7) + 2$$
$$y = -14 + 2$$
$$y = -12$$
(7, −12)

2. Let $y = 2x$ in the 1st equation.
$$x - y = -3$$
$$x - 2x = -3$$
$$-x = -3$$
$$x = 3$$
$$y = 2x$$
$$y = 2(3)$$
$$y = 6$$
(3, 6)

3. $-2x = 6y + 12$
$$x = -3y - 6$$

Let $x = -3y - 6$ in the 1st equation.
$$x + 3y = 6$$
$$-3y - 6 + 3y = 6$$
$$-6 = 6$$
False

No solution

4. $-x + y = 4$
$$y = x + 4$$

Let $y = x + 4$ in the 2nd equation.
$$5x - y = -4$$
$$5x - (x + 4) = -4$$
$$5x - x - 4 = -4$$
$$4x - 4 = -4$$
$$4x = 0$$
$$x = 0$$
$$y = x + 4$$
$$y = 0 + 4$$
$$y = 4$$
(0, 4)

5. $x - y = 1$
$$x = y + 1$$

Let $x = y + 1$ in the 1st equation.
$$3x - 2y = 0$$
$$3(y + 1) - 2y = 0$$
$$3y + 3 - 2y = 0$$
$$y + 3 = 0$$
$$y = -3$$
$$x = y + 1$$
$$x = -3 + 1$$
$$x = -2$$
(−2, −3)

6. $x + 4y = -2$
$$x = -4y - 2$$

Let $x = -4y - 2$ in the 1st equation.

$$\frac{1}{2}x + 2y = -1$$
$$\frac{1}{2}(-4y - 2) + 2y = -1$$
$$-2y - 1 + 2y = -1$$
$$-1 = -1$$
True

Infinite number of solutions

7. $x - 3y = 2$
$$x = 3y + 2$$

Let $x = 3y + 2$ in the 1st equation.

$$8x + 7y = -15$$
$$8(3y + 2) + 7y = -15$$
$$24y + 16 + 7y = -15$$
$$31y + 16 = -15$$
$$31y = -31$$
$$y = -1$$

$$x = 3y + 2$$
$$x = 3(-1) + 2$$
$$x = -1$$

$(-1, -1)$

8. $4x = 6y - 4$
$$x = \frac{3}{2}y - 1$$

Let $x = \frac{3}{2}y - 1$ in the 1st equation.

$$2x + 3y = 0$$
$$2\left(\frac{3}{2}y - 1\right) + 3y = 0$$
$$3y - 2 + 3y = 0$$
$$6y - 2 = 0$$
$$6y = 2$$
$$y = \frac{1}{3}$$

$$x = \frac{3}{2}y - 1$$
$$x = \frac{3}{2}\left(\frac{1}{3}\right) - 1$$
$$x = -\frac{1}{2}$$

$\left(-\frac{1}{2}, \frac{1}{3}\right)$

9. $3x + 9y = 18$
$$3x = -9y + 18$$
$$x = -3y + 6$$

Let $x = -3y + 6$ in the 2nd equation.

$$4x + 12y = 24$$
$$4(-3y + 6) + 12y = 24$$
$$-12y + 24 + 12y = 24$$
$$24 = 24$$
True

Infinite number of solutions

10. $-5x + y = 3$
$$y = 5x + 3$$

Let $y = 5x + 3$ in the 2nd equation.

$$7x - 2y = -3$$
$$7x - 2(5x + 3) = -3$$
$$7x - 10x - 6 = -3$$
$$-3x - 6 = -3$$
$$-3x = 3$$
$$x = -1$$

$$y = 5x + 3$$
$$y = 5(-1) + 3$$
$$y = -2$$

$(-1, -2)$

11. $5x = 10y - 15$
$$x = 2y - 3$$

Let $x = 2y - 3$ in the 2nd equation.

$$2x - 4y = -6$$
$$2(2y - 3) - 4y = -6$$
$$4y - 6 - 4y = -6$$
$$-6 = -6$$
True

Infinite number of solutions

12. Let $y = 3x$ in the 2nd equation.

$$x + y = 1$$
$$x + 3x = 1$$
$$4x = 1$$
$$x = \frac{1}{4}$$

$y = 3x$
$y = 3\left(\frac{1}{4}\right)$
$y = \frac{3}{4}$

$\left(\frac{1}{4}, \frac{3}{4}\right)$

13. $x + y = 28$
$x = -y + 28$

Let $x = -y + 28$ in the 1st equation.

$2x - y = -4$
$2(-y + 28) - y = -4$
$-2y + 56 - y = -4$
$-3y + 56 = -4$
$-3y = -60$
$y = 20$

$x = -y + 28$
$x = -20 + 28$
$x = 8$

(8, 20)

14. Let $x = -y - 1$ in the 1st equation.

$6x + 3y = -1$
$6(-y - 1) + 3y = -1$
$-6y - 6 + 3y = -1$
$-3y - 6 = -1$
$-3y = 5$
$y = -\frac{5}{3}$

$x = -y - 1$
$x = -\left(-\frac{5}{3}\right) - 1$
$x = \frac{2}{3}$

$\left(\frac{2}{3}, -\frac{5}{3}\right)$

15. $6x - y = 0$
$6x = y$

Let $y = 6x$ in the 2nd equation.

$-9x + 4y = 0$
$-9x + 4(6x) = 0$
$-9x + 24x = 0$
$15x = 0$
$x = 0$

$y = 6x$
$y = 6(0)$
$y = 0$

(0, 0)

8.3 SOLUTIONS TO EXERCISES

1. $2x - y = -5$
$\underline{3x + y = 10}$
$5x = 5$
$x = 1$

$3x + y = 10$
$3(1) + y = 10$
$3 + y = 10$
$y = 7$

(1, 7)

2. $x - 4y = 24$
$\underline{-x + 3y = -20}$
$-y = 4$
$y = -4$

$x - 4y = 24$
$x - 4(-4) = 24$
$x + 16 = 24$
$x = 8$

(8, -4)

3. $5x + 2y = -6$
$\underline{7x - 2y = 6}$
$12x = 0$
$x = 0$

$5x + 2y = -6$
$5(0) + 2y = -6$
$2y = -6$
$y = -3$

(0, -3)

4. $-7x + 3y = 8$
$\underline{7x + 4y = -22}$
$7y = -14$
$y = -2$

$$\begin{aligned} 7x + 4y &= -22 \\ 7x + 4(-2) &= -22 \\ 7x - 8 &= -22 \\ 7x &= -14 \\ x &= -2 \end{aligned}$$

(-2, -2)

5. $$\begin{cases} -2(x + 3y) = -2(16) \\ 2x - 5y = -23 \end{cases}$$

$$\begin{aligned} -2x - 6y &= -32 \\ \underline{2x - 5y} &\underline{= -23} \\ -11y &= -55 \\ y &= 5 \end{aligned}$$

$$\begin{aligned} x + 3y &= 16 \\ x + 3(5) &= 16 \\ x + 15 &= 16 \\ x &= 1 \end{aligned}$$

(1, 5)

6. $$\begin{cases} -2(3x + 6y) = -2(18) \\ 3(2x + 4y) = 3(5) \end{cases}$$

$$\begin{aligned} -6x - 12y &= -36 \\ \underline{6x + 12y} &\underline{= 15} \\ 0 &= -21 \end{aligned}$$

False

No solution

7. $$\begin{cases} -3(2x + 4y) = -3(2) \\ 2(3x + 6y) = 2(3) \end{cases}$$

$$\begin{aligned} -6x - 12y &= -6 \\ \underline{6x + 12y} &\underline{= 6} \\ 0 &= 0 \end{aligned}$$

True

Infinite number of solutions

8. $$\begin{cases} 8x - 32y = 16 \\ 2\left(-\frac{1}{2}x + 2y\right) = 2(-1) \end{cases}$$

$$\begin{cases} 8x - 32y = 16 \\ -x + 4y = -2 \end{cases}$$

$$\begin{cases} 8x - 32y = 16 \\ 8(-x + 4y) = 8(-2) \end{cases}$$

$$\begin{aligned} 8x - 32y &= 16 \\ \underline{-8x + 32y} &\underline{= -16} \\ 0 &= 0 \end{aligned}$$

True

Infinite number of solutions

9. $$\begin{cases} 6\left(\frac{x}{3} - \frac{y}{6}\right) = 6(1) \\ 36\left(-\frac{x}{9} + \frac{y}{4}\right) = 36(2) \end{cases}$$

$$\begin{cases} 2x - y = 6 \\ -4x + 9y = 72 \end{cases}$$

$$\begin{cases} 2(2x - y) = 2(6) \\ -4x + 9y = 72 \end{cases}$$

$$\begin{aligned} 4x - 2y &= 12 \\ \underline{-4x + 9y} &\underline{= 72} \\ 7y &= 84 \\ y &= 12 \end{aligned}$$

$$\begin{aligned} 2x - y &= 6 \\ 2x - 12 &= 6 \\ 2x &= 18 \\ x &= 9 \end{aligned}$$

(9, 12)

10. $$\begin{cases} -4(3x - 6y) = -4(0) \\ 12x + 28y = 13 \end{cases}$$

$$\begin{aligned} -12x + 24y &= 0 \\ \underline{12x + 28y} &\underline{= 13} \\ 52y &= 13 \\ y &= \frac{1}{4} \end{aligned}$$

$$12x + 28y = 13$$
$$12x + 28\left(\frac{1}{4}\right) = 13$$
$$12x + 7 = 13$$
$$12x = 6$$
$$x = \frac{1}{2}$$

$\left(\frac{1}{2}, \frac{1}{4}\right)$

11. $\begin{cases} 2(-6x + 5y) = 2(-42) \\ 3(4x + 2y) = 3(28) \end{cases}$

$$\begin{array}{r} -12x + 10y = -84 \\ \underline{12x + 6y = 84} \\ 16y = 0 \\ y = 0 \end{array}$$

$$4x + 2y = 28$$
$$4x + 2(0) = 28$$
$$4x = 28$$
$$x = 7$$

(7, 0)

12. $\begin{cases} -7(2x + 10y) = -7(5) \\ 2(7x + 35y) = 2(3) \end{cases}$

$$\begin{array}{r} -14x - 70y = -35 \\ \underline{14x + 70y = 6} \\ 0 = -29 \end{array}$$

False

No solution

13. $\begin{cases} 20x + 30y = -9 \\ -4(5x + 10y) = -4(-2) \end{cases}$

$$\begin{array}{r} 20x + 30y = -9 \\ \underline{-20x - 40y = 8} \\ -10y = -1 \\ y = \frac{1}{10} \end{array}$$

$$5x + 10y = -2$$
$$5x + 10\left(\frac{1}{10}\right) = -2$$
$$5x + 1 = -2$$
$$5x = -3$$
$$x = -\frac{3}{5}$$

$\left(-\frac{3}{5}, \frac{1}{10}\right)$

14. $\begin{cases} 7\left(\frac{x}{7} - y\right) = 7(-5) \\ 6\left(x + \frac{y}{6}\right) = 6(8) \end{cases}$

$\begin{cases} x - 7y = -35 \\ 6x + y = 48 \end{cases}$

$\begin{cases} x - 7y = -35 \\ 7(6x + y) = 7(48) \end{cases}$

$$\begin{array}{r} x - 7y = -35 \\ \underline{42x + 7y = 336} \\ 43x = 301 \\ x = 7 \end{array}$$

$$6x + y = 48$$
$$6(7) + y = 48$$
$$42 + y = 48$$
$$y = 6$$

(7, 6)

15. $\begin{cases} 2(8x - 3y) = 2(-14) \\ 3(-12x + 2y) = 3(11) \end{cases}$

$$\begin{array}{r} 16x - 6y = -28 \\ \underline{-36x + 6y = 33} \\ -20x = 5 \\ x = -\frac{1}{4} \end{array}$$

$$8x - 3y = -14$$
$$8\left(-\frac{1}{4}\right) - 3y = -14$$
$$-2 - 3y = -14$$
$$-3y = -12$$
$$y = 4$$
$$\left(-\frac{1}{4}, 4\right)$$

8.4 SOLUTIONS TO EXERCISES

1. x = 1st number
   y = 2nd number

   $$\begin{cases} x + y = 28 \\ x - y = 14 \end{cases}$$

2. x = amount in smaller account
   y = amount in larger account

   $$\begin{cases} x + y = 9300 \\ y = 1200 + x \end{cases}$$

3. x = 1st number
   y = 2nd number

   $$\begin{cases} x + y = 31 \\ x + (7 + 4y) = 71 \end{cases}$$

   or $$\begin{cases} x + y = 31 \\ x + 4y = 64 \end{cases}$$

4. x = 1st number
   y = 2nd number

   $$\begin{cases} x + y = 89 \\ x - y = 33 \end{cases}$$

   $$\begin{array}{r} x + y = 89 \\ \underline{x - y = 33} \\ 2x \quad = 122 \\ x = 61 \end{array}$$

   $$x + y = 89$$
   $$61 + y = 89$$
   $$y = 28$$
   The numbers are 61 and 28.

5. x = 1st number
   y = 2nd number

   $$\begin{cases} x = 5 + 2y \\ x + y = -1 \end{cases}$$

   $$x + y = -1$$
   $$(5 + 2y) + y = -1$$
   $$5 + 3y = -1$$
   $$3y = -6$$
   $$y = -2$$

   $$x + y = -1$$
   $$x + (-2) = -1$$
   $$x = 1$$

   The numbers are 1 and −2.

6. x = 1st number
   y = 2nd number

   $$\begin{cases} x + 3y = 43 \\ 4x + y = 40 \end{cases}$$

   $$\begin{cases} -4(x + 3y) = -4(43) \\ 4x + y = 40 \end{cases}$$

   $$\begin{array}{r} -4x - 12y = -172 \\ \underline{4x + \quad y = 40} \\ -11y = -132 \\ y = 12 \end{array}$$

   $$x + 3y = 43$$
   $$x + 3(12) = 43$$
   $$x + 36 = 43$$
   $$x = 7$$

   The numbers are 7 and 12.

7. x = price of each cassette
   y = price of each compact disc

   $$\begin{cases} 4x + 3y = 84 \\ 6x + 2y = 86 \end{cases}$$

   $$\begin{cases} -2(4x + 3y) = -2(84) \\ 3(6x + 2y) = 3(86) \end{cases}$$

$$\begin{array}{r} -8x - 6y = -168 \\ \underline{18x + 6y = 258} \\ 10x \quad\quad = 90 \\ x = 9 \end{array}$$

$$\begin{aligned} 4x + 3y &= 84 \\ 4(9) + 3y &= 84 \\ 36 + 3y &= 84 \\ 3y &= 48 \\ y &= 16 \end{aligned}$$

The price of each cassette is \$9 and the price of each compact disc is \$16.

8. d = number of dimes
   q = number of quarters

$$\begin{cases} d + q = 75 \\ 10d + 25q = 1380 \end{cases}$$

$$\begin{cases} -10(d + q) = -10(75) \\ 10d + 25q = 1380 \end{cases}$$

$$\begin{array}{r} -10d - 10q = -750 \\ \underline{10d + 25q = 1380} \\ 15q = 630 \\ q = 42 \end{array}$$

$$\begin{aligned} d + q &= 75 \\ d + 42 &= 75 \\ d &= 33 \end{aligned}$$

She has 33 dimes and 42 quarters.

9. x = number of 32 cent stamps
   y = number of 19 cent stamps

$$\begin{cases} x + y = 66 \\ 32x + 19y = 1878 \end{cases}$$

$$\begin{cases} -19(x + y) = -19(66) \\ 32x + 19y = 1878 \end{cases}$$

$$\begin{array}{r} -19x - 19y = -1254 \\ \underline{32x + 19y = 1878} \\ 13x \quad\quad = 624 \\ x = 48 \end{array}$$

$$\begin{aligned} x + y &= 66 \\ 48 + y &= 66 \\ y &= 18 \end{aligned}$$

Judi purchased 48 32-cent stamps and 18 19-cent stamps.

10. x = Alan's rate in still water
    y = rate of current

	rate ·	time =	distance
downriver	$x + y$	3	$3(x + y)$
upriver	$x - y$	12	$12(x - y)$

distance downriver = 24
$3(x + y) = 24$

distance upriver = 24
$12(x - y) = 24$

$$\begin{cases} 3(x + y) = 24 \\ 12(x - y) = 24 \end{cases}$$

$$\begin{cases} 3x + 3y = 24 \\ 12x - 12y = 24 \end{cases}$$

$$\begin{cases} 4(3x + 3y) = 4(24) \\ 12x - 12y = 24 \end{cases}$$

$$\begin{array}{r} 12x + 12y = 96 \\ \underline{12x - 12y = 24} \\ 24x \quad\quad = 120 \\ x = 5 \end{array}$$

$$\begin{aligned} 3x + 3y &= 24 \\ 3(5) + 3y &= 24 \\ 15 + 3y &= 24 \\ 3y &= 9 \\ y &= 3 \end{aligned}$$

Alan's rate in still water is 5 mph and the current's rate is 3 mph.

11. x = speed of plane in still air
y = speed of the wind

$$15 \text{ min} = \frac{15}{60}\, hr = \frac{1}{4} \text{ hr}$$

$$3 \text{ hr } 15 \text{ min} = 3\frac{1}{4} \text{ hr} = 3.25 \text{ hr}$$

	rate ·	time =	distance
with wind	$x + y$	3.25	$3.25(x + y)$
against	$x - y$	4	$4(x - y)$

distance each way = 1686

$$\begin{cases} 3.25(x + y) = 1686 \\ 4(x - y) = 1686 \end{cases}$$

$$\begin{cases} 3.25x + 3.25y = 1686 \\ 4x - 4y = 1686 \end{cases}$$

$$\begin{aligned} 4x - 4y &= 1686 \\ 4x &= 4y + 1686 \\ x &= y + 421.5 \end{aligned}$$

Substituting,

$$\begin{aligned} 3.25x + 3.25y &= 1686 \\ 3.25(y + 421.5) + 3.25y &= 1686 \\ 3.25y + 1369.875 + 3.25y &= 1686 \\ 6.5y + 1369.875 &= 1686 \\ 6.5y &= 316.125 \\ y &= 48.6 \end{aligned}$$

$$\begin{aligned} 4x - 4y &= 1686 \\ 4x - 4(48.6) &= 1686 \\ 4x - 194.4 &= 1686 \\ 4x &= 1880.4 \\ x &= 470.1 \end{aligned}$$

The speed of the plane in still air is 470.1 mph and the speed of the wind is 48.6 mph.

12. x = amount of 5% solution
y = amount of 14% solution

	concentration rate	ounces of Solution	ounces of pure acid
1st solution	5%	x	$0.05x$
2nd solution	14%	y	$0.14y$
mixture	12.2%	15	0.122(15)

$$\begin{cases} x + y = 15 \\ 0.05x + 0.14y = 0.122(15) \end{cases}$$

$$\begin{cases} x + y = 15 \\ 100(0.05x + 0.14y) = 100[0.122(15)] \end{cases}$$

$$\begin{cases} x + y = 15 \\ 5x + 14y = 183 \end{cases}$$

$$\begin{cases} -5(x + y) = -5(15) \\ 5x + 14y = 183 \end{cases}$$

$$\begin{aligned} -5x - 5y &= -75 \\ \underline{5x + 14y} &= \underline{183} \\ 9y &= 108 \\ y &= 12 \end{aligned}$$

$$\begin{aligned} x + y &= 15 \\ x + 12 &= 15 \\ x &= 3 \end{aligned}$$

She should use 3 oz. of the 5% solution and 12 oz. of the 14% solution.

13. x = number of pounds of high-quality coffee
y = number of pounds of cheaper coffee

	number of lbs ·	price per lb	= total cost
high quality	x	8.25	$8.25x$
cheaper quality	y	4.75	$4.75y$
mixture needed	150	5.68	5.68(150)

$$\begin{cases} x + y = 150 \\ 8.25x + 4.75y = 5.68(150) \end{cases}$$

$$\begin{aligned} x + y &= 150 \\ x &= 150 - y \end{aligned}$$

Substituting,

$$\begin{aligned} 8.25x + 4.75y &= 5.68(150) \\ 8.25(150 - y) + 4.75y &= 852 \\ 1237.5 - 8.25y + 4.75y &= 852 \\ 1237.5 - 3.5y &= 852 \\ -3.5y &= -385.5 \\ y &= 110 \quad \text{(Rounded to nearest lb)} \end{aligned}$$

$$\begin{aligned} x &= 150 - y \\ x &= 150 - 110 = 40 \end{aligned}$$

He should use 40 lbs of the high quality bean and 110 lbs of the cheaper bean.

14. x = number of adults
    y = number of children

	number of people $\cdot$	price per person $=$	total dollars
adults	x	6.25	$6.25x$
children	y	3.75	$3.75y$
total	342		1932.50

$$\begin{cases} x + y = 342 \\ 6.25x + 3.75y = 1932.50 \end{cases}$$

$$\begin{cases} -3.75(x + y) = -3.75(342) \\ 6.25x + 3.75y = 1932.50 \end{cases}$$

$$\begin{aligned} -3.75x - 3.75y &= -1282.50 \\ 6.25x + 3.75y &= 1932.50 \\ \hline 2.5x &= 650 \\ x &= 260 \end{aligned}$$

$$\begin{aligned} x + y &= 342 \\ 260 + y &= 342 \\ y &= 82 \end{aligned}$$

There were 260 adults and 82 children.

15.

river

x x

y

$$\begin{cases} x + x + y = 122 \\ x = \frac{1}{2}y - 5 \end{cases}$$

$$\begin{cases} 2x + y = 122 \\ x = \frac{1}{2}y - 5 \end{cases}$$

Substituting,

$$2x + y = 122$$

$$2\left(\frac{1}{2}y - 5\right) + y = 122$$

$$\begin{aligned} y - 10 + y &= 122 \\ 2y - 10 &= 122 \\ 2y &= 132 \\ y &= 66 \end{aligned}$$

$$x = \frac{1}{2}y - 5$$

$$x = \frac{1}{2}(66) - 5$$

$$x = 28$$

The dimensions are 28 ft. by 66 ft.

8.5 SOLUTIONS TO EXERCISES

In Exercises 1–15, the solution set to the system is where the shadings overlap.

1. $y \geq x + 2$ (solid line)

$y = x + 2$

x	y
0	2
1	3
2	4

Test point: (0, 0)
$y \geq x + 2$
$0 \geq 0 + 2$
$0 \geq 2$
False

Shade half-plane not containing (0, 0).

$y \geq 4 - x$ (solid line)

$y = 4 - x$

x	y
0	4
1	3
2	2

Test point: (0, 0)
$y \geq 4 - x$
$0 \geq 4 - 0$
$0 \geq 4$
False

Shade half-plane not containing (0, 0).

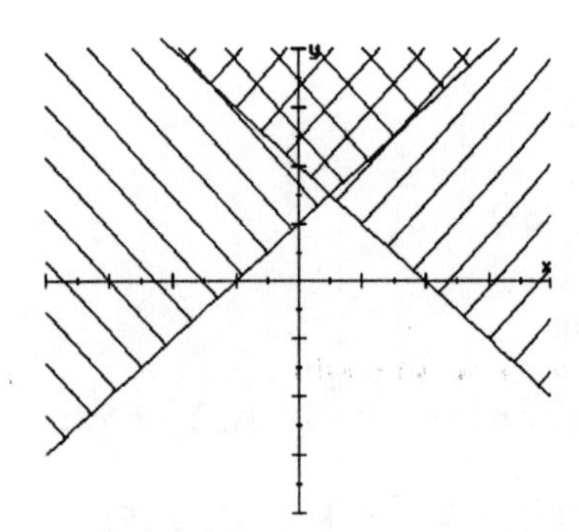

$y \geq -3 - x$ (solid line)

$y = -3 - x$

x	y
0	-3
1	-4
2	-5

Test point: (0, 0)
$y \geq -3 - x$
$0 \geq -3 - 0$
$0 \geq -3$
True

Shade half-plane containing (0, 0).

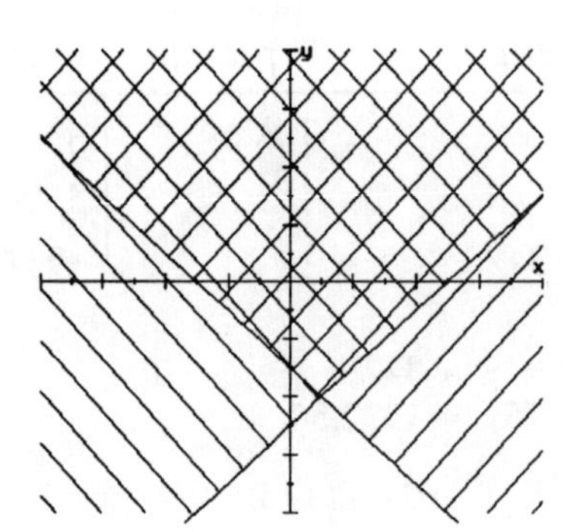

2. $y \geq x - 5$ (solid line)

$y = x - 5$

x	y
0	-5
1	-4
2	-3

Test point: (0, 0)
$y \geq x - 5$
$0 \geq 0 - 5$
$0 \geq -5$
True

Shade half-plane containing (0, 0).

3. $y < 4x - 3$ (dashed line)

$y = 4x - 3$

x	y
0	-3
1	1
2	5

Test point: (0, 0)
$y < 4x - 3$
$0 < 4(0) - 3$
$0 < -3$
False

Shade half-plane not containing (0, 0).

$y \le x + 3$ (solid line)

$y = x + 3$

x	y
0	3
1	4
2	5

Test point: (0, 0)
$y \le x + 3$
$0 \le 0 + 3$
$0 \le 3$
True

Shade half-plane containing (0, 0).

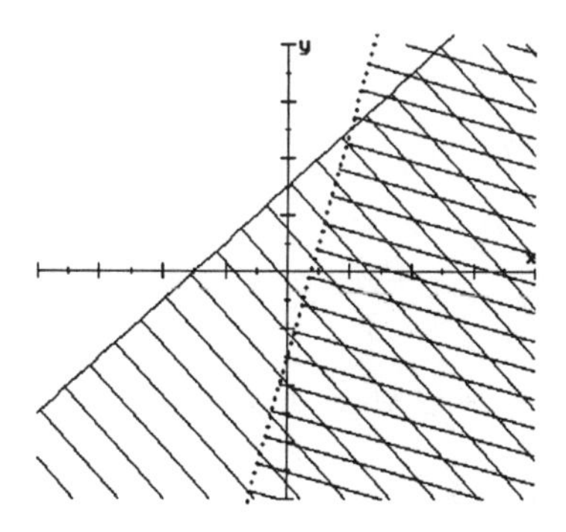

$y > x + 4$ (dashed line)

$y = x + 4$

x	y
0	4
1	5
2	6

Test point: (0, 0)
$y > x + 4$
$0 > 0 + 4$
$0 > 4$
False

Shade half-plane not containing (0, 0).

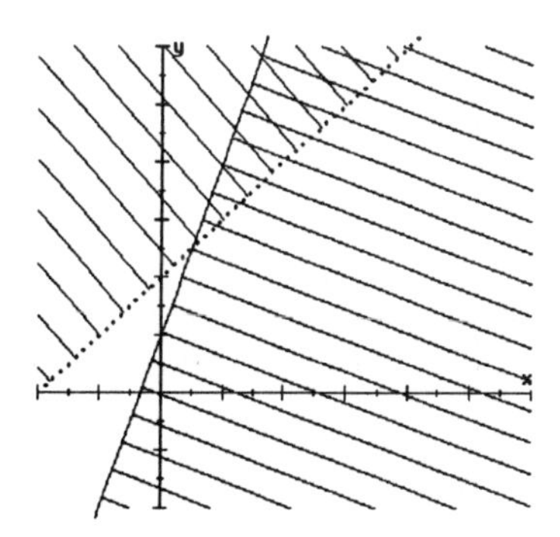

4. $y \le 3x + 2$ (solid line)

$y = 3x + 2$

x	y
0	2
1	5
2	8

Test point: (0, 0)
$y \le 3x + 2$
$0 \le 3(0) + 2$
$0 \le 2$
True

Shade half-plane containing (0, 0).

5. $y \le -5x - 5$ (solid line)

$y = -5x - 5$

x	y
0	-5
-1	0
-2	5

Test point: (0, 0)
$y \le -5x - 5$
$0 \le -5(0) - 5$
$0 \le -5$
False

Shade half-plane not containing (0, 0).

$y \geq x + 6$ (solid line)

$y = x + 6$

x	y
-2	4
-1	5
0	6

Test point: (0, 0)
$y \geq x + 6$
$0 \geq 0 + 6$
$0 \geq 6$
False

Shade half-plane not containing (0, 0).

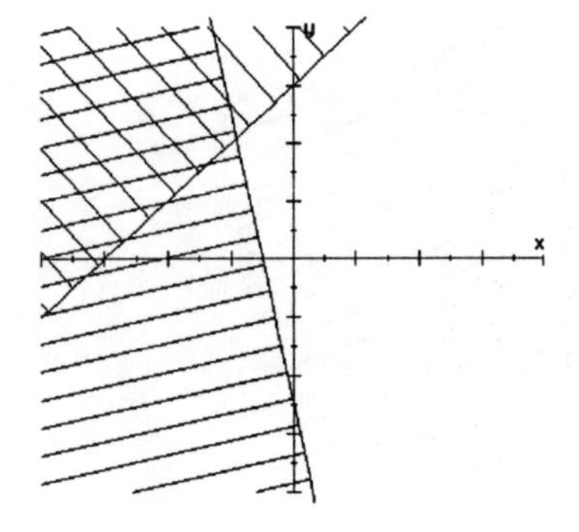

$y \geq -x - 7$ (solid line)

$y = -x - 7$

x	y
-2	-5
-1	-6
0	-7

Test point: (0, 0)
$y \geq -x - 7$
$0 \geq 0 - 7$
$0 \geq -7$
True

Shade half-plane containing (0, 0).

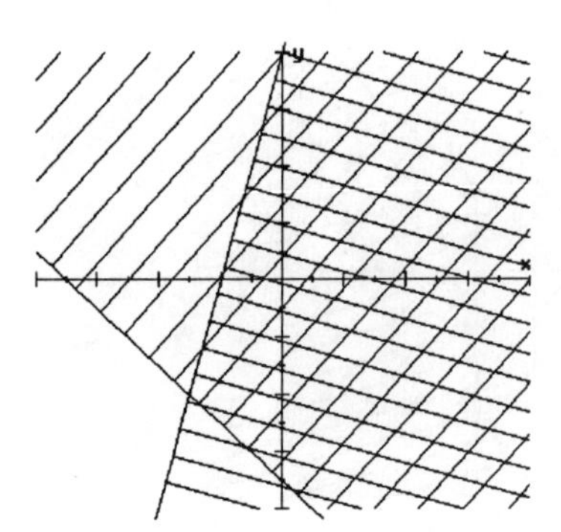

6. $y \leq 4x + 8$ (solid line)

$y = 4x + 8$

x	y
-2	0
-1	4
0	8

Test point: (0, 0)
$y \leq 4x + 8$
$0 \leq 4(0) + 8$
$0 \leq 8$
True

Shade half-plane containing (0, 0).

7. $y \geq -x + 1$ (solid line)

$y = -x + 1$

x	y
-1	2
0	1
1	0

Test point: (0, 0)
$y \geq -x + 1$
$0 \geq 0 + 1$
$0 \geq 1$
False

Shade half-plane not containing (0, 0).

$y \le 4x + 1$ (solid line)

$y = 4x + 1$

x	y
-1	-3
0	1
1	5

Test point: (0, 0)
$y \le 4x + 1$
$0 \le 4(0) + 1$
$0 \le 1$
True

Shade half-plane containing (0, 0).

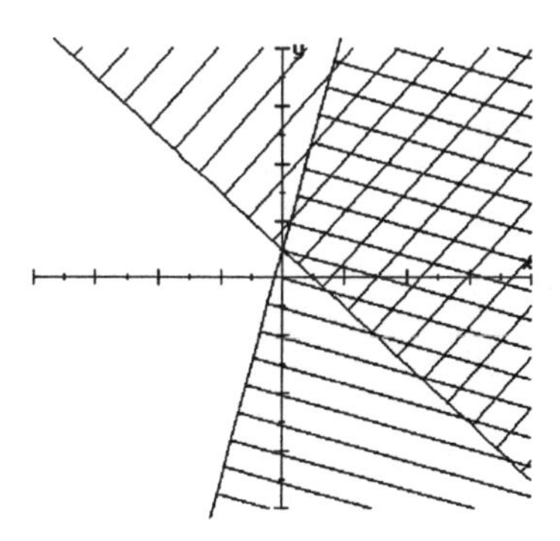

$y \le -2x + 2$ (solid line)

$y = -2x + 2$

x	y
0	2
1	0
2	-2

Test point: (0, 0)
$y \le -2x + 2$
$0 \le -2(0) + 2$
$0 \le 2$
True

Shade half-plane containing (0, 0).

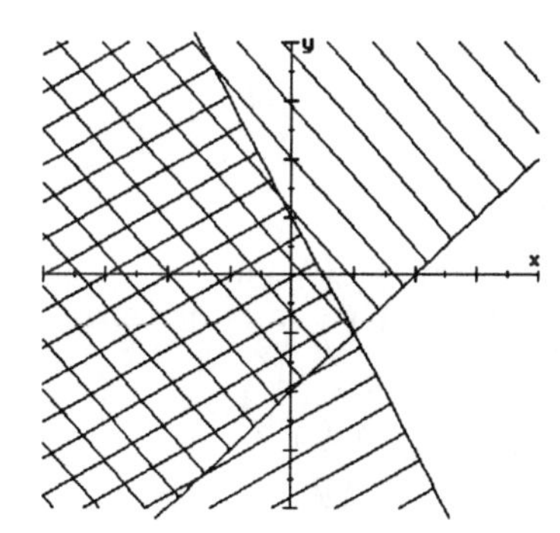

8. $y \ge x - 4$ (solid line)

$y = x - 4$

x	y
0	-4
1	-3
2	-2

Test point: (0, 0)
$y \ge x - 4$
$0 \ge 0 - 4$
$0 \ge -4$
True

Shade half-plane containing (0, 0).

9. $x \ge 4y$ (solid line)

$x = 4y$

x	y
0	0
4	1
-4	-1

Test point: (1, 0)
$x \ge 4y$
$1 \ge 4(0)$
$1 \ge 0$
True

Shade half-plane containing (1, 0).

$x + 2y \le 8$ (solid line)

$x + 2y = 8$

x	y
0	4
-2	5
2	3

Test point: (0, 0)

$x + 2y \le 8$
$0 + 2(0) \le 8$
$0 \le 8$
True

Shade half-plane containing (0, 0).

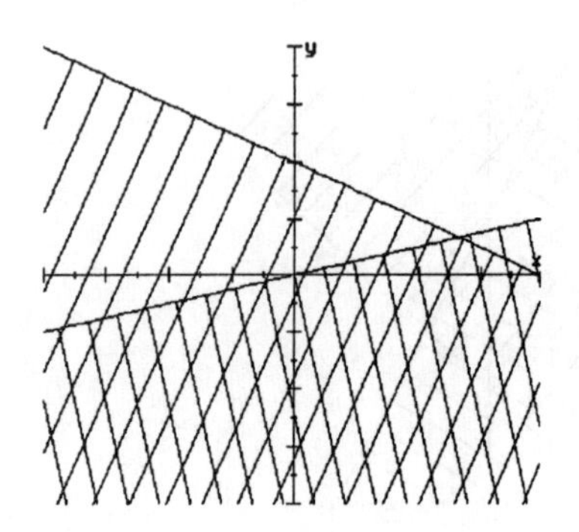

$2x + y < 5$ (dashed line)

$2x + y = 5$

x	y
0	5
1	3
2	1

Test point: (0, 0)

$2x + y < 5$
$2(0) + 0 < 5$
$0 < 5$
True

Shade half-plane containing (0, 0).

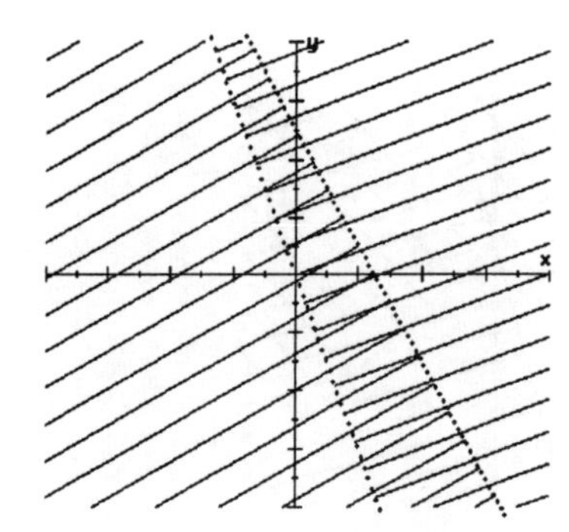

10. $-3x < y$ (dashed line)

$-3x = y$

x	y
-1	3
0	0
1	-3

Test point: (1, 0)

$-3x < y$
$-3(1) < 0$
$-3 < 0$
True

Shade half-plane containing (1, 0).

11. $y + 4x \ge 0$ (solid line)

$y + 4x = 0$

x	y
0	0
-1	4
1	-4

Test point: (1, 0)

$y + 4x \ge 0$
$0 + 4(1) \ge 0$
$4 \ge 0$
True

Shade half-plane containing (1, 0).

$7x - 2y \le 14$ (solid line)

$7x - 2y = 14$

x	y
0	-7
2	0
4	7

Test point: (0, 0)

$7x - 2y \le 14$
$7(0) - 2(0) \le 14$
$0 \le 14$
True

Shade half-plane containing (0, 0).

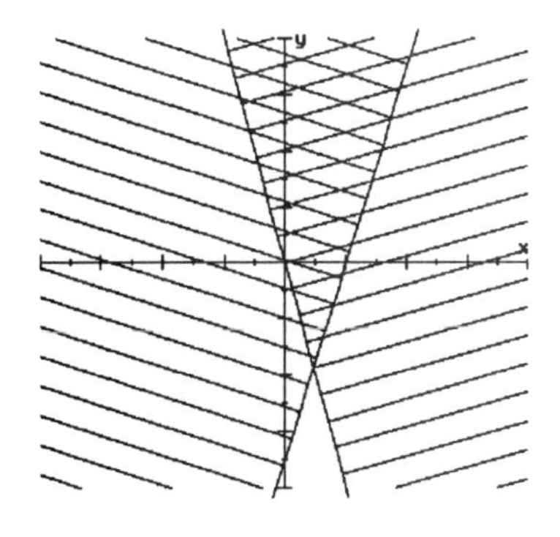

12. $x \le 3$ (solid vertical line)
Test point: (0, 0)
$0 \le 3$
True
Shade half-plane containing (0, 0).

$y \ge -4$ (solid horizontal line)
Test point: (0, 0)
$0 \ge -4$
True
Shade half-plane containing (0, 0).

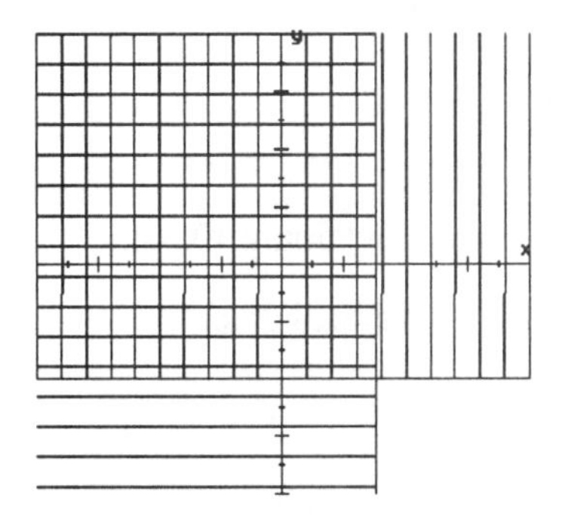

13. $x > -5$ (dashed vertical line)
Test point: (0, 0)
$0 > -5$
True
Shade half-plane containing (0, 0).

$y > -1$ (dashed horizontal line)
Test point: (0, 0)
$0 > -1$
True
Shade half-plane containing (0, 0).

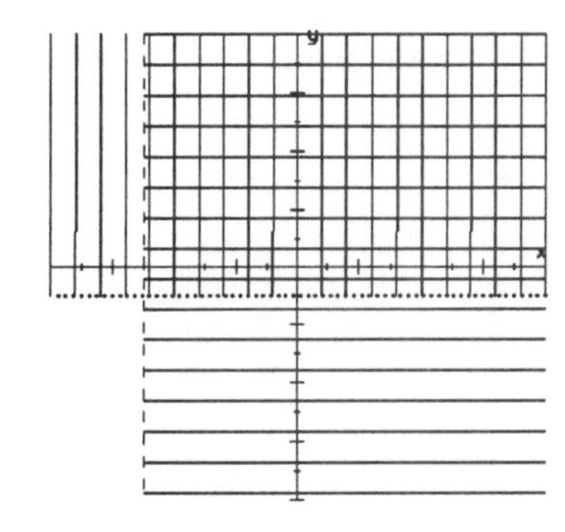

14. $x + y \le 4$ (solid line)

$x + y = 4$

x	y
0	4
1	3
2	2

Test point: (0, 0)
$x + y \le 4$
$0 + 0 \le 4$
$0 \le 4$
True
Shade half-plane containing (0, 0).

$x < 3$ (dashed vertical line)
Test point: (0, 0)
$0 < 3$
True
Shade half-plane containing (0, 0).

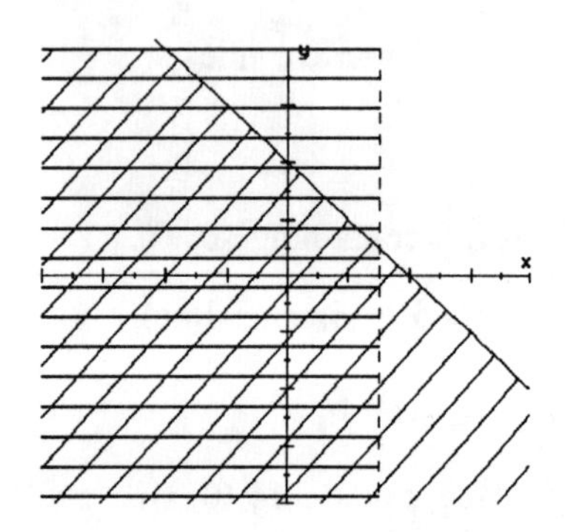

$y \le 4$ (solid horizontal line)

Test point: (0, 0)
$y \le 4$
$0 \le 4$
True
Shade half-plane containing (0, 0).

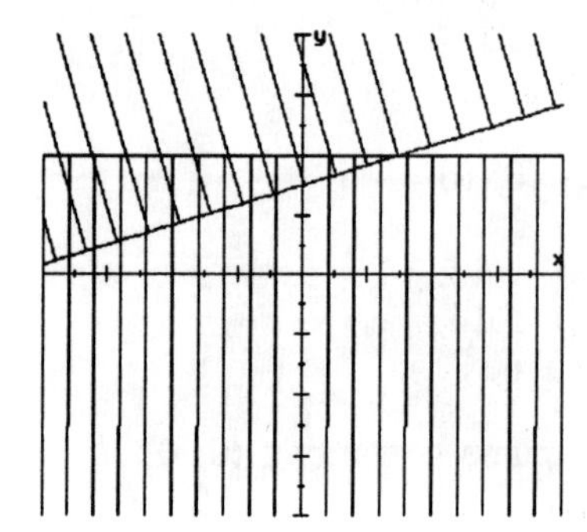

15. $y \ge \frac{1}{3}x + 3$ (solid line)

$y = \frac{1}{3}x + 3$

x	y
-3	2
0	3
3	4

Test point: (0, 0)
$y \ge \frac{1}{3}x + 3$

$0 \ge \frac{1}{3}(0) + 3$

$0 \ge 3$
False

Shade half-plane not containing (0, 0).

CHAPTER 8 PRACTICE TEST SOLUTIONS

1. $3(2) + 4(-1) = 2$
   $6 - 4 = 2$
   $2 = 2$
   True

   $5(2) - (-1) = 9$
   $10 + 1 = 9$
   $11 = 9$
   False

   It is not a solution.

2. $7(-2) - 2(5) = -24$
   $-14 - 10 = -24$
   $-24 = -24$
   True

   $7(-2) + 3(5) = 1$
   $-14 + 15 = 1$
   $1 = 1$
   True

   It is a solution.

3. $y - x = 4$
   $y = x + 4$

x	y
-1	3
0	4
1	5

   $y + 3x = 16$
   $y = -3x + 16$

x	y
2	10
3	7
4	4

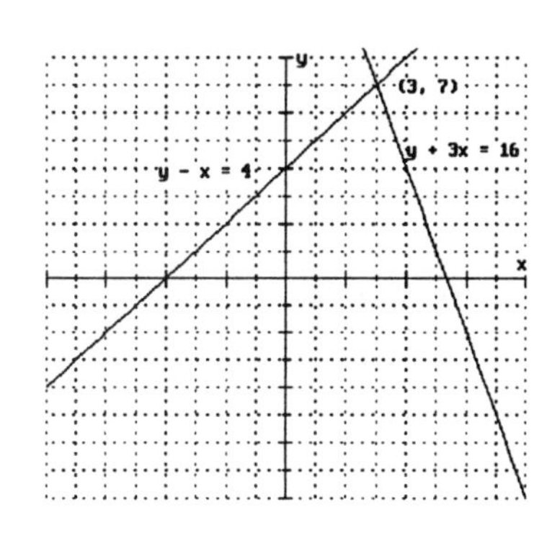

4. $x - 2y = 4$
   $x = 2y + 4$

   Substituting into other equation:

   $4(2y + 4) + y = -11$
   $8y + 16 + y = -11$
   $9y + 16 = -11$
   $9y = -27$
   $y = -3$
   $x = 2(-3) + 4 = -2$

   (-2, -3)

5. $6x = -y$
   $-6x = y$

   $\frac{1}{3}x + 3y = -\frac{53}{3}$

   $3\left(\frac{1}{3}x + 3y\right) = 3\left(-\frac{53}{3}\right)$

   $x + 9y = -53$
   $x + 9(-6x) = -53$
   $x - 54x = -53$
   $-53x = -53$
   $x = 1$
   $y = -6(1) = -6$

   (1, -6)

6. $\begin{cases} 2x + 7y = 23 \\ -3x + 4y = 9 \end{cases}$

   $\begin{cases} 3(2x + 7y) = 3(23) \\ 2(-3x + 4y) = 2(9) \end{cases}$

   $6x + 21y = 69$
   $\underline{-6x + 8y = 18}$
   $29y = 87$
   $y = 3$

   $2x + 7(3) = 23$
   $2x + 21 = 23$
   $2x = 2$
   $x = 1$

   (1, 3)

7. $\begin{cases} 9x - 7y = 46 \\ 8x - 5y = 36 \end{cases}$

$$\begin{cases} -5(9x - 7y) = -5(46) \\ 7(8x - 5y) = 7(36) \end{cases}$$

$$\begin{aligned} -45x + 35y &= -230 \\ \underline{56x - 35y} &\underline{= 252} \\ 11x &= 22 \\ x &= 2 \end{aligned}$$

$$\begin{aligned} 9(2) - 7y &= 46 \\ 18 - 7y &= 46 \\ -7y &= 28 \\ y &= -4 \end{aligned}$$

(2, -4)

8. $4x - y = 22$

$$\begin{aligned} -y &= -4x + 22 \\ y &= 4x - 22 \end{aligned}$$

$$\begin{aligned} 5x + 3(4x - 22) &= 53 \\ 5x + 12x - 66 &= 53 \\ 17x - 66 &= 53 \\ 17x &= 119 \\ x &= 7 \\ y &= 4(7) - 22 = 6 \end{aligned}$$

(7, 6)

9. $2(3x + y) = 5x - 3$

$$\begin{aligned} 6x + 2y &= 5x - 3 \\ x + 2y &= -3 \\ x &= -2y - 3 \end{aligned}$$

$$\begin{aligned} 2(-2y - 3) - y &= -1 \\ -4y - 6 - y &= -1 \\ -5y - 6 &= -1 \\ -5y &= 5 \\ y &= -1 \\ x &= -2(-1) - 3 = -1 \end{aligned}$$

(-1, -1)

10. $\dfrac{x - 11}{2} = \dfrac{2 - 7y}{6}$

$$6\left(\frac{x - 11}{2}\right) = 6\left(\frac{2 - 7y}{6}\right)$$

$$3(x - 11) = 2 - 7y$$

$$\begin{aligned} 3x - 33 &= 2 - 7y \\ 3x + 7y &= 35 \end{aligned}$$

$$\frac{45 - 4x}{18} = \frac{y}{2}$$

$$18\left(\frac{45 - 4x}{18}\right) = 18\left(\frac{y}{2}\right)$$

$$\begin{aligned} 45 - 4x &= 9y \\ 45 &= 4x + 9y \end{aligned}$$

$$\begin{aligned} 3x + 7y &= 35 \\ 4x + 9y &= 45 \end{aligned}$$

$$\begin{aligned} -4(3x + 7y) &= -4(35) \\ 3(4x + 9y) &= 3(45) \end{aligned}$$

$$\begin{aligned} -12x - 28y &= -140 \\ \underline{12x + 27y} &\underline{= 135} \\ -y &= -5 \\ y &= 5 \end{aligned}$$

$$\begin{aligned} 3x + 7(5) &= 35 \\ 3x + 35 &= 35 \\ 3x &= 0 \\ x &= 0 \end{aligned}$$

(0, 5)

11. x = number of \$1 bills
y = number of \$5 bills

$$\begin{aligned} x + y &= 41 \\ 1x + 5y &= 117 \end{aligned}$$

$$\begin{aligned} -x - y &= -41 \\ \underline{x + 5y} &\underline{= 117} \\ 4y &= 76 \\ y &= 19 \end{aligned}$$

$$\begin{aligned} x + 19 &= 41 \\ x &= 22 \end{aligned}$$

22 \$1 bills; 19 \$5 bills

12. x = amount invested at 4%
y = amount invested at 7%

$$\begin{cases} x + y = 6000 \\ 0.04x + 0.07y = 375 \end{cases}$$

$$x + y = 6000$$
$$4x + 7y = 37500$$

$$-4(x + y) = -4(6000)$$
$$4x + 7y = 37500$$

$$-4x - 4y = -24000$$
$$4x + 7y = 37500$$
$$3y = 13500$$
$$y = 4500$$

$$x + 4500 = 6000$$
$$x = 1500$$

$1500 @ 4% ; $4500 @ 7%

13. $y - 3x \le 6$ (solid line)
$y - 3x = 6$

x	y
-2	0
-1	3
0	6

Test point: (0, 0)
$y - 3x \le 6$
$0 - 3(0) \le 6$
$0 \le 6$
True

Shade half-plane containing (0, 0).

$y \ge 4$
$y = 4$ (Solid horizontal line)

Test point: (0, 0)

$y \ge 4$
$0 \ge 4$
False

Shade half-plane not containing (0, 0).

The solution set is where the two shadings overlap.

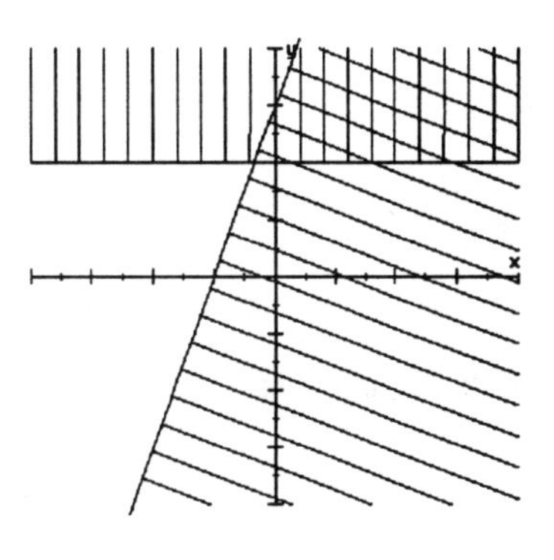

14. $2y + x \ge -1$ (Solid line)
$2y + x = -1$

x	y
-1	0
0	-1/2
1	-1

Test point: (0, 0)
$2y + x \ge -1$
$2(0) + 0 \ge -1$
$0 \ge -1$
True

Shade half-plane containing (0, 0).

$x - y \ge 3$ (Solid line)
$x - y = 3$

x	y
0	-3
1	-2
2	-1

Test point: (0, 0)
$x - y \geq 3$
$0 - 0 \geq 3$
$0 \geq 3$
False

Shade half-plane not containing (0, 0).

The solution set is where the two shadings overlap.

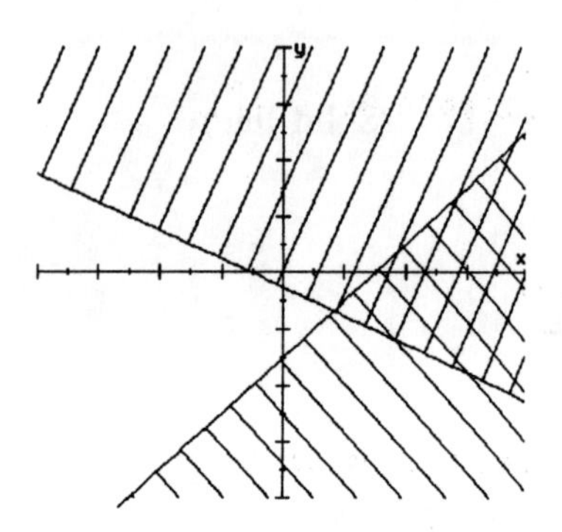

9.1 SOLUTIONS TO EXERCISES

1. $\sqrt{121} = 11$, because $11^2 = 121$ and 11 is positive.

2. $\sqrt{\dfrac{4}{49}} = \dfrac{2}{7}$, because $\left(\dfrac{2}{7}\right)^2 = \dfrac{4}{49}$ and $\dfrac{2}{7}$ is positive.

3. $-\sqrt{169} = -13$; the negative sign in front of the radical indicates the negative square root.

4. $\sqrt[3]{27} = 3$, because $3^3 = 27$.

5. $-\sqrt[3]{64} = -4$; the negative sign in front of the radical indicates the opposite of $\sqrt[3]{64}$.

6. $\sqrt[5]{\dfrac{1}{243}} = \dfrac{1}{3}$, because $\left(\dfrac{1}{3}\right)^5 = \dfrac{1}{243}$.

7. $\sqrt[4]{0} = 0$, because $0^4 = 0$.

8. $-\sqrt[4]{-16}$ is not a real number since the radicand is negative and the index is even.

9. $-\sqrt[4]{256} = -4$; the negative sign in front of the radical indicates the negative fourth root.

10. $\sqrt{a^8} = a^4$, because $(a^4)^2 = a^8$.

11. $\sqrt{25x^4} = 5x^2$, because $(5x^2)^2 = 25x^4$.

12. $\sqrt[3]{b^{15}} = b^5$, because $(b^5)^3 = b^{15}$.

13. $-\sqrt[3]{8m^6n^9} = -2m^2n^3$; the negative sign in front of the radical indicates the opposite of $\sqrt[3]{8m^6n^9}$.

14. $\sqrt{x^{10}y^2} = x^5y$, because $(x^5y)^2 = x^{10}y^2$.

15. $\sqrt[3]{-c^{12}d^9} = -c^4d^3$, because $(-c^4d^3)^3 = -c^{12}d^9$.

16. $-\sqrt{144x^6} = -12x^3$; the negative sign in front of the radical indicates the negative square root.

17. $\sqrt{-144x^6}$ is not a real number since the radicand is negative and the index is even.

18. $-\sqrt[3]{-27b^3} = -(-3b) = 3b$; the negative sign in front of the radical indicates the opposite of $\sqrt[3]{-27b^3}$.

9.2 SOLUTIONS TO EXERCISES

1. $\sqrt{28} = \sqrt{4 \cdot 7}$
$= \sqrt{4} \cdot \sqrt{7}$
$= 2\sqrt{7}$

2. $\sqrt{108} = \sqrt{36 \cdot 3}$
$= \sqrt{36} \cdot \sqrt{3}$
$= 6\sqrt{3}$

3. $\sqrt[3]{32} = \sqrt[3]{8 \cdot 4}$
$= \sqrt[3]{8} \cdot \sqrt[3]{4}$
$= 2\sqrt[3]{4}$

4. $\sqrt[3]{-54} = \sqrt[3]{-27 \cdot 2}$
$= \sqrt[3]{-27} \cdot \sqrt[3]{2}$
$= -3\sqrt[3]{2}$

5. $\sqrt{40} = \sqrt{4 \cdot 10}$
$= \sqrt{4} \cdot \sqrt{10}$
$= 2\sqrt{10}$

6. $-\sqrt[3]{40} = -\sqrt[3]{8 \cdot 5}$
$= -\sqrt[3]{8} \cdot \sqrt[3]{5}$
$= -2\sqrt[3]{5}$

7. $\sqrt{\dfrac{8}{18}} = \sqrt{\dfrac{4}{9}}$
$= \dfrac{\sqrt{4}}{\sqrt{9}}$
$= \dfrac{2}{3}$

8. $-\sqrt{\dfrac{180}{125}} = -\sqrt{\dfrac{36}{25}}$
$= -\dfrac{\sqrt{36}}{\sqrt{25}}$
$= -\dfrac{6}{5}$

9. $\sqrt[3]{\dfrac{9}{8}} = \dfrac{\sqrt[3]{9}}{\sqrt[3]{8}}$
$= \dfrac{\sqrt[3]{9}}{2}$

10. $\sqrt{x^{11}} = \sqrt{x^{10} \cdot x}$
$= \sqrt{x^{10}} \cdot \sqrt{x}$
$= x^5\sqrt{x}$

11. $\sqrt{a^5b^4} = \sqrt{a^4 \cdot a \cdot b^4}$
$= \sqrt{a^4b^4 \cdot a}$
$= \sqrt{a^4b^4} \cdot \sqrt{a}$
$= a^2b^2\sqrt{a}$

12. $-\sqrt{m^{16}n^7} = -\sqrt{m^{16} \cdot n^6 \cdot n}$
$= -\sqrt{m^{16}n^6} \cdot \sqrt{n}$
$= -m^8n^3\sqrt{n}$

13. $\sqrt[3]{y^{17}} = \sqrt[3]{y^{15} \cdot y^2}$
$= \sqrt[3]{y^{15}} \cdot \sqrt[3]{y^2}$
$= y^5\sqrt[3]{y^2}$

14. $-\sqrt[3]{b^7} = -\sqrt[3]{b^6 \cdot b}$

$= -\sqrt[3]{b^6} \cdot \sqrt[3]{b}$

$= -b^2 \sqrt[3]{b}$

15. $\sqrt[3]{8a^{10}b^{20}} = \sqrt[3]{8a^9 \cdot a \cdot b^{18} \cdot b^2}$

$= \sqrt[3]{8a^9b^{18} \cdot ab^2}$

$= \sqrt[3]{8a^9b^{18}} \cdot \sqrt[3]{ab^2}$

$= 2a^3b^6 \sqrt[3]{ab^2}$

16. $\sqrt{\dfrac{4x}{y^6}} = \dfrac{\sqrt{4x}}{\sqrt{y^6}}$

$= \dfrac{\sqrt{4} \cdot \sqrt{x}}{\sqrt{y^6}}$

$= \dfrac{2\sqrt{x}}{y^3}$

17. $\sqrt{\dfrac{12m^6}{n^8}} = \dfrac{\sqrt{12m^6}}{\sqrt{n^8}}$

$= \dfrac{\sqrt{4 \cdot 3 \cdot m^6}}{\sqrt{n^8}}$

$= \dfrac{\sqrt{4m^6 \cdot 3}}{\sqrt{n^8}}$

$= \dfrac{\sqrt{4m^6} \cdot \sqrt{3}}{\sqrt{n^8}}$

$= \dfrac{2m^3\sqrt{3}}{n^4}$

18. $\sqrt[3]{\dfrac{b^5c^9}{8d^3}} = \dfrac{\sqrt[3]{b^5c^9}}{\sqrt[3]{8d^3}}$

$= \dfrac{\sqrt[3]{b^3 \cdot b^2 \cdot c^9}}{\sqrt[3]{8d^3}}$

$= \dfrac{\sqrt[3]{b^3c^9}\sqrt[3]{b^2}}{\sqrt[3]{8d^3}}$

$= \dfrac{bc^3\sqrt[3]{b^2}}{2d}$

19. $\sqrt[4]{80} = \sqrt[4]{16 \cdot 5}$

$= \sqrt[4]{16} \cdot \sqrt[4]{5}$

$= 2\sqrt[4]{5}$

20. $\sqrt[4]{162a^6b^8} = \sqrt[4]{81 \cdot 2 \cdot a^4 \cdot a^2 \cdot b^8}$

$= \sqrt[4]{81a^4b^8 \cdot 2a^2}$

$= \sqrt[4]{81a^4b^8} \cdot \sqrt[4]{2a^2}$

$= 3ab^2\sqrt[4]{2a^2}$

9.3 SOLUTIONS TO EXERCISES

1. $9 - 5\sqrt{6} + 12 - 2\sqrt{6} = (9 + 12) + (-5 - 2)\sqrt{6}$
$= 21 - 7\sqrt{6}$

2. $4\sqrt[3]{5} + 8\sqrt[3]{5} - 6\sqrt{5} = (4 + 8)\sqrt[3]{5} - 6\sqrt{5}$
$= 12\sqrt[3]{5} - 6\sqrt{5}$

3. $4\sqrt{12} - 6\sqrt{27} = 4\sqrt{4 \cdot 3} - 6\sqrt{9 \cdot 3}$
$= 4\sqrt{4} \cdot \sqrt{3} - 6\sqrt{9} \cdot \sqrt{3}$
$= 4 \cdot 2 \cdot \sqrt{3} - 6 \cdot 3 \cdot \sqrt{3}$
$= 8\sqrt{3} - 18\sqrt{3}$
$= -10\sqrt{3}$

4. $\sqrt{200} + 3\sqrt{18} = \sqrt{100 \cdot 2} + 3\sqrt{9 \cdot 2}$
$= \sqrt{100} \cdot \sqrt{2} + 3\sqrt{9} \cdot \sqrt{2}$
$= 10\sqrt{2} + 3 \cdot 3 \cdot \sqrt{2}$
$= 10\sqrt{2} + 9\sqrt{2}$
$= 19\sqrt{2}$

5. $2\sqrt[3]{9} - 8\sqrt[3]{243}$

$= 2\sqrt[3]{9} - 8\sqrt[3]{27 \cdot 9}$

$= 2\sqrt[3]{9} - 8\sqrt[3]{27} \cdot \sqrt[3]{9}$

$= 2\sqrt[3]{9} - 8 \cdot 3 \cdot \sqrt[3]{9}$

$= 2\sqrt[3]{9} - 24\sqrt[3]{9}$

$= -22\sqrt[3]{9}$

6. $14x + 6\sqrt{x} - 5\sqrt{x^2} = 14x + 6\sqrt{x} - 5x$
$= 9x + 6\sqrt{x}$

7. $\sqrt{25x} - 5\sqrt{x^3} = \sqrt{25x} - 5\sqrt{x^2 \cdot x}$
$= \sqrt{25} \cdot \sqrt{x} - 5\sqrt{x^2} \cdot \sqrt{x}$
$= 5\sqrt{x} - 5x\sqrt{x}$

8. $19 - 3\sqrt{2} - \sqrt{8} = 19 - 3\sqrt{2} - \sqrt{4 \cdot 2}$
$= 19 - 3\sqrt{2} - \sqrt{4} \cdot \sqrt{2}$
$= 19 - 3\sqrt{2} - 2\sqrt{2}$
$= 19 - 5\sqrt{2}$

9. $\sqrt{\frac{7}{25}} + \sqrt{\frac{7}{64}} = \frac{\sqrt{7}}{\sqrt{25}} + \frac{\sqrt{7}}{\sqrt{64}}$

$= \frac{\sqrt{7}}{5} + \frac{\sqrt{7}}{8}$

$= \frac{\sqrt{7} \cdot 8}{5 \cdot 8} + \frac{\sqrt{7} \cdot 5}{8 \cdot 5}$

$= \frac{8\sqrt{7}}{40} + \frac{5\sqrt{7}}{40}$

$= \frac{8\sqrt{7} + 5\sqrt{7}}{40}$

$= \frac{13\sqrt{7}}{40}$

10. $\sqrt{9x^3} + 2\sqrt{81x^3} - 5\sqrt{x}$

$= \sqrt{9 \cdot x^2 \cdot x} + 2\sqrt{81 \cdot x^2 \cdot x} - 5\sqrt{x}$

$= \sqrt{9x^2} \cdot \sqrt{x} + 2\sqrt{81x^2} \cdot \sqrt{x} - 5\sqrt{x}$

$= 3x\sqrt{x} + 2 \cdot 9x\sqrt{x} - 5\sqrt{x}$

$= 3x\sqrt{x} + 18x\sqrt{x} - 5\sqrt{x}$

$= 21x\sqrt{x} - 5\sqrt{x}$

11. $x\sqrt{36x^2} + \sqrt{4x^4} = x \cdot 6x + 2x^2$
$= 6x^2 + 2x^2$
$= 8x^2$

12. $-10a\sqrt{98b} - 4\sqrt{2a^2b}$

$= -10a\sqrt{49 \cdot 2b} - 4\sqrt{a^2 \cdot 2b}$

$= -10a\sqrt{49} \cdot \sqrt{2b} - 4\sqrt{a^2} \cdot \sqrt{2b}$

$= -10a \cdot 7 \cdot \sqrt{2b} - 4a \cdot \sqrt{2b}$

$= -70a\sqrt{2b} - 4a\sqrt{2b}$

$= -74a\sqrt{2b}$

13. $\sqrt{49y^2} + 2\sqrt[3]{49y^2} + \sqrt{9x^2}$

$= 7y + 2\sqrt[3]{49y^2} + 3x$

14. $5\sqrt{6} - \sqrt{16} + 3\sqrt{24} + \sqrt{121}$
$= 5\sqrt{6} - \sqrt{16} + 3\sqrt{4 \cdot 6} + \sqrt{121}$
$= 5\sqrt{6} - \sqrt{16} + 3 \cdot \sqrt{4} \cdot \sqrt{6} + \sqrt{121}$
$= 5\sqrt{6} - 4 + 3 \cdot 2\sqrt{6} + 11$
$= 5\sqrt{6} - 4 + 6\sqrt{6} + 11$
$= 11\sqrt{6} + 7$

15. $\sqrt{\frac{10}{45}} - \sqrt{\frac{8}{225}} = \sqrt{\frac{2}{9}} - \sqrt{\frac{8}{225}}$
$= \frac{\sqrt{2}}{\sqrt{9}} - \frac{\sqrt{8}}{\sqrt{225}}$
$= \frac{\sqrt{2}}{\sqrt{9}} - \frac{\sqrt{4 \cdot 2}}{\sqrt{225}}$
$= \frac{\sqrt{2}}{3} - \frac{2\sqrt{2}}{15}$
$= \frac{5\sqrt{2}}{15} - \frac{2\sqrt{2}}{15}$
$= \frac{3\sqrt{2}}{15}$
$= \frac{\sqrt{2}}{5}$

16. $\sqrt{15} - \sqrt{60} = \sqrt{15} - \sqrt{4 \cdot 15}$
$= \sqrt{15} - \sqrt{4} \cdot \sqrt{15}$
$= \sqrt{15} - 2\sqrt{15}$
$= -\sqrt{15}$

17. $\sqrt[3]{27} - \sqrt[3]{135} + 6 = \sqrt[3]{27} - \sqrt[3]{27 \cdot 5} + 6$
$= \sqrt[3]{27} - \sqrt[3]{27} \cdot \sqrt[3]{5} + 6$
$= 3 - 3\sqrt[3]{5} + 6$
$= 9 - 3\sqrt[3]{5}$

18. $\sqrt{20x} + \sqrt[3]{189x^4} - 2\sqrt{5x} - \sqrt[3]{7x^4}$
$= \sqrt{4 \cdot 5x} + \sqrt[3]{27 \cdot 7 \cdot x^3 \cdot x} - 2\sqrt{5x} - \sqrt[3]{7 \cdot x^3 \cdot x}$
$= \sqrt{4} \cdot \sqrt{5x} + \sqrt[3]{27x^3} \cdot \sqrt[3]{7x} - 2\sqrt{5x} - \sqrt[3]{x^3} \cdot \sqrt[3]{7x}$
$= 2\sqrt{5x} + 3x\sqrt[3]{7x} - 2\sqrt{5x} - x\sqrt[3]{7x}$
$= 2x\sqrt[3]{7x}$

9.4 SOLUTIONS TO EXERCISES

1. $2\sqrt{3} \cdot 4\sqrt{12} = 2 \cdot 4\sqrt{3 \cdot 12}$
$= 8\sqrt{36}$
$= 8 \cdot 6$
$= 48$

2. $\sqrt[3]{20} \cdot \sqrt[3]{6} = \sqrt[3]{20 \cdot 6}$
$= \sqrt[3]{120}$
$= \sqrt[3]{8 \cdot 15}$
$= \sqrt[3]{8} \cdot \sqrt[3]{15}$
$= 2\sqrt[3]{15}$

3. $\sqrt{14}(\sqrt{7} - \sqrt{2})$
$= \sqrt{14} \cdot \sqrt{7} - \sqrt{14} \cdot \sqrt{2}$
$= \sqrt{98} - \sqrt{28}$
$= \sqrt{49 \cdot 2} - \sqrt{4 \cdot 7}$
$= \sqrt{49} \cdot \sqrt{2} - \sqrt{4} \cdot \sqrt{7}$
$= 7\sqrt{2} - 2\sqrt{7}$

4. $\sqrt{20}(2\sqrt{5} + \sqrt{2})$
$= \sqrt{20} \cdot 2\sqrt{5} + \sqrt{20} \cdot \sqrt{2}$
$= 2\sqrt{20 \cdot 5} + \sqrt{20 \cdot 2}$
$= 2\sqrt{100} + \sqrt{40}$
$= 2\sqrt{100} + \sqrt{4 \cdot 10}$
$= 2\sqrt{100} + \sqrt{4} \cdot \sqrt{10}$
$= 2 \cdot 10 + 2 \cdot \sqrt{10}$
$= 20 + 2\sqrt{10}$

5. $(2\sqrt{6} - \sqrt{5})(3 + 4\sqrt{5})$

$= 2\sqrt{6} \cdot 3 + 2\sqrt{6} \cdot 4\sqrt{5} - \sqrt{5} \cdot 3 - \sqrt{5} \cdot 4\sqrt{5}$

$= 6\sqrt{6} + 8\sqrt{30} - 3\sqrt{5} - 4\sqrt{25}$

$= 6\sqrt{6} + 8\sqrt{30} - 3\sqrt{5} - 4 \cdot 5$

$= 6\sqrt{6} + 8\sqrt{30} - 3\sqrt{5} - 20$

6. $(\sqrt{2} + 3)^2 = (\sqrt{2})^2 + 2(\sqrt{2})(3) + 3^2$
$= 2 + 6\sqrt{2} + 9$
$= 11 + 6\sqrt{2}$

7. $(3 + \sqrt{x})(3 - \sqrt{x}) = 3^2 - (\sqrt{x})^2$
$= 9 - x$

8. $(\sqrt{5} + 3\sqrt{y})(\sqrt{5} - 3\sqrt{y}) = (\sqrt{5})^2 - (3\sqrt{y})^2$
$= 5 - 9y$

9. $(\sqrt{a} + 2\sqrt{b})^2 = (\sqrt{a})^2 + 2(\sqrt{a})(2\sqrt{b}) + (2\sqrt{b})^2$
$= a + 4\sqrt{ab} + 4b$

10. $(9\sqrt{x})^2 = 9^2(\sqrt{x})^2$
$= 81x$

11. $\dfrac{\sqrt{50}}{\sqrt{2}} = \sqrt{\dfrac{50}{2}}$

$= \sqrt{25}$

$= 5$

12. $\dfrac{\sqrt{56}}{\sqrt{7}} = \sqrt{\dfrac{56}{7}}$

$= \sqrt{8}$

$= \sqrt{4 \cdot 2}$

$= \sqrt{4} \cdot \sqrt{2}$

$= 2\sqrt{2}$

13. $\dfrac{\sqrt{125a^5}}{\sqrt{5a}} = \sqrt{\dfrac{125a^5}{5a}}$

$= \sqrt{25a^4}$

$= 5a^2$

14. $\sqrt{\dfrac{2}{5}} = \dfrac{\sqrt{2}}{\sqrt{5}}$

$= \dfrac{\sqrt{2} \cdot \sqrt{5}}{\sqrt{5} \cdot \sqrt{5}}$

$= \dfrac{\sqrt{10}}{5}$

15. $\dfrac{1}{\sqrt{7x}} = \dfrac{1 \cdot \sqrt{7x}}{\sqrt{7x} \cdot \sqrt{7x}}$

$= \dfrac{\sqrt{7x}}{7x}$

16. $\sqrt{\dfrac{3}{80x}} = \dfrac{\sqrt{3}}{\sqrt{80x}}$

$= \dfrac{\sqrt{3}}{\sqrt{16 \cdot 5x}}$

$= \dfrac{\sqrt{3}}{4\sqrt{5x}}$

$= \dfrac{\sqrt{3} \cdot \sqrt{5x}}{4\sqrt{5x} \cdot \sqrt{5x}}$

$= \dfrac{\sqrt{15x}}{4 \cdot 5x}$

$= \dfrac{\sqrt{15x}}{20x}$

17. $\frac{7}{\sqrt[3]{4}} = \frac{7 \cdot \sqrt[3]{2}}{\sqrt[3]{4} \cdot \sqrt[3]{2}}$

$= \frac{7\sqrt[3]{2}}{\sqrt[3]{8}}$

$= \frac{7\sqrt[3]{2}}{2}$

18. $\sqrt[3]{\frac{6}{25x}} = \frac{\sqrt[3]{6}}{\sqrt[3]{25x}}$

$= \frac{\sqrt[3]{6} \cdot \sqrt[3]{5x^2}}{\sqrt[3]{25x} \cdot \sqrt[3]{5x^2}}$

$= \frac{\sqrt[3]{30x^2}}{\sqrt[3]{125x^3}}$

$= \frac{\sqrt[3]{30x^2}}{5x}$

19. $\frac{2}{\sqrt{3} - 1} = \frac{2}{\sqrt{3} - 1} \cdot \frac{\sqrt{3} + 1}{\sqrt{3} + 1}$

$= \frac{2(\sqrt{3} + 1)}{(\sqrt{3})^2 - 1^2}$

$= \frac{2(\sqrt{3} + 1)}{3 - 1}$

$= \frac{2(\sqrt{3} + 1)}{2}$

$= \sqrt{3} + 1$

20. $\frac{\sqrt{2} + 3}{\sqrt{2} - 3} = \frac{(\sqrt{2} + 3)(\sqrt{2} + 3)}{(\sqrt{2} - 3)(\sqrt{2} + 3))}$

$= \frac{(\sqrt{2})^2 + 2(\sqrt{2})(3) + 3^2}{(\sqrt{2})^2 - 3^2}$

$= \frac{2 + 6\sqrt{2} + 9}{2 - 9}$

$= \frac{11 + 6\sqrt{2}}{-7}$

$= -\frac{11 + 6\sqrt{2}}{7}$

9.5 SOLUTIONS TO EXERCISES

1. $\sqrt{x} = 11$
$(\sqrt{x})^2 = 11^2$
$x = 121$

Check: $\sqrt{x} = 11$
$\sqrt{121} = 11$
$11 = 11$ True

$\{121\}$

2. $\sqrt{x - 5} = 6$
$(\sqrt{x - 5})^2 = 6^2$
$x - 5 = 36$
$x = 41$

Check: $\sqrt{x - 5} = 6$
$\sqrt{41 - 5} = 6$
$\sqrt{36} = 6$
$6 = 6$ True

$\{41\}$

3. $\sqrt{x} - 6 = 3$
$\sqrt{x} = 9$
$(\sqrt{x})^2 = 9^2$
$x = 81$

Check: $\sqrt{x} - 6 = 3$
$\sqrt{81} - 6 = 3$
$9 - 6 = 3$
$3 = 3$ True

$\{81\}$

4. $$\begin{aligned} \sqrt{3x+4} &= 5 \\ \left(\sqrt{3x+4}\right)^2 &= 5^2 \\ 3x+4 &= 25 \\ 3x &= 21 \\ x &= 7 \end{aligned}$$

Check: $$\begin{aligned} \sqrt{3x+4} &= 5 \\ \sqrt{31(7)+4} &= 5 \\ \sqrt{25} &= 5 \\ 5 &= 5 \quad \text{True} \end{aligned}$$

$\{7\}$

5. $$\begin{aligned} \sqrt{2x+7} &= \sqrt{x-8} \\ \left(\sqrt{2x+7}\right)^2 &= \left(\sqrt{x-8}\right)^2 \\ 2x+7 &= x-8 \\ x+7 &= -8 \\ x &= -15 \end{aligned}$$

Check: $$\begin{aligned} \sqrt{2x+7} &= \sqrt{x-8} \\ \sqrt{2(-15)+7} &= \sqrt{-15-8} \\ \sqrt{-23} &= \sqrt{-23} \end{aligned}$$

Not a real number

$\emptyset$

6. $$\begin{aligned} \sqrt{7x+3} &= \sqrt{9x-1} \\ \left(\sqrt{7x+3}\right)^2 &= \left(\sqrt{9x-1}\right)^2 \\ 7x+3 &= 9x-1 \\ 3 &= 2x-1 \\ 4 &= 2x \\ 2 &= x \end{aligned}$$

Check: $$\begin{aligned} \sqrt{7x+3} &= \sqrt{9x-1} \\ \sqrt{7(2)+3} &= \sqrt{9(2)-1} \\ \sqrt{17} &= \sqrt{17} \quad \text{True} \end{aligned}$$

$\{2\}$

7. $$\begin{aligned} \sqrt{5x} - \sqrt{x+2} &= 0 \\ \sqrt{5x} &= \sqrt{x+2} \\ \left(\sqrt{5x}\right)^2 &= \left(\sqrt{x+2}\right)^2 \\ 5x &= x+2 \\ 4x &= 2 \\ x &= \frac{1}{2} \end{aligned}$$

Check: $$\begin{aligned} \sqrt{5x} - \sqrt{x+2} &= 0 \\ \sqrt{5\left(\frac{1}{2}\right)} - \sqrt{\frac{1}{2}+2} &= 0 \\ \sqrt{\frac{5}{2}} - \sqrt{\frac{5}{2}} &= 0 \\ 0 &= 0 \quad \text{True} \end{aligned}$$

$\left\{\frac{1}{2}\right\}$

8. $$\begin{aligned} \sqrt{10-x} + 8 &= 3 \\ \sqrt{10-x} &= -5 \end{aligned}$$

The principal root is not negative, hence it cannot equal -5.

$\emptyset$

9. $$\begin{aligned} \sqrt{2x^2+8x+15} &= x \\ \left(\sqrt{2x^2+8x+15}\right)^2 &= x^2 \\ 2x^2+8x+15 &= x^2 \\ x^2+8x+15 &= 0 \\ (x+5)(x+3) &= 0 \end{aligned}$$

$$x+5 = 0 \quad \text{or} \quad x+3 = 0$$
$$x = -5 \quad \text{or} \quad x = -3$$

Check: Let $x = -5$

$$\begin{aligned} \sqrt{2x^2+8x+15} &= x \\ \sqrt{2(-5)^2+8(-5)+15} &= -5 \end{aligned}$$

The principal square root cannot be negative, hence $x = -5$ is not a solution.

Let $x = -3$

$$\begin{aligned} \sqrt{2x^2+8x+15} &= x \\ \sqrt{2(-3)^2+8(-3)+15} &= -3 \end{aligned}$$

The principal square root cannot be negative, hence $x = -3$ is not a solution.

$\emptyset$

10. $\sqrt{2x - 4} = x - 6$

$(\sqrt{2x - 4})^2 = (x - 6)^2$

$2x - 4 = x^2 - 12x + 36$

$0 = x^2 - 14x + 40$

$0 = (x - 4)(x - 10)$

$x - 4 = 0$ or $x - 10 = 0$

$x = 4$ or $x = 10$

Check: Let $x = 4$

$\sqrt{2x - 4} = x - 6$

$\sqrt{2(4) - 4} = 4 - 6$

$\sqrt{4} = -2$

$2 = -2$ False

Let $x = 10$

$\sqrt{2x - 4} = x - 6$

$\sqrt{2(10) - 4} = 10 - 6$

$\sqrt{16} = 4$

$4 = 4$ True

$\{10\}$

11. $\sqrt{21 - 2x} + 3 = x$

$\sqrt{21 - 2x} = x - 3$

$(\sqrt{21 - 2x})^2 = (x - 3)^2$

$21 - 2x = x^2 - 6x + 9$

$0 = x^2 - 4x - 12$

$0 = (x - 6)(x + 2)$

$x - 6 = 0$ or $x + 2 = 0$

$x = 6$ or $x = -2$

Check: Let $x = 6$

$\sqrt{21 - 2x} + 3 = x$

$\sqrt{21 - 2(6)} + 3 = 6$

$\sqrt{9} + 3 = 6$

$3 + 3 = 6$

$6 = 6$ True

Let $x = -2$

$\sqrt{21 - 2x} + 3 = x$

$\sqrt{21 - 2(-2)} + 3 = -2$

$\sqrt{25} + 3 = -2$

$5 + 3 = -2$

$8 = -2$ False

$\{6\}$

12. $2 = \sqrt{12x^2 - 20x + 7}$

$2^2 = (\sqrt{12x^2 - 20x + 7})^2$

$4 = 12x^2 - 20x + 7$

$0 = 12x^2 - 20x + 3$

$0 = (2x - 3)(6x - 1)$

$0 = 2x - 3$ or $0 = 6x - 1$

$3 = 2x$ or $1 = 6x$

$\frac{3}{2} = x$ or $\frac{1}{6} = x$

Check: Let $x = \frac{3}{2}$

$2 = \sqrt{12x^2 - 20x + 7}$

$2 = \sqrt{12\left(\frac{3}{2}\right)^2 - 20\left(\frac{3}{2}\right) + 7}$

$2 = \sqrt{12\left(\frac{9}{4}\right) - 30 + 7}$

$2 = \sqrt{27 - 30 + 7}$

$2 = \sqrt{4}$

$2 = 2$ True

Let $x = \frac{1}{6}$

$2 = \sqrt{12x^2 - 20x + 7}$

$2 = \sqrt{12\left(\frac{1}{6}\right)^2 - 20\left(\frac{1}{6}\right) + 7}$

$2 = \sqrt{12\left(\frac{1}{36}\right) - \frac{10}{3} + 7}$

$2 = \sqrt{\frac{1}{3} - \frac{10}{3} + 7}$

$2 = \sqrt{\frac{12}{3}}$

$2 = \sqrt{4}$

$2 = 2$ True

$\left\{\frac{3}{2}, \frac{1}{6}\right\}$

13. $8\sqrt{x} + 7 = 2$

$$8\sqrt{x} = -5$$
$$\sqrt{x} = -\frac{5}{8}$$

The principal square root cannot be negative, hence it cannot equal $-\frac{5}{8}$.

$\emptyset$

14. x = unknown number

$$x = 20 + \sqrt{x}$$
$$x - 20 = \sqrt{x}$$
$$(x - 20)^2 = (\sqrt{x})^2$$
$$x^2 - 40x + 400 = x$$
$$x^2 - 41x + 400 = 0$$
$$(x - 25)(x - 16) = 0$$
$$x - 25 = 0 \quad \text{or} \quad x - 16 = 0$$
$$x = 25 \quad \text{or} \quad x = 16$$

Check: Let $x = 25$

$$x = 20 + \sqrt{x}$$
$$25 = 20 + \sqrt{25}$$
$$25 = 20 + 5$$
$$25 = 25 \quad \text{True}$$

Let $x = 16$

$$x = 20 + \sqrt{x}$$
$$16 = 20 + \sqrt{16}$$
$$16 = 20 + 4$$
$$16 = 24 \quad \text{False}$$

The number is 25.

15. x = unknown number

$$x = 2\sqrt{3x}$$
$$x^2 = (2\sqrt{3x})^2$$
$$x^2 = 4(3x)$$
$$x^2 = 12x$$
$$x^2 - 12x = 0$$
$$x(x - 12) = 0$$
$$x = 0 \quad \text{or} \quad x - 12 = 0$$
$$x = 12$$

Check: Let $x = 0$

$$x = 2\sqrt{3x}$$
$$0 = 2\sqrt{3(0)}$$
$$0 = 2\sqrt{0}$$
$$0 = 2(0)$$
$$0 = 0 \quad \text{True}$$

Let $x = 12$

$$x = 2\sqrt{3x}$$
$$12 = 2\sqrt{3(12)}$$
$$12 = 2\sqrt{36}$$
$$12 = 2(6)$$
$$12 = 12 \quad \text{True}$$

The number is 0 or 12.

9.6 SOLUTIONS TO EXERCISES

1. $a^2 + b^2 = c^2$

$$15^2 + 20^2 = c^2$$
$$225 + 400 = c^2$$
$$625 = c^2$$
$$\sqrt{625} = \sqrt{c^2}$$
$$25 = c$$

2. $a^2 + b^2 = c^2$

$$21^2 + b^2 = 35^2$$
$$441 + b^2 = 1225$$
$$b^2 = 784$$
$$\sqrt{b^2} = \sqrt{784}$$
$$b = 28$$

3. $a^2 + b^2 = c^2$

$$a^2 + 2^2 = 6^2$$
$$a^2 + 4 = 36$$
$$a^2 = 32$$
$$\sqrt{a^2} = \sqrt{32}$$
$$a = 4\sqrt{2}$$

4. $a^2 + b^2 = c^2$

$$(\sqrt{3})^2 + 2^2 = c^2$$
$$3 + 4 = c^2$$
$$7 = c^2$$
$$\sqrt{7} = \sqrt{c^2}$$
$$\sqrt{7} = c$$

5. $d = \sqrt{(x_2 - x_1)^2 + (y_2 - y_1)^2}$
$= \sqrt{(3 - 5)^2 + (-2 - 7)^2}$
$= \sqrt{(-2)^2 + (-9)^2}$
$= \sqrt{4 + 81}$
$= \sqrt{85}$

6. $d = \sqrt{(x_2 - x_1)^2 + (y_2 - y_1)^2}$
$= \sqrt{(-9 - 0)^2 + (-1 - 4)^2}$
$= \sqrt{(-9)^2 + (-5)^2}$
$= \sqrt{81 + 25}$
$= \sqrt{106}$

7. $d = \sqrt{(x_2 - x_1)^2 + (y_2 - y_1)^2}$
$= \sqrt{[-6 - (-2)]^2 + [-4 - (-8)]^2}$
$= \sqrt{(-4)^2 + (4)^2}$
$= \sqrt{16 + 16}$
$= \sqrt{32}$
$= 4\sqrt{2}$

8. $d = \sqrt{(x_2 - x_1)^2 + (y_2 - y_1)^2}$
$= \sqrt{(0 - 6)^2 + (-8 - 0)^2}$
$= \sqrt{(-6)^2 + (-8)^2}$
$= \sqrt{36 + 64}$
$= \sqrt{100}$
$= 10$

9. $d = \sqrt{(x_2 - x_1)^2 + (y_2 - y_1)^2}$
$= \sqrt{\left(4 - \frac{3}{2}\right)^2 + (-1 - 1)^2}$
$= \sqrt{\left(\frac{5}{2}\right)^2 + (-2)^2}$
$= \sqrt{\frac{25}{4} + 4}$
$= \sqrt{\frac{41}{4}}$
$= \frac{\sqrt{41}}{2}$

10. $d = \sqrt{(x_2 - x_1)^2 + (y_2 - y_1)^2}$
$= \sqrt{\left(\frac{4}{3} - \frac{1}{3}\right)^2 + (-7 - 3)^2}$
$= \sqrt{1^2 + (-10)^2}$
$= \sqrt{1 + 100}$
$= \sqrt{101}$

11. $b = \sqrt{\frac{3V}{H}}$

$8 = \sqrt{\frac{3V}{6}}$

$8^2 = \left(\sqrt{\frac{3V}{6}}\right)^2$

$64 = \frac{3V}{6}$

$384 = 3V$
$128 = V$

Check: $8 = \sqrt{\frac{3V}{6}}$

$8 = \sqrt{\frac{3(128)}{6}}$

$8 = \sqrt{64}$
$8 = 8$ True

The volume is 128 cubic feet.

12.

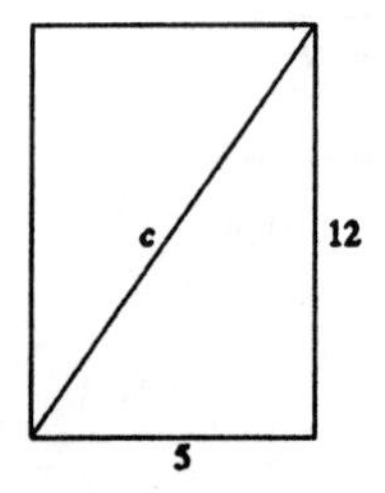

c = length of brace

$$a^2 + b^2 = c^2$$
$$5^2 + 12^2 = c^2$$
$$25 + 144 = c^2$$
$$169 = c^2$$
$$\sqrt{169} = \sqrt{c^2}$$
$$13 = c$$

The brace is 13 feet long.

13. $s = \sqrt{30fd}$
$$s = \sqrt{30(0.35)(320)}$$
$$s = \sqrt{3360} \approx 58$$

The car was traveling 58 mph.

14. $45 \text{ min} = \dfrac{45}{60} \text{ hr} = \dfrac{3}{4} \text{ hr}$

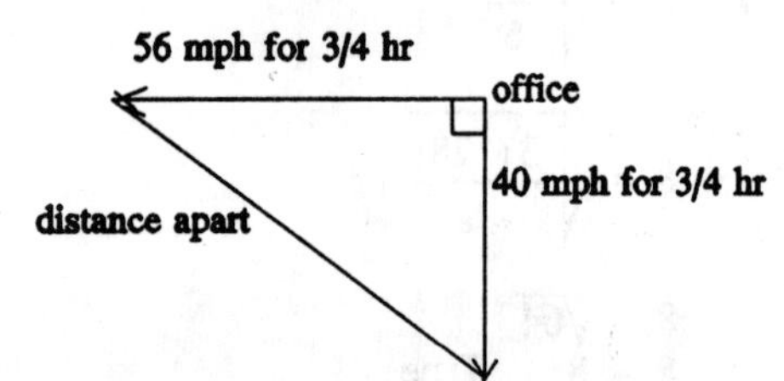

$d = rt$

Dave's distance $= (40)\left(\dfrac{3}{4}\right) = 30$

Dan's distance $= (56)\left(\dfrac{3}{4}\right) = 42$

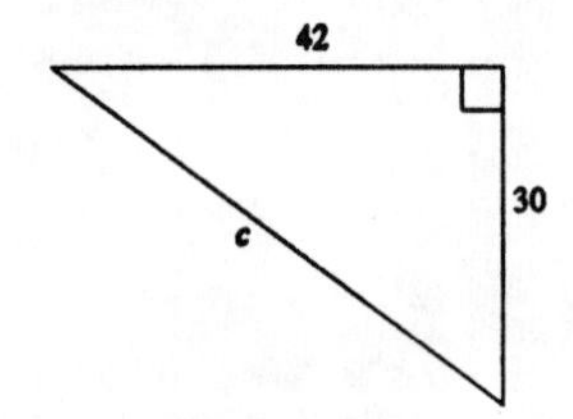

$$c^2 = 42^2 + 30^2$$
$$c^2 = 1764 + 900$$
$$c^2 = 2664$$
$$\sqrt{c^2} = \sqrt{2664}$$
$$c = \sqrt{2664} \approx 52$$

They are 52 miles apart.

15. Look at smaller right triangle first.

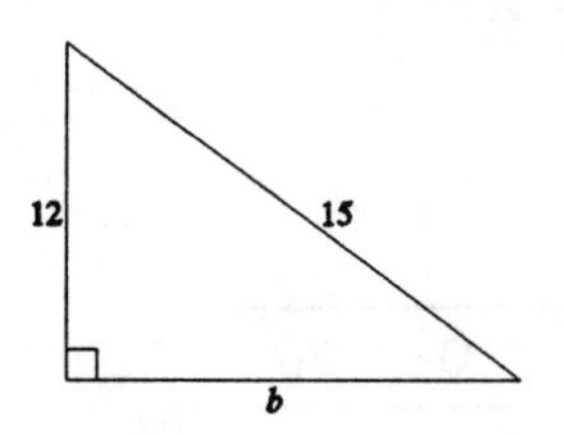

$$12^2 + b^2 = 15^2$$
$$144 + b^2 = 225$$
$$b^2 = 81$$
$$b = 9$$

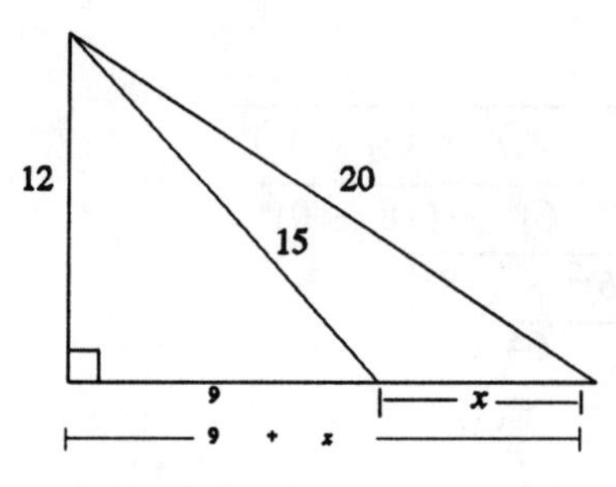

$$12^2 + (9 + x)^2 = 20^2$$
$$144 + 81 + 18x + x^2 = 400$$
$$x^2 + 18x + 225 = 400$$
$$x^2 + 18x - 175 = 0$$
$$(x + 25)(x - 7) = 0$$
$$x + 25 = 0 \quad \text{or} \quad x - 7 = 0$$
$$x = -25 \quad \text{or} \quad x = 7$$

Since x represents a length, it must be positive. Hence $x = 7$.

9.7 SOLUTIONS TO EXERCISES

1. $8^{4/3} = \left(\sqrt[3]{8}\right)^4$
$$= 2^4$$
$$= 16$$

2. $100^{3/2} = (\sqrt{100})^3$
$= 10^3$
$= 1000$

3. $64^{2/3} = (\sqrt[3]{64})^2$
$= 4^2$
$= 16$

4. $-49^{1/2} = -\sqrt{49}$
$= -7$

5. $16^{-3/4} = \dfrac{1}{16^{3/4}}$

$= \dfrac{1}{(\sqrt[4]{16})^3}$

$= \dfrac{1}{2^3}$

$= \dfrac{1}{8}$

6. $-32^{-3/5} = -\dfrac{1}{32^{3/5}}$

$= -\dfrac{1}{(\sqrt[5]{32})^3}$

$= -\dfrac{1}{2^3}$

$= -\dfrac{1}{8}$

7. $\left(\dfrac{36}{25}\right)^{-1/2} = \left(\dfrac{25}{36}\right)^{1/2}$

$= \sqrt{\dfrac{25}{36}}$

$= \dfrac{5}{6}$

8. $\left(\dfrac{64}{125}\right)^{-2/3} = \left(\dfrac{125}{64}\right)^{2/3}$

$= \left(\sqrt[3]{\dfrac{125}{64}}\right)^2$

$= \left(\dfrac{5}{4}\right)^2$

$= \dfrac{25}{16}$

9. $\dfrac{x^{1/7}}{x^{5/7}} = x^{1/7-5/7}$

$= x^{-4/7}$

$= \dfrac{1}{x^{4/7}}$

10. $(y^{3/4})^{20} = y^{(3/4)(20)}$

$= y^{15}$

11. $x^{2/5} \cdot x^{7/5} = x^{2/5+7/5}$

$= x^{9/5}$

12. $y^{3/4} \cdot y^{1/3} = y^{3/4+1/3}$

$= y^{9/12+4/12}$

$= y^{13/12}$

13. $\dfrac{x^{5/6}}{x^{-2/3}} = x^{5/6-(-2/3)}$

$= x^{5/6+2/3}$

$= x^{5/6+4/6}$

$= x^{9/6}$

$= x^{3/2}$

14. $\left(\dfrac{x^{3/4}}{y^{1/6}}\right)^2 = \dfrac{(x^{3/4})^2}{(y^{1/6})^2}$

$= \dfrac{x^{(3/4)(2)}}{y^{(1/6)(2)}}$

$= \dfrac{x^{3/2}}{y^{1/3}}$

15. $\left(\dfrac{x^{2/5}}{y^{7/5}}\right)^{10} = \dfrac{(x^{2/5})^{10}}{(y^{7/5})^{10}}$

$= \dfrac{x^{(2/5)(10)}}{y^{(7/5)(10)}}$

$= \dfrac{x^4}{y^{14}}$

16. $\dfrac{7^{-2/9}}{7^{5/9}} = 7^{-2/9-5/9}$

$= 7^{-7/9}$

$= \dfrac{1}{7^{7/9}}$

17. $(x^{3/8})^{1/4} = x^{(3/8)(1/4)}$

$= x^{3/32}$

18. $3^{2/5} \cdot 3^{8/5} = 3^{2/5+8/5}$

$= 3^{10/5}$

$= 3^2$
$= 9$

CHAPTER 9 PRACTICE TEST SOLUTIONS

1. $\sqrt{49} = \sqrt{7^2} = 7$

2. $\sqrt[3]{-125} = \sqrt[3]{(-5)^3} = -5$

3. $81^{1/4} = \sqrt[4]{81} = \sqrt[4]{3^4} = 3$

4. $\left(\frac{16}{9}\right)^{3/2} = \frac{16^{3/2}}{9^{3/2}} = \frac{(\sqrt{16})^3}{(\sqrt{9})^3} = \frac{4^3}{3^3} = \frac{64}{27}$

5. This is not a real number since the radicand, −256, is negative and the index, 4, is even.

6. $64^{-3/2} = \frac{1}{64^{3/2}}$
$= \frac{1}{(\sqrt{64})^3}$
$= \frac{1}{8^3}$
$= \frac{1}{512}$

7. $\sqrt{112} = \sqrt{16 \cdot 7}$
$= \sqrt{16}\sqrt{7}$
$= 4\sqrt{7}$

8. $\sqrt{12x^3y^9} = \sqrt{4 \cdot 3 \cdot x^2 \cdot x \cdot y^8 \cdot y}$
$= \sqrt{4x^2y^8} \cdot \sqrt{3xy}$
$= 2xy^4\sqrt{3xy}$

9. $\sqrt[3]{27x^{12}y^7} = \sqrt[3]{27x^{12}y^6} \cdot \sqrt[3]{y}$
$= 3x^4y^2\sqrt[3]{y}$

10. $\sqrt{\frac{8}{50}} = \sqrt{\frac{4}{25}}$
$= \frac{\sqrt{4}}{\sqrt{25}}$
$= \frac{2}{5}$

11. $\sqrt{3x^2} + \sqrt[3]{56} - 2x\sqrt{12}$
$= \sqrt{3}\sqrt{x^2} + \sqrt[3]{8}\sqrt[3]{7} - 2x\sqrt{4} \cdot \sqrt{3}$
$= x\sqrt{3} + 2\sqrt[3]{7} - 2x(2)\sqrt{3}$
$= x\sqrt{3} + 2\sqrt[3]{7} - 4x\sqrt{3}$
$= -3x\sqrt{3} + 2\sqrt[3]{7}$

12. $\sqrt{\frac{10}{7}} = \frac{\sqrt{10}}{\sqrt{7}}$
$= \frac{\sqrt{10}}{\sqrt{7}} \cdot \frac{\sqrt{7}}{\sqrt{7}}$
$= \frac{\sqrt{70}}{7}$

13. $\sqrt[3]{\frac{11}{9x}} = \frac{\sqrt[3]{11}}{\sqrt[3]{3^2x}}$
$= \frac{\sqrt[3]{11}}{\sqrt[3]{3^2x}} \cdot \frac{\sqrt[3]{3x^2}}{\sqrt[3]{3x^2}}$
$= \frac{\sqrt[3]{33x^2}}{3x}$

14. $\frac{4}{\sqrt{6}-2} = \frac{4}{\sqrt{6}-2} \cdot \frac{\sqrt{6}+2}{\sqrt{6}+2}$

$= \frac{4(\sqrt{6}+2)}{6-4}$

$= \frac{4(\sqrt{6}+2)}{2}$

$= 2(\sqrt{6}+2)$
$= 2\sqrt{6}+4$

15. $\sqrt{x} + 7 = 19$
$\sqrt{x} = 12$
$(\sqrt{x})^2 = 12^2$
$x = 144$

Check: $\sqrt{144} + 7 = 19$
$12 + 7 = 19$
$19 = 19$
True

{144}

16. $\sqrt{2x-4} = \sqrt{x+1}$
$(\sqrt{2x-4})^2 = (\sqrt{x+1})^2$
$2x - 4 = x + 1$
$x - 4 = 1$
$x = 5$

Check: $\sqrt{2(5)-4} = \sqrt{5+1}$
$\sqrt{6} = \sqrt{6}$
True

{5}

17. unknown leg: x

$a^2 + b^2 = c^2$
$9^2 + x^2 = 15^2$
$81 + x^2 = 225$
$x^2 = 144$
$x = \sqrt{144}$
$x = 12$

12 inches

18. $d = \sqrt{(x_2 - x_1)^2 + (y_2 - y_1)^2}$
$= \sqrt{(-8-2)^2 + (3-6)^2}$
$= \sqrt{(-10)^2 + (-3)^2}$
$= \sqrt{100+9}$
$= \sqrt{109}$

19. $25^{-3/2} \cdot 25^{-1/2} = 25^{-3/2+(-1/2)}$
$= 25^{-2}$

$= \frac{1}{25^2}$

$= \frac{1}{625}$

20. $(x^{3/7})^4 = x^{(3/7)(4)} = x^{12/7}$

10.1 SOLUTIONS TO EXERCISES

1. $x^2 = \frac{1}{25}$

$x = \pm\sqrt{\frac{1}{25}}$

$x = \pm\frac{1}{5}$

$\left\{\pm\frac{1}{5}\right\}$

2. $y^2 = 8$

$y = \pm\sqrt{8}$

$y = \pm 2\sqrt{2}$

$\{\pm 2\sqrt{2}\}$

3. $2k^2 - 98 = 0$

$2k^2 = 98$

$k^2 = 49$

$k = \pm\sqrt{49}$

$k = \pm 7$

$\{\pm 7\}$

4. $5x^2 - 5 = 0$

$5x^2 = 5$

$x^2 = 1$

$x = \pm\sqrt{1}$

$x = \pm 1$

$\{\pm 1\}$

5. $n^2 + 16 = 0$

$n^2 = -16$

$n = \pm\sqrt{-16}$

No real solution.

6. $(x + 4)^2 = 16$

$x + 4 = \pm\sqrt{16}$

$x + 4 = \pm 4$

$x = -4 \pm 4$

$x = -4 + 4$ or $x = -4 - 4$

$x = 0$ or $x = -8$

$\{-8, 0\}$

7. $(y - 2)^2 = 36$

$y - 2 = \pm\sqrt{36}$

$y - 2 = \pm 6$

$y = 2 \pm 6$

$y = 2 + 6$ or $y = 2 - 6$

$y = 8$ or $y = -4$

$\{-4, 8\}$

8. $(m - 7)^2 = 100$

$m - 7 = \pm\sqrt{100}$

$m - 7 = \pm 10$

$m = 7 \pm 10$

$m = 7 + 10$ or $m = 7 - 10$

$m = 17$ or $m = -3$

$\{-3, 17\}$

9. $(z + 8)^2 = -100$

$z + 8 = \pm\sqrt{-100}$

No real solution.

10. $p^2 = 28$

$p = \pm\sqrt{28}$

$p = \pm 2\sqrt{7}$

$\{\pm 2\sqrt{7}\}$

11. $(2x + 7)^2 = 48$

$2x + 7 = \pm\sqrt{48}$

$2x + 7 = \pm 4\sqrt{3}$

$2x = -7 \pm 4\sqrt{3}$

$x = \frac{-7 \pm 4\sqrt{3}}{2}$

$\left\{\frac{-7 + 4\sqrt{3}}{2}, \frac{-7 - 4\sqrt{3}}{2}\right\}$

12. $(6y - 13)^2 = 12$

$6y - 13 = \pm\sqrt{12}$

$6y - 13 = \pm 2\sqrt{3}$

$6y = 13 \pm 2\sqrt{3}$

$y = \dfrac{13 \pm 2\sqrt{3}}{6}$

$\left\{\dfrac{13 + 2\sqrt{3}}{6}, \dfrac{13 - 2\sqrt{3}}{6}\right\}$

13. $(5m + 4)^2 = 18$

$5m + 4 = \pm\sqrt{18}$

$5m + 4 = \pm 3\sqrt{2}$

$5m = -4 \pm 3\sqrt{2}$

$m = \dfrac{-4 \pm 3\sqrt{2}}{5}$

$\left\{\dfrac{-4 + 3\sqrt{2}}{5}, \dfrac{-4 - 3\sqrt{2}}{5}\right\}$

14. $(3 + 2y)^2 = 24$

$3 + 2y = \pm\sqrt{24}$

$3 + 2y = \pm 2\sqrt{6}$

$2y = -3 \pm 2\sqrt{6}$

$y = \dfrac{-3 \pm 2\sqrt{6}}{2}$

$\left\{\dfrac{-3 + 2\sqrt{6}}{2}, \dfrac{-3 - 2\sqrt{6}}{2}\right\}$

15. $4x^2 = 169$

$x^2 = \dfrac{169}{4}$

$x = \pm\sqrt{\dfrac{169}{4}}$

$x = \pm\dfrac{13}{2}$

$\left\{\pm\dfrac{13}{2}\right\}$

16. $-3a^2 = 27$

$a^2 = -9$

$a = \pm\sqrt{-9}$

No real solution.

10.2 SOLUTIONS TO EXERCISES

1. $\left[\dfrac{1}{2}(20)\right]^2 = 10^2 = 100$

$x^2 + 20x + 100 = (x + 10)^2$

2. $\left[\dfrac{1}{2}(-18)\right]^2 = (-9)^2 = 81$

$y^2 - 18y + 81 = (y - 9)^2$

3. $\left[\dfrac{1}{2}(-7)\right]^2 = \left(-\dfrac{7}{2}\right)^2 = \dfrac{49}{4}$

$x^2 - 7x + \dfrac{49}{4} = \left(x - \dfrac{7}{2}\right)^2$

4. $\left[\dfrac{1}{2}(13)\right]^2 = \left(\dfrac{13}{2}\right)^2 = \dfrac{169}{4}$

$m^2 + 13m + \dfrac{169}{4} = \left(m + \dfrac{13}{2}\right)^2$

5. $x^2 + 4x = 8$

$\left[\dfrac{1}{2}(4)\right]^2 = 2^2 = 4$

$x^2 + 4x + 4 = 8 + 4$

$(x + 2)^2 = 12$

$x + 2 = \pm\sqrt{12}$

$x + 2 = \pm 2\sqrt{3}$

$x = -2 \pm 2\sqrt{3}$

$\{-2 + 2\sqrt{3}, -2 - 2\sqrt{3}\}$

6. $y^2 - 2y = 0$

$$\left[\frac{1}{2}(-2)\right]^2 = (-1)^2 = 1$$

$$y^2 - 2y + 1 = 0 + 1$$
$$(y - 1)^2 = 1$$
$$y - 1 = \pm\sqrt{1}$$
$$y - 1 = \pm 1$$
$$y = 1 \pm 1$$
$$y = 1 + 1 \quad \text{or} \quad y = 1 - 1$$
$$y = 2 \quad \text{or} \quad y = 0$$

$\{0, 2\}$

7. $m^2 - 6m - 7 = 0$

$$m^2 - 6m = 7$$

$$\left[\frac{1}{2}(-6)\right]^2 = (-3)^2 = 9$$

$$m^2 - 6m + 9 = 7 + 9$$
$$(m - 3)^2 = 16$$
$$m - 3 = \pm\sqrt{16}$$
$$m - 3 = \pm 4$$
$$m = 3 \pm 4$$
$$m = 3 + 4 \quad \text{or} \quad m = 3 - 4$$
$$m = 7 \quad \text{or} \quad m = -1$$

$\{-1, 7\}$

8. $x^2 + 14x + 24 = 0$

$$x^2 + 14x = -24$$

$$\left[\frac{1}{2}(14)\right]^2 = 7^2 = 49$$

$$x^2 + 14x + 49 = -24 + 49$$
$$(x + 7)^2 = 25$$
$$x + 7 = \pm\sqrt{25}$$
$$x + 7 = \pm 5$$
$$x = -7 \pm 5$$
$$x = -7 + 5 \quad \text{or} \quad x = -7 - 5$$
$$x = -2 \quad \text{or} \quad x = -12$$

$\{-12, -2\}$

9. $z^2 + 8z + 1 = 0$

$$z^2 + 8z = -1$$

$$\left[\frac{1}{2}(8)\right]^2 = 4^2 = 16$$

$$z^2 + 8z + 16 = -1 + 16$$
$$(z + 4)^2 = 15$$
$$z + 4 = \pm\sqrt{15}$$
$$z = -4 \pm \sqrt{15}$$

$\{-4 + \sqrt{15}, -4 - \sqrt{15}\}$

10. $y^2 = 10y + 3$

$$y^2 - 10y = 3$$

$$\left[\frac{1}{2}(-10)\right]^2 = (-5)^2 = 25$$

$$y^2 - 10y + 25 = 3 + 25$$
$$(y - 5)^2 = 28$$
$$y - 5 = \pm\sqrt{28}$$
$$y - 5 = \pm 2\sqrt{7}$$
$$y = 5 \pm 2\sqrt{7}$$

$\{5 + 2\sqrt{7}, 5 - 2\sqrt{7}\}$

11. $2x^2 - 14x + 5 = 0$

$$x^2 - 7x + \frac{5}{2} = 0$$

$$x^2 - 7x = -\frac{5}{2}$$

$$\left[\frac{1}{2}(-7)\right]^2 = \left(-\frac{7}{2}\right)^2 = \frac{49}{4}$$

$$x^2 - 7x + \frac{49}{4} = -\frac{5}{2} + \frac{49}{4}$$

$$\left(x - \frac{7}{2}\right)^2 = \frac{39}{4}$$

$$x - \frac{7}{2} = \pm\sqrt{\frac{39}{4}}$$

$$x - \frac{7}{2} = \pm\frac{\sqrt{39}}{2}$$

$$x = \frac{7}{2} \pm \frac{\sqrt{39}}{2} = \frac{7 \pm \sqrt{39}}{2}$$

$$\left\{\frac{7 + \sqrt{39}}{2}, \frac{7 - \sqrt{39}}{2}\right\}$$

12.

$$3x^2 - 17x = -10$$

$$x^2 - \frac{17}{3}x = -\frac{10}{3}$$

$$\left[\frac{1}{2}\left(-\frac{17}{3}\right)\right]^2 = \left(-\frac{17}{6}\right)^2 = \frac{289}{36}$$

$$x^2 - \frac{17}{3}x + \frac{289}{36} = -\frac{10}{3} + \frac{289}{36}$$

$$\left(x - \frac{17}{6}\right)^2 = \frac{169}{36}$$

$$x - \frac{17}{6} = \pm\sqrt{\frac{169}{36}}$$

$$x - \frac{17}{6} = \pm\frac{13}{6}$$

$$x = \frac{17}{6} \pm \frac{13}{6}$$

$$x = \frac{17}{6} + \frac{13}{6} \quad \text{or} \quad x = \frac{17}{6} - \frac{13}{6}$$

$$x = 5 \quad \text{or} \quad x = \frac{2}{3}$$

$$\left\{\frac{2}{3}, 5\right\}$$

13.

$$4y^2 = 12y - 9$$

$$y^2 = 3y - \frac{9}{4}$$

$$y^2 - 3y = -\frac{9}{4}$$

$$\left[\frac{1}{2}(-3)\right]^2 = \left(-\frac{3}{2}\right)^2 = \frac{9}{4}$$

$$y^2 - 3y + \frac{9}{4} = -\frac{9}{4} + \frac{9}{4}$$

$$\left(y - \frac{3}{2}\right)^2 = 0$$

$$y - \frac{3}{2} = \pm\sqrt{0}$$

$$y - \frac{3}{2} = 0$$

$$y = \frac{3}{2}$$

$$\left\{\frac{3}{2}\right\}$$

14.

$$5m^2 - 2 = 10m$$

$$m^2 - \frac{2}{5} = 2m$$

$$m^2 - 2m = \frac{2}{5}$$

$$\left[\frac{1}{2}(-2)\right]^2 = (-1)^2 = 1$$

$$m^2 - 2m + 1 = \frac{2}{5} + 1$$

$$(m - 1)^2 = \frac{7}{5}$$

$$m - 1 = \pm\sqrt{\frac{7}{5}}$$

$$m - 1 = \pm\frac{\sqrt{35}}{5}$$

$$m = 1 \pm \frac{\sqrt{35}}{5} = \frac{5 \pm \sqrt{35}}{5}$$

$$\left\{\frac{5 + \sqrt{35}}{5}, \frac{5 - \sqrt{35}}{5}\right\}$$

15. $z^2 + z + 8 = 0$

$$z^2 + z = -8$$

$$\left[\frac{1}{2}(1)\right]^2 = \left(\frac{1}{2}\right)^2 = \frac{1}{4}$$

$$z^2 + z + \frac{1}{4} = -8 + \frac{1}{4}$$

$$\left(z + \frac{1}{2}\right)^2 = -\frac{31}{4}$$

No real solution.

16. $x(x - 1) = 6$

$$x^2 - x = 6$$

$$\left[\frac{1}{2}(-1)\right]^2 = \left(-\frac{1}{2}\right)^2 = \frac{1}{4}$$

$$x^2 - x + \frac{1}{4} = 6 + \frac{1}{4}$$

$$\left(x - \frac{1}{2}\right)^2 = \frac{25}{4}$$

$$x - \frac{1}{2} = \pm\sqrt{\frac{25}{4}}$$

$$x - \frac{1}{2} = \pm\frac{5}{2}$$

$$x = \frac{1}{2} \pm \frac{5}{2}$$

$$x = \frac{1}{2} + \frac{5}{2} \quad \text{or} \quad x = \frac{1}{2} - \frac{5}{2}$$

$$x = 3 \quad \text{or} \quad x = -2$$

$\{-2, 3\}$

10.3 SOLUTIONS TO EXERCISES

1. $x^2 - 10x + 24 = 0$
   $a = 1, b = -10, c = 24$

$$x = \frac{-b \pm \sqrt{b^2 - 4ac}}{2a}$$

$$= \frac{-(-10) \pm \sqrt{(-10)^2 - 4(1)(24)}}{2(1)}$$

$$= \frac{10 \pm \sqrt{100 - 96}}{2}$$

$$= \frac{10 \pm \sqrt{4}}{2} = \frac{10 \pm 2}{2}$$

$$x = \frac{10 + 2}{2} = 6 \quad \text{or} \quad x = \frac{10 - 2}{2} = 4$$

$\{4, 6\}$

2. $x^2 - 18x + 81 = 0$
   $a = 1, b = -18, c = 81$

$$x = \frac{-b \pm \sqrt{b^2 - 4ac}}{2a}$$

$$= \frac{-(-18) \pm \sqrt{(-18)^2 - 4(1)(81)}}{2(1)}$$

$$= \frac{18 \pm \sqrt{324 - 324}}{2}$$

$$= \frac{18 \pm \sqrt{0}}{2} = \frac{18}{2} = 9$$

$\{9\}$

3. $49x^2 - 10 = 0$
   $a = 49, b = 0, c = -10$

$$x = \frac{-b \pm \sqrt{b^2 - 4ac}}{2a}$$

$$= \frac{-0 \pm \sqrt{0^2 - 4(49)(-10)}}{2(49)}$$

$$= \frac{\pm\sqrt{1960}}{98} = \frac{\pm 14\sqrt{10}}{98} = \frac{\pm\sqrt{10}}{7}$$

$$\left\{\frac{\sqrt{10}}{7}, -\frac{\sqrt{10}}{7}\right\}$$

4. $3x^2 + 7x + 4 = 0$
$a = 3, b = 7, c = 4$

$$x = \frac{-b \pm \sqrt{b^2 - 4ac}}{2a}$$

$$= \frac{-7 \pm \sqrt{7^2 - 4(3)(4)}}{2(3)}$$

$$= \frac{-7 \pm \sqrt{49 - 48}}{6}$$

$$= \frac{-7 \pm \sqrt{1}}{6} = \frac{-7 \pm 1}{6}$$

$$x = \frac{-7 + 1}{6} = -1 \quad \text{or} \quad x = \frac{-7 - 1}{6} = -\frac{4}{3}$$

$$\left\{-1, -\frac{4}{3}\right\}$$

5. $y^2 + 6y - 8 = 0$
$a = 1, b = 6, c = -8$

$$y = \frac{-b \pm \sqrt{b^2 - 4ac}}{2a}$$

$$= \frac{-6 \pm \sqrt{6^2 - 4(1)(-8)}}{2(1)}$$

$$= \frac{-6 \pm \sqrt{36 + 32}}{2}$$

$$= \frac{-6 \pm \sqrt{68}}{2}$$

$$= \frac{-6 \pm 2\sqrt{17}}{2}$$

$$= \frac{2(-3 \pm \sqrt{17})}{2} = -3 \pm \sqrt{17}$$

$$\{-3 + \sqrt{17}, -3 - \sqrt{17}\}$$

6. $2z^2 + 3z - 1 = 0$
$a = 2, b = 3, c = -1$

$$z = \frac{-b \pm \sqrt{b^2 - 4ac}}{2a}$$

$$= \frac{-3 \pm \sqrt{3^2 - 4(2)(-1)}}{2(2)}$$

$$= \frac{-3 \pm \sqrt{9 + 8}}{4}$$

$$= \frac{-3 \pm \sqrt{17}}{4}$$

$$\left\{\frac{-3 + \sqrt{17}}{4}, \frac{-3 - \sqrt{17}}{4}\right\}$$

7. $7m^2 + 8m = 2$
$7m^2 + 8m - 2 = 0$
$a = 7, b = 8, c = -2$

$$m = \frac{-b \pm \sqrt{b^2 - 4ac}}{2a}$$

$$= \frac{-8 \pm \sqrt{8^2 - 4(7)(-2)}}{2(7)}$$

$$= \frac{-8 \pm \sqrt{64 + 56}}{14}$$

$$= \frac{-8 \pm \sqrt{120}}{14}$$

$$= \frac{-8 \pm 2\sqrt{30}}{14}$$

$$= \frac{2(-4 \pm \sqrt{30})}{14} = \frac{-4 \pm \sqrt{30}}{7}$$

$$\left\{\frac{-4 + \sqrt{30}}{7}, \frac{-4 - \sqrt{30}}{7}\right\}$$

8. $5 - x^2 = x$

$$0 = x^2 + x - 5$$

$a = 1,\ b = 1,\ c = -5$

$$x = \frac{-b \pm \sqrt{b^2 - 4ac}}{2a}$$

$$= \frac{-1 \pm \sqrt{1^2 - 4(1)(-5)}}{2(1)}$$

$$= \frac{-1 \pm \sqrt{1 + 20}}{2}$$

$$= \frac{-1 \pm \sqrt{21}}{2}$$

$$\left\{\frac{-1 + \sqrt{21}}{2}, \frac{-1 - \sqrt{21}}{2}\right\}$$

9. $2x^2 + 3x + 5 = 0$

$a = 2,\ b = 3,\ c = 5$

$$x = \frac{-b \pm \sqrt{b^2 - 4ac}}{2a}$$

$$= \frac{-3 \pm \sqrt{3^2 - 4(2)(5)}}{2(2)}$$

$$= \frac{-3 \pm \sqrt{9 - 40}}{4}$$

$$= \frac{-3 \pm \sqrt{-31}}{4}$$

No real solution.

10. $4y^2 = 5y + 8$

$$4y^2 - 5y - 8 = 0$$

$a = 4,\ b = -5,\ c = -8$

$$y = \frac{-b \pm \sqrt{b^2 - 4ac}}{2a}$$

$$= \frac{-(-5) \pm \sqrt{(-5)^2 - 4(4)(-8)}}{2(4)}$$

$$= \frac{5 \pm \sqrt{25 + 128}}{8}$$

$$= \frac{5 \pm \sqrt{153}}{8} = \frac{5 \pm 3\sqrt{17}}{8}$$

$$\left\{\frac{5 + 3\sqrt{17}}{8}, \frac{5 - 3\sqrt{17}}{8}\right\}$$

11. $4d^2 - \frac{1}{2}d - 1 = 0$

$$8d^2 - d - 2 = 0$$

$a = 8,\ b = -1,\ c = -2$

$$d = \frac{-b \pm \sqrt{b^2 - 4ac}}{2a}$$

$$= \frac{-(-1) \pm \sqrt{(-1)^2 - 4(8)(-2)}}{2(8)}$$

$$= \frac{1 \pm \sqrt{1 + 64}}{16}$$

$$= \frac{1 \pm \sqrt{65}}{16}$$

$$\left\{\frac{1 + \sqrt{65}}{16}, \frac{1 - \sqrt{65}}{16}\right\}$$

12. $$\frac{x^2}{5} = 2x + 3$$

$$x^2 = 10x + 15$$
$$x^2 - 10x - 15 = 0$$
$a = 1$, $b = -10$, $c = -15$

$$x = \frac{-b \pm \sqrt{b^2 - 4ac}}{2a}$$
$$= \frac{-(-10) \pm \sqrt{(-10)^2 - 4(1)(-15)}}{2(1)}$$
$$= \frac{10 \pm \sqrt{100 + 60}}{2}$$
$$= \frac{10 \pm \sqrt{160}}{2}$$
$$= \frac{10 \pm 4\sqrt{10}}{2}$$
$$= \frac{2(5 \pm 2\sqrt{10})}{2}$$
$$= 5 \pm 2\sqrt{10}$$

$\{5 + 2\sqrt{10}, 5 - 2\sqrt{10}\}$

13. $10y^2 + y - 5 = 0$
$a = 10$, $b = 1$, $c = -5$

$$y = \frac{-b \pm \sqrt{b^2 - 4ac}}{2a}$$
$$= \frac{-1 \pm \sqrt{1^2 - 4(10)(-5)}}{2(10)}$$
$$= \frac{-1 \pm \sqrt{1 + 200}}{20}$$
$$= \frac{-1 \pm \sqrt{201}}{20}$$

$$\left\{\frac{-1 + \sqrt{201}}{20}, \frac{-1 - \sqrt{201}}{20}\right\}$$

14. $p^2 - 6p + 18 = 0$
$a = 1$, $b = -6$, $c = 18$

$$p = \frac{-b \pm \sqrt{b^2 - 4ac}}{2a}$$
$$= \frac{-(-6) \pm \sqrt{(-6)^2 - 4(1)(18)}}{2(1)}$$
$$= \frac{6 \pm \sqrt{36 - 72}}{2}$$
$$= \frac{6 \pm \sqrt{-36}}{2}$$

No real solution.

15. $$\frac{1}{4}x^2 + 2x - \frac{3}{4} = 0$$

$$x^2 + 8x - 3 = 0$$
$a = 1$, $b = 8$, $c = -3$

$$x = \frac{-b \pm \sqrt{b^2 - 4ac}}{2a}$$
$$= \frac{-8 \pm \sqrt{8^2 - 4(1)(-3)}}{2(1)}$$
$$= \frac{-8 \pm \sqrt{64 + 12}}{2}$$
$$= \frac{-8 \pm \sqrt{76}}{2}$$
$$= \frac{-8 \pm 2\sqrt{19}}{2}$$
$$= -4 \pm \sqrt{19}$$

$\{-4 + \sqrt{19}, -4 - \sqrt{19}\}$

16. $1 - 3x - x^2 = 0$
$-x^2 - 3x + 1 = 0$
$a = -1,\ b = -3,\ c = 1$

$$x = \frac{-b \pm \sqrt{b^2 - 4ac}}{2a}$$

$$= \frac{-(-3) \pm \sqrt{(-3)^2 - 4(-1)(1)}}{2(-1)}$$

$$= \frac{3 \pm \sqrt{9 + 4}}{-2}$$

$$= \frac{3 \pm \sqrt{13}}{-2} = \frac{-3 \pm \sqrt{13}}{2}$$

$$\left\{\frac{-3 + \sqrt{13}}{2}, \frac{-3 - \sqrt{13}}{2}\right\}$$

10.4 SOLUTIONS TO EXERCISES

1. $6x^2 + 11x + 5 = 0$
$(6x + 5)(x + 1) = 0$
$6x + 5 = 0$ or $x + 1 = 0$
$6x = -5$ or $x = -1$
$x = -\frac{5}{6}$

$$\left\{-1, -\frac{5}{6}\right\}$$

2. $b^2 = 44$
$b = \pm\sqrt{44}$
$b = \pm 2\sqrt{11}$

$\{-2\sqrt{11}, 2\sqrt{11}\}$

3. $y^2 + 5 = 3y$
$y^2 - 3y + 5 = 0$
$a = 1,\ b = -3,\ c = 5$

$$y = \frac{-b \pm \sqrt{b^2 - 4ac}}{2a}$$

$$= \frac{-(-3) \pm \sqrt{(-3)^2 - 4(1)(5)}}{2(1)}$$

$$= \frac{3 \pm \sqrt{9 - 20}}{2}$$

$$= \frac{3 \pm \sqrt{-11}}{2}$$

No real solution.

4. $16x^2 + 24x + 9 = 0$
$(4x + 3)^2 = 0$
$4x + 3 = \pm\sqrt{0}$
$4x + 3 = 0$
$4x = -3$
$x = -\frac{3}{4}$

$$\left\{-\frac{3}{4}\right\}$$

5. $d^2 - 7d = 2$
$d^2 - 7d - 2 = 0$
$a = 1,\ b = -7,\ c = -2$

$$d = \frac{-b \pm \sqrt{b^2 - 4ac}}{2a}$$

$$= \frac{-(-7) \pm \sqrt{(-7)^2 - 4(1)(-2)}}{2(1)}$$

$$= \frac{7 \pm \sqrt{49 + 8}}{2}$$

$$= \frac{7 \pm \sqrt{57}}{2}$$

$$\left\{\frac{7 + \sqrt{57}}{2}, \frac{7 - \sqrt{57}}{2}\right\}$$

6. $14x = x^2 - 4$

$0 = x^2 - 14x - 4$

$a = 1,\ b = -14,\ c = -4$

$$x = \frac{-b \pm \sqrt{b^2 - 4ac}}{2a}$$

$$= \frac{-(-14) \pm \sqrt{(-14)^2 - 4(1)(-4)}}{2(1)}$$

$$= \frac{14 \pm \sqrt{196 + 16}}{2}$$

$$= \frac{14 \pm \sqrt{212}}{2}$$

$$= \frac{14 \pm 2\sqrt{53}}{2}$$

$$= \frac{2(7 \pm \sqrt{53})}{2}$$

$$= 7 \pm \sqrt{53}$$

$\{7 + \sqrt{53},\ 7 - \sqrt{53}\}$

7. $(9b - 7)^2 = 48$

$9b - 7 = \pm\sqrt{48}$

$9b - 7 = \pm 4\sqrt{3}$

$9b = 7 \pm 4\sqrt{3}$

$$b = \frac{7 \pm 4\sqrt{3}}{9}$$

$$\left\{\frac{7 + 4\sqrt{3}}{9},\ \frac{7 - 4\sqrt{3}}{9}\right\}$$

8. $x^3 + x^2 - 6x = 0$

$x(x^2 + x - 6) = 0$

$x(x + 3)(x - 2) = 0$

$x = 0$ or $x + 3 = 0$ or $x - 2 = 0$

$x = -3$ or $x = 2$

$\{-3, 0, 2\}$

9. $y^2 - \frac{7}{4}y - \frac{1}{4} = 0$

$4y^2 - 7y - 1 = 0$

$a = 4,\ b = -7,\ c = -1$

$$y = \frac{-b \pm \sqrt{b^2 - 4ac}}{2a}$$

$$= \frac{-(-7) \pm \sqrt{(-7)^2 - 4(4)(-1)}}{2(4)}$$

$$= \frac{7 \pm \sqrt{49 + 16}}{8}$$

$$= \frac{7 \pm \sqrt{65}}{8}$$

$$\left\{\frac{7 + \sqrt{65}}{8},\ \frac{7 - \sqrt{65}}{8}\right\}$$

10. $(8 - z)^2 = 16$

$8 - z = \pm\sqrt{16}$

$8 - z = \pm 4$

$-z = -8 \pm 4$

$z = 8 \pm 4$

$z = 8 + 4$ or $z = 8 - 4$

$z = 12$ or $z = 4$

$\{4, 12\}$

11. $8x^2 + 7x - 6 = 0$

$a = 8,\ b = 7,\ c = -6$

$$x = \frac{-b \pm \sqrt{b^2 - 4ac}}{2a}$$

$$= \frac{-7 \pm \sqrt{7^2 - 4(8)(-6)}}{2(8)}$$

$$= \frac{-7 \pm \sqrt{49 + 192}}{16}$$

$$= \frac{-7 \pm \sqrt{241}}{16}$$

$$\left\{\frac{-7 + \sqrt{241}}{16},\ \frac{-7 - \sqrt{241}}{16}\right\}$$

12. $2b^2 - 200 = 0$
$$2b^2 = 200$$
$$b^2 = 100$$
$$b = \pm\sqrt{100}$$
$$b = \pm 10$$

$\{-10, 10\}$

13. $\frac{2}{3}x^2 - \frac{5}{6}x + 1 = 0$

$$4x^2 - 5x + 6 = 0$$
$$a = 4, b = -5, c = 6$$

$$x = \frac{-b \pm \sqrt{b^2 - 4ac}}{2a}$$

$$= \frac{-(-5) \pm \sqrt{(-5)^2 - 4(4)(6)}}{2(4)}$$

$$= \frac{5 \pm \sqrt{25 - 96}}{8}$$

$$= \frac{5 \pm \sqrt{-71}}{8}$$

No real solution.

14. $\frac{3}{2}y^2 - \frac{1}{2}y - 1 = 0$

$$3y^2 - y - 2 = 0$$
$$(3y + 2)(y - 1) = 0$$
$3y + 2 = 0$ or $y - 1 = 0$
$3y = -2$ or $y = 1$

$$y = -\frac{2}{3}$$

$\left\{-\frac{2}{3}, 1\right\}$

15. $x^2 + 4 = 16x$
$$x^2 - 16x + 4 = 0$$
$$a = 1, b = -16, c = 4$$

$$x = \frac{-b \pm \sqrt{b^2 - 4ac}}{2a}$$

$$= \frac{-(-16) \pm \sqrt{(-16)^2 - 4(1)(4)}}{2(1)}$$

$$= \frac{16 \pm \sqrt{256 - 16}}{2}$$

$$= \frac{16 \pm \sqrt{240}}{2}$$

$$= \frac{16 \pm 4\sqrt{15}}{2}$$

$$= \frac{2\left(8 \pm 2\sqrt{15}\right)}{2}$$

$$= 8 \pm 2\sqrt{15}$$

$\left\{8 + 2\sqrt{15}, 8 - 2\sqrt{15}\right\}$

16. $m^2 - 10m = -25$
$$m^2 - 10m + 25 = 0$$
$$(m - 5)^2 = 0$$
$$m - 5 = \pm\sqrt{0}$$
$$m - 5 = 0$$
$$m = 5$$

$\{5\}$

10.5 SOLUTIONS TO EXERCISES

1. $\sqrt{-144} = \sqrt{-1 \cdot 144} = \sqrt{-1} \cdot \sqrt{144} = i \cdot 12 = 12i$

2. $\sqrt{-75} = \sqrt{-1 \cdot 75} = \sqrt{-1} \cdot \sqrt{75} = i \cdot 5\sqrt{3} = 5i\sqrt{3}$

3. $\sqrt{-19} = \sqrt{-1 \cdot 19} = \sqrt{-1} \cdot \sqrt{19} = i\sqrt{19}$

4. $(3 + i) + (-4 + 9i) = (3 - 4) + (i + 9i)$
$$= -1 + 10i$$

5. $(7 - 6i) - (-10 + 6i)$
$= 7 - 6i + 10 - 6i$
$= (7 + 10) + (-6i - 6i)$
$= 17 - 12i$

6. $3i(5 - 9i) = 15i - 27i^2$
$= 15i - 27(-1)$
$= 15i + 27$
$= 27 + 15i$

7. $(8 + 7i)(5 + 2i)$
$= 40 + 16i + 35i + 14i^2$
$= 40 + 51i + 14(-1)$
$= 40 + 51i - 14$
$= 26 + 51i$

8. $\frac{6 - 5i}{2i} = \frac{6 - 5i}{2i} \cdot \frac{-2i}{-2i}$

$= \frac{-12i + 10i^2}{-4i^2}$

$= \frac{-12i + 10(-1)}{-4(-1)}$

$= \frac{-12i - 10}{4}$

$= -\frac{12}{4}i - \frac{10}{4}$

$= -3i - \frac{5}{2} = -\frac{5}{2} - 3i$

9. $\frac{3 + 4i}{8 - i} = \frac{3 + 4i}{8 - i} \cdot \frac{8 + i}{8 + i}$

$= \frac{24 + 3i + 32i + 4i^2}{64 - i^2}$

$= \frac{24 + 35i + 4(-1)}{64 - (-1)}$

$= \frac{24 + 35i - 4}{65}$

$= \frac{20 + 35i}{65}$

$= \frac{20}{65} + \frac{35}{65}i$

$= \frac{4}{13} + \frac{7}{13}i$

10. $(x - 5)^2 = -36$
$x - 5 = \pm\sqrt{-36}$

$x - 5 = \pm 6i$
$x = 5 \pm 6i$

$\{5 - 6i, 5 + 6i\}$

11. $(3y + 7)^2 = -24$
$3y + 7 = \pm\sqrt{-24}$
$3y + 7 = \pm 2i\sqrt{6}$
$3y = -7 \pm 2i\sqrt{6}$

$y = \frac{-7 \pm 2i\sqrt{6}}{3}$

$\left\{\frac{-7 + 2i\sqrt{6}}{3}, \frac{-7 - 2i\sqrt{6}}{3}\right\}$

12. $z^2 + 2z + 6 = 0$
$a = 1, b = 2, c = 6$

$z = \frac{-b \pm \sqrt{b^2 - 4ac}}{2a}$

$= \frac{-2 \pm \sqrt{2^2 - 4(1)(6)}}{2(1)}$

$= \frac{-2 \pm \sqrt{4 - 24}}{2}$

$$= \frac{-2 \pm \sqrt{-20}}{2}$$

$$= \frac{-2 \pm 2i\sqrt{5}}{2}$$

$$= \frac{2(-1 \pm i\sqrt{5})}{2}$$

$$= -1 \pm i\sqrt{5}$$

$$\{-1 + i\sqrt{5}, -1 - i\sqrt{5}\}$$

13. $3d^2 - d + 5 = 0$
$a = 3,\ b = -1,\ c = 5$

$$d = \frac{-b \pm \sqrt{b^2 - 4ac}}{2a}$$

$$= \frac{-(-1) \pm \sqrt{(-1)^2 - 4(3)(5)}}{2(3)}$$

$$= \frac{1 \pm \sqrt{1 - 60}}{6}$$

$$= \frac{1 \pm \sqrt{-59}}{6}$$

$$= \frac{1 \pm i\sqrt{59}}{6}$$

$$\left\{\frac{1 + i\sqrt{59}}{6}, \frac{1 - i\sqrt{59}}{6}\right\}$$

14. $9x^2 + 49 = 0$

$$9x^2 = -49$$

$$x^2 = \frac{-49}{9}$$

$$x = \pm\sqrt{-\frac{49}{9}}$$

$$x = \pm\frac{7}{3}i$$

$$\left\{-\frac{7}{3}i, \frac{7}{3}i\right\}$$

15. $y^2 + 4y + 8 = 0$
$a = 1,\ b = 4,\ c = 8$

$$y = \frac{-b \pm \sqrt{b^2 - 4ac}}{2a}$$

$$= \frac{-4 \pm \sqrt{4^2 - 4(1)(8)}}{2(1)}$$

$$= \frac{-4 \pm \sqrt{16 - 32}}{2}$$

$$= \frac{-4 \pm \sqrt{-16}}{2}$$

$$= \frac{-4 \pm 4i}{2}$$

$$= -\frac{4}{2} \pm \frac{4}{2}i = -2 \pm 2i$$

$$\{-2 + 2i, -2 - 2i\}$$

10.6 SOLUTIONS TO EXERCISES

1. $y = 3x^2$

x	y
-1	$3(-1)^2 = 3$
0	$3(0)^2 = 0$
1	$3(1)^2 = 3$

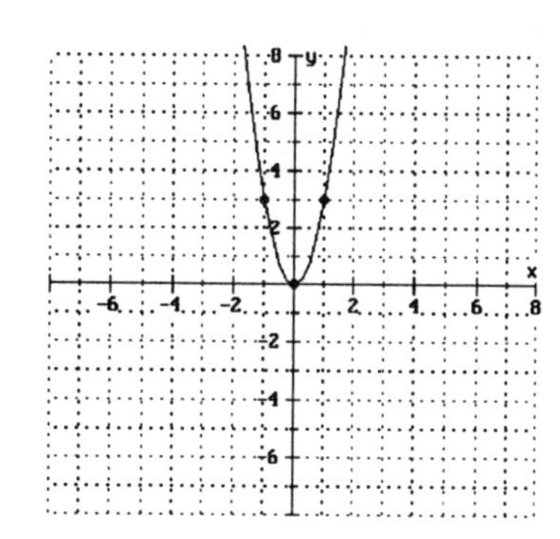

Vertex, x- and y-intercept: (0, 0)

2. $y = -3x^2$

x	y
-1	$-3(-1)^2 = -3$
0	$-3(0)^2 = 0$
1	$-3(1)^2 = -3$

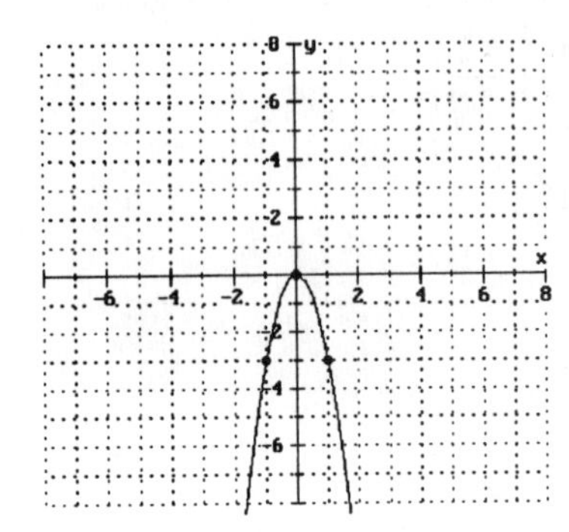

Vertex, x- and y-intercept:
(0, 0)

3. $y = \frac{1}{2}x^2$

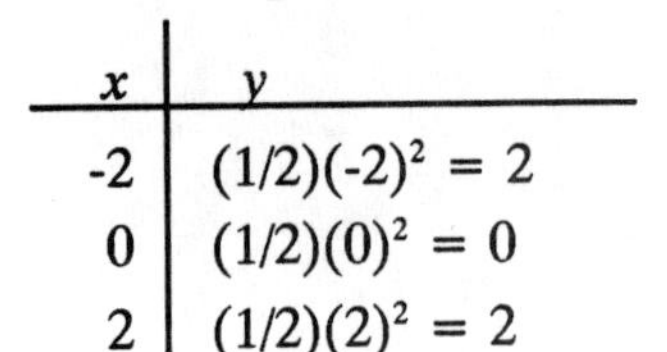

x	y
-2	$(1/2)(-2)^2 = 2$
0	$(1/2)(0)^2 = 0$
2	$(1/2)(2)^2 = 2$

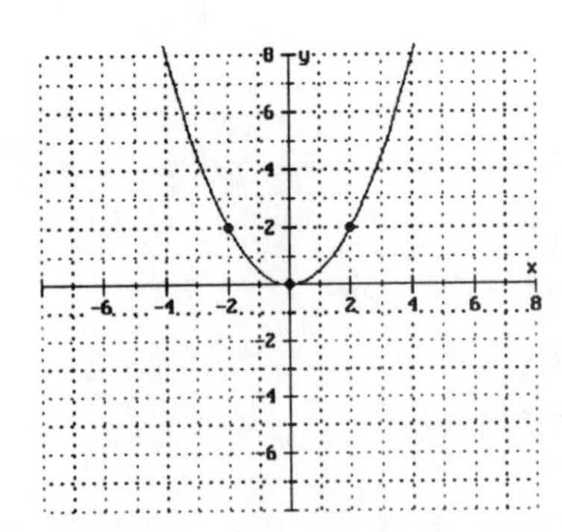

Vertex, x- and y-intercept:
(0, 0)

4. $y = (x - 2)^2$

Vertex: (2, 0)
y-intercept: Let $x = 0$
$y = (0 - 2)^2 = 4$

x-intercepts: Let $y = 0$

$$0 = (x - 2)^2$$
$$\pm\sqrt{0} = x - 2$$
$$0 = x - 2$$
$$2 = x$$

x	y
1	1
3	1

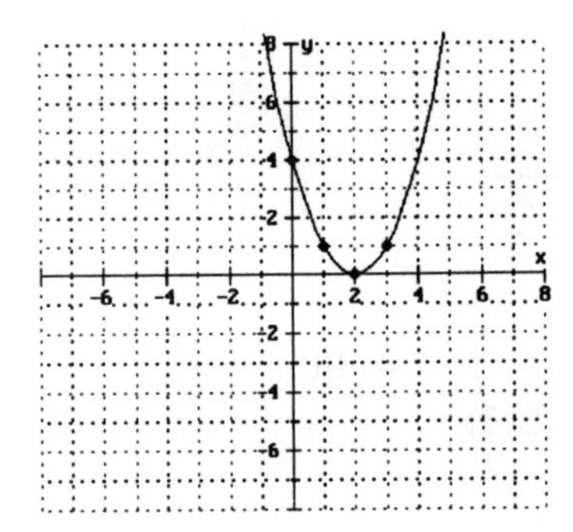

5. $y = -x^2 + 3$

Vertex: $x = \frac{-b}{2a} = \frac{-0}{2(-1)} = 0$

$y = -0^2 + 3 = 3$
(0, 3) (this is also the y-intercept)

x-intercepts: Let $y = 0$

$$0 = -x^2 + 3$$
$$x^2 = 3$$
$$x = \pm\sqrt{3}$$

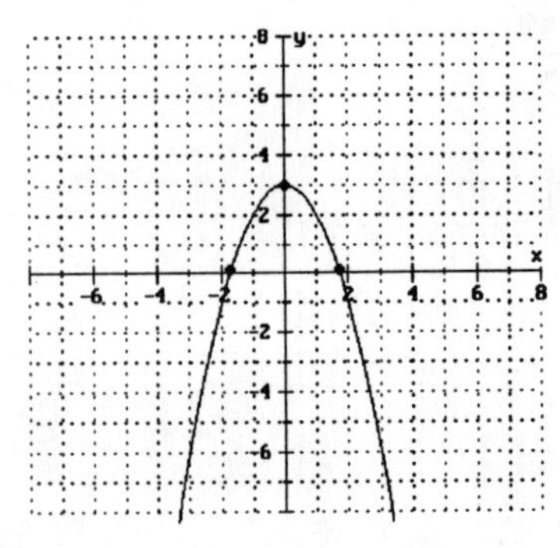

6. $y = 9 - x^2$

Vertex: $x = \frac{-b}{2a} = \frac{-0}{2(-1)} = 0$

$y = 9 - 0^2 = 9$
(0, 9) (this is also the y-intercept)

x-intercepts: Let $y = 0$
$0 = 9 - x^2$
$x^2 = 9$
$x = \pm\sqrt{9} = \pm 3$

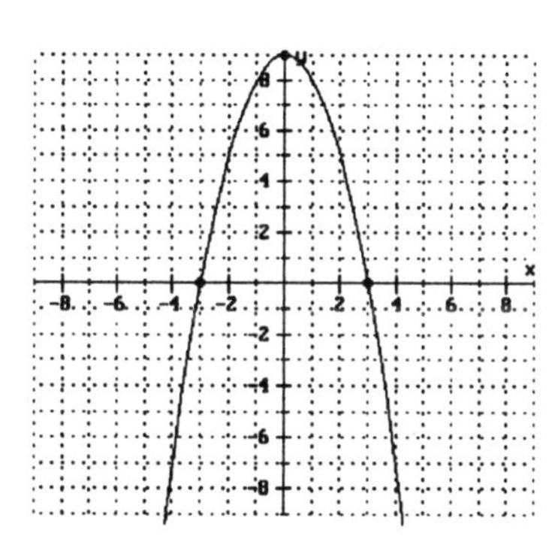

7. $y = -2(x + 1)^2$

Vertex: $(-1, 0)$ (this is also the x-intercept)

y-intercept: Let $x = 0$
$y = -2(0 + 1)^2 = -2$

x	y
-2	-2

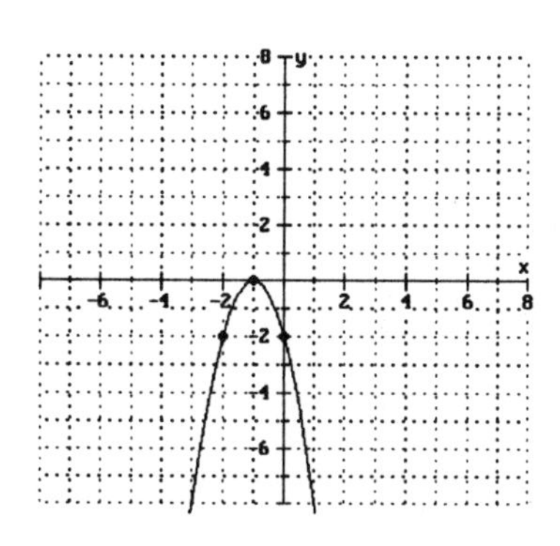

8. $y = -x^2 - 5$

Vertex: $x = \frac{-b}{2a} = \frac{-0}{2(-1)} = 0$

$y = -0^2 - 5 = -5$
(0, -5) (this is also the y-intercept)

x-intercepts: Let $y = 0$
$0 = -x^2 - 5$
$x^2 = -5$
No x-intercepts

x	y
-1	-6
1	-6

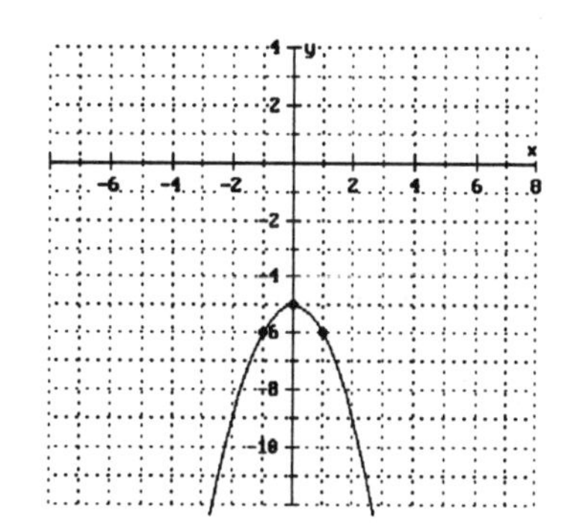

9. $y = 2(x - 3)^2 - 8$

Vertex: $(3, -8)$

y-intercept: Let $x = 0$
$y = 2(0 - 3)^2 - 8 = 10$

x-intercepts: Let $y = 0$
$0 = 2(x - 3)^2 - 8$
$8 = 2(x - 3)^2$
$4 = (x - 3)^2$
$\pm\sqrt{4} = x - 3$
$\pm 2 = x - 3$
$3 \pm 2 = x$
$x = 3 + 2 = 5$ or $x = 3 - 2 = 1$

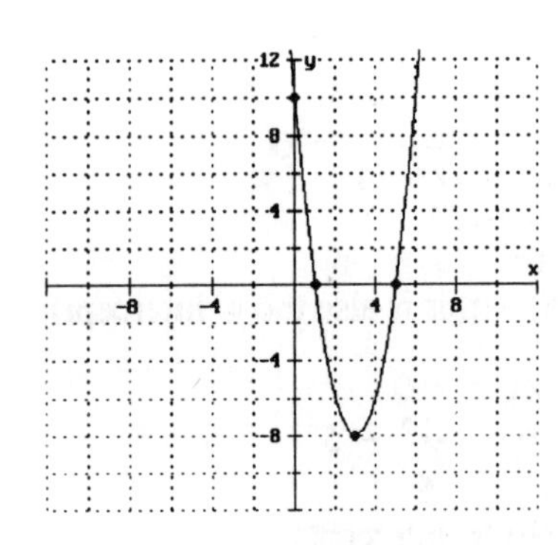

10. $y = -(x - 3)^2 + 1$

Vertex: (3, 1)

y-intercept: Let $x = 0$

$$y = -(0 - 3)^2 + 1 = -8$$

x-intercepts: Let $y = 0$

$$0 = -(x - 3)^2 + 1$$
$$(x - 3)^2 = 1$$
$$x - 3 = \pm\sqrt{1}$$
$$x - 3 = \pm 1$$
$$x = 3 \pm 1$$
$$x = 3 + 1 = 4 \quad \text{or} \quad x = 3 - 1 = 2$$

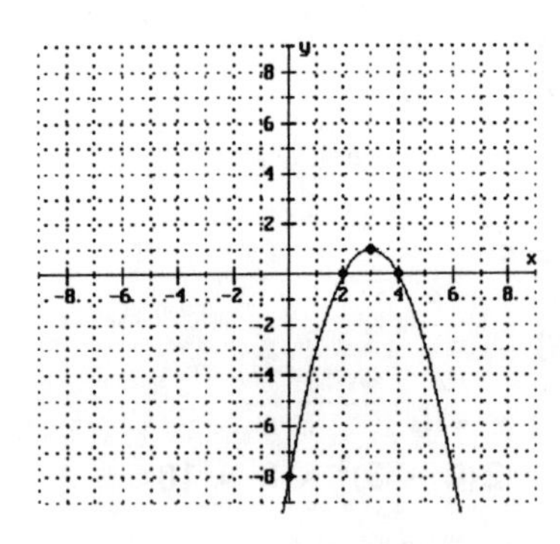

11. $y = \frac{1}{2}(x + 2)^2 + 2$

Vertex: (-2, 2)

y-intercept: Let $x = 0$

$$y = \frac{1}{2}(0 + 2)^2 + 2 = 4$$

x-intercepts: Let $y = 0$

$$0 = \frac{1}{2}(x + 2)^2 + 2$$

$$0 = (x + 2)^2 + 4$$
$$-4 = (x + 2)^2$$

No x-intercepts.

x	y
-4	4

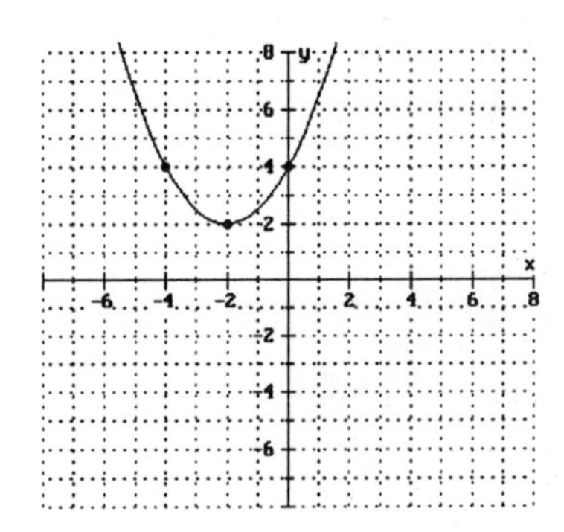

12. $y = -\frac{1}{9}(x - 6)^2 - 2$

Vertex: (6, -2)

y-intercept: Let $x = 0$

$$y = -\frac{1}{9}(0 - 6)^2 - 2 = -6$$

x-intercepts: Let $y = 0$

$$0 = -\frac{1}{9}(x - 6)^2 - 2$$

$$\frac{1}{9}(x - 6)^2 = -2$$

No x-intercepts.

x	y
12	-6

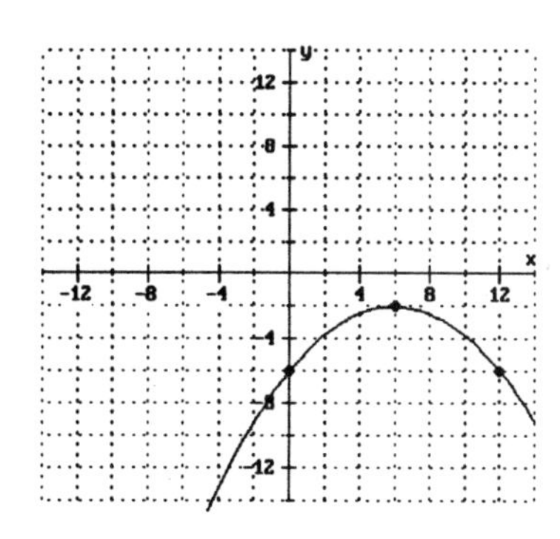

13. $y = x^2 + 4x$

Vertex: $x = \dfrac{-b}{2a} = \dfrac{-4}{2(1)} = -2$

$y = (-2)^2 + 4(-2) = -4$
$(-2, -4)$

y-intercept: Let $x = 0$
$y = 0^2 + 4(0) = 0$

x-intercepts: Let $y = 0$
$0 = x^2 + 4x$
$0 = x(x + 4)$
$x = 0$ or $x + 4 = 0$
$x = -4$

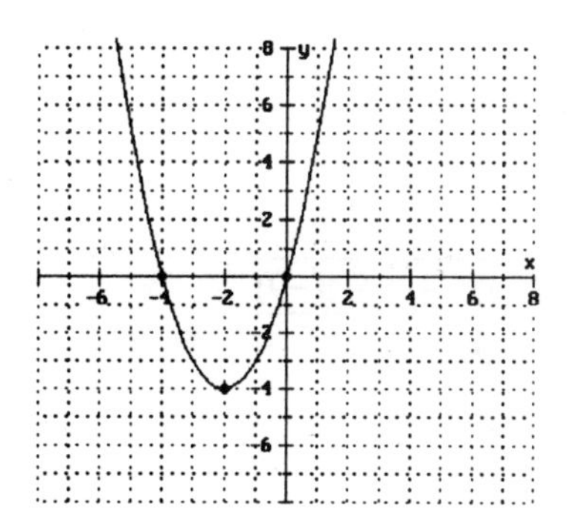

14. $y = x^2 - 6x$

Vertex: $x = \dfrac{-b}{2a} = \dfrac{-(-6)}{2(1)} = 3$

$y = 3^2 - 6(3) = -9$
$(3, -9)$

y-intercept: Let $x = 0$
$y = 0^2 - 6(0) = 0$

x-intercepts: Let $y = 0$
$0 = x^2 - 6x$
$0 = x(x - 6)$
$x = 0$ or $x - 6 = 0$
$x = 6$

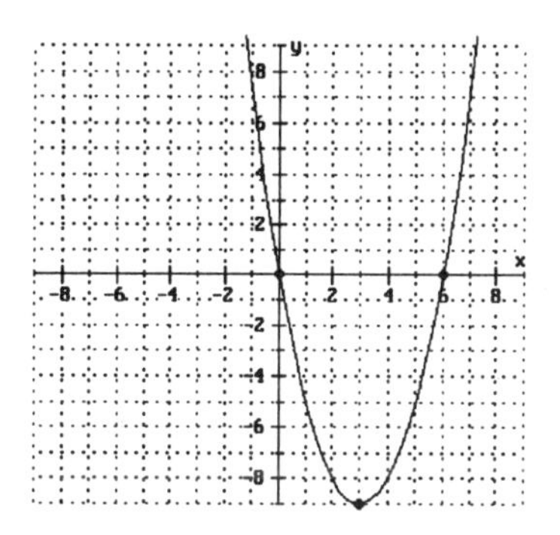

15. $y = -x^2 + 4x - 3$

Vertex: $x = \dfrac{-b}{2a} = \dfrac{-4}{2(-1)} = 2$

$y = -2^2 + 4(2) - 3 = 1$
$(2, 1)$

y-intercept: Let $x = 0$
$y = -0^2 + 4(0) - 3 = -3$

x-intercepts: Let $y = 0$
$0 = -x^2 + 4x - 3$
$0 = x^2 - 4x + 3$
$0 = (x - 3)(x - 1)$
$x - 3 = 0$ or $x - 1 = 0$
$x = 3$ or $x = 1$

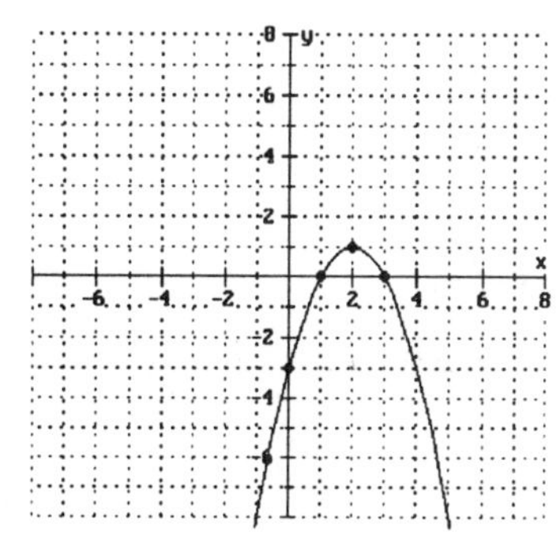

CHAPTER 10 PRACTICE TEST SOLUTIONS

1. $3x^2 - x = 10$

$3x^2 - x - 10 = 0$

$(3x + 5)(x - 2) = 0$

$3x + 5 = 0$ or $x - 2 = 0$

$3x = -5$ or $x = 2$

$x = -\frac{5}{3}$

$\left\{-\frac{5}{3}, 2\right\}$

2. $x^5 - 7x^4 + 12x^3 = 0$

$x^3(x^2 - 7x + 12) = 0$

$x^3(x - 4)(x - 3) = 0$

$x^3 = 0$ or $x - 4 = 0$ or $x - 3 = 0$

$x = 0$ or $x = 4$ or $x = 3$

$\{0, 4, 3\}$

3. $6k^2 = 150$

$k^2 = 25$

$k = \pm\sqrt{25}$

$k = \pm 5$

$\{\pm 5\}$

4. $(2m - 7)^2 = 5$

$2m - 7 = \pm\sqrt{5}$

$2m = 7 \pm \sqrt{5}$

$m = \frac{7 \pm \sqrt{5}}{2}$

$\left\{\frac{7 \pm \sqrt{5}}{2}\right\}$

5. $x^2 - 14x + 48 = 0$

$x^2 - 14x = -48$

$x^2 - 14x + 49 = -48 + 49$

$(x - 7)^2 = 1$

$x - 7 = \pm\sqrt{1}$

$x - 7 = \pm 1$

$x = 7 \pm 1$

$x = 7 + 1 = 8$ or $x = 7 - 1 = 6$

$\{8, 6\}$

6. $3x^2 + 5x = 4$

$x^2 + \frac{5}{3}x = \frac{4}{3}$

$x^2 + \frac{5}{3}x + \frac{25}{36} = \frac{4}{3} + \frac{25}{36}$

$\left(x + \frac{5}{6}\right)^2 = \frac{73}{36}$

$x + \frac{5}{6} = \pm\sqrt{\frac{73}{36}}$

$x + \frac{5}{6} = \pm\frac{\sqrt{73}}{6}$

$x = -\frac{5}{6} \pm \frac{\sqrt{73}}{6}$

$x = \frac{-5 \pm \sqrt{73}}{6}$

$\left\{\frac{-5 \pm \sqrt{73}}{6}\right\}$

7. $x^2 - 4x - 9 = 0$

$a = 1, b = -4, c = -9$

$x = \frac{-b \pm \sqrt{b^2 - 4ac}}{2a}$

$= \frac{-(-4) \pm \sqrt{(-4)^2 - 4(1)(-9)}}{2(1)}$

$= \frac{4 \pm \sqrt{16 + 36}}{2}$

$= \frac{4 \pm \sqrt{52}}{2}$

$$= \frac{4 \pm 2\sqrt{13}}{2}$$

$$= \frac{2(2 \pm \sqrt{13})}{2}$$

$$= 2 \pm \sqrt{13}$$

$$\{2 \pm \sqrt{13}\}$$

8. $p^2 - \frac{7}{6}p - \frac{1}{6} = 0$

$$6\left(p^2 - \frac{7}{6}p - \frac{1}{6}\right) = 6(0)$$

$$6p^2 - 7p - 1 = 0$$

$a = 6,\ b = -7,\ c = -1$

$$p = \frac{-b \pm \sqrt{b^2 - 4ac}}{2a}$$

$$= \frac{-(-7) \pm \sqrt{(-7)^2 - 4(6)(-1)}}{2(6)}$$

$$= \frac{7 \pm \sqrt{49 + 24}}{12}$$

$$= \frac{7 \pm \sqrt{73}}{12}$$

$$\left\{\frac{7 \pm \sqrt{73}}{12}\right\}$$

9. $(2x - 5)(x + 3) = -5$

$2x^2 + x - 15 = -5$

$2x^2 + x - 10 = 0$

$(2x + 5)(x - 2) = 0$

$2x + 5 = 0$ or $x - 2 = 0$

$2x = -5$ or $x = 2$

$x = -\frac{5}{2}$ or $x = 2$

$$\left\{-\frac{5}{2}, 2\right\}$$

10. $(7x + 6)^2 = 49$

$7x + 6 = \pm\sqrt{49}$

$7x + 6 = \pm 7$

$7x = -6 \pm 7$

$$x = \frac{-6 \pm 7}{7}$$

$x = \frac{-6 + 7}{7} = \frac{1}{7}$ or $x = \frac{-6 - 7}{7} = -\frac{13}{7}$

$$\left\{\frac{1}{7}, -\frac{13}{7}\right\}$$

11. $8x^3 = 2x$

$8x^3 - 2x = 0$

$2x(4x^2 - 1) = 0$

$2x(2x + 1)(2x - 1) = 0$

$2x = 0$ or $2x + 1 = 0$ or $2x - 1 = 0$

$x = 0$ or $2x = -1$ or $2x = 1$

$2x = -\frac{1}{2}$ or $x = \frac{1}{2}$

$$\left\{0, -\frac{1}{2}, \frac{1}{2}\right\}$$

12. $x^4 + x^3 - 56x^2 = 0$

$x^2(x^2 + x - 56) = 0$

$x^2(x + 8)(x - 7) = 0$

$x^2 = 0$ or $x + 8 = 0$ or $x - 7 = 0$

$x = 0$ or $x = -8$ or $x = 7$

$\{0, -8, 7\}$

13. $\sqrt{-81} = \sqrt{-1 \cdot 9^2}$

$= 9i$

14. $(9 - 7i) + (4 - i) = (9 + 4) + (-7 - 1)i$

$= 13 - 8i$

15. $(8 + 3i) + (8 - 3i) = (8 + 8) + (3 - 3)i$

$= 16 + 0i$

$= 16$

16. $(11 - 2i) - (-9 - 7i) = (11 - 2i) + (9 + 7i)$

$= (11 + 9) + (-2 + 7)i$

$= 20 + 5i$

17. $(1 + 3i)(5 - 4i) = 5 - 4i + 15i - 12i^2$
$= 5 + 11i - 12(-1)$
$= 17 + 11i$

18. $(2 - 5i)(2 + 5i) = 2^2 - (5i)^2$
$= 4 - 25i^2$
$= 4 - 25(-1)$
$= 29$

19. $\frac{4 - i}{5 + i} = \frac{4 - i}{5 + i} \cdot \frac{5 - i}{5 - i}$
$= \frac{20 - 4i - 5i + i^2}{5^2 - i^2}$
$= \frac{20 - 9i - 1}{25 - (-1)}$
$= \frac{19 - 9i}{26}$
$= \frac{19}{26} - \frac{9}{26}i$

20. $\frac{3 + 2i}{6 - 2i} = \frac{3 + 2i}{6 - 2i} \cdot \frac{6 + 2i}{6 + 2i}$
$= \frac{18 + 6i + 12i + 4i^2}{6^2 - (2i)^2}$
$= \frac{18 + 18i + 4(-1)}{36 - 4i^2}$
$= \frac{14 + 18i}{36 - 4(-1)}$
$= \frac{14 + 18i}{40}$
$= \frac{14}{40} + \frac{18}{40}i$
$= \frac{7}{20} + \frac{9}{20}i$

21. $y = (x - 2)^2 + 1$
Vertex: $(2, 1)$

Let $x = 0$: $y = (0 - 2)^2 + 1 = 5$
y-intercept: $(0, 5)$

Let $y = 0$: $0 = (x - 2)^2 + 1$
$-1 = (x - 2)^2$
$\pm\sqrt{-1} = x - 2$

No x-intercepts

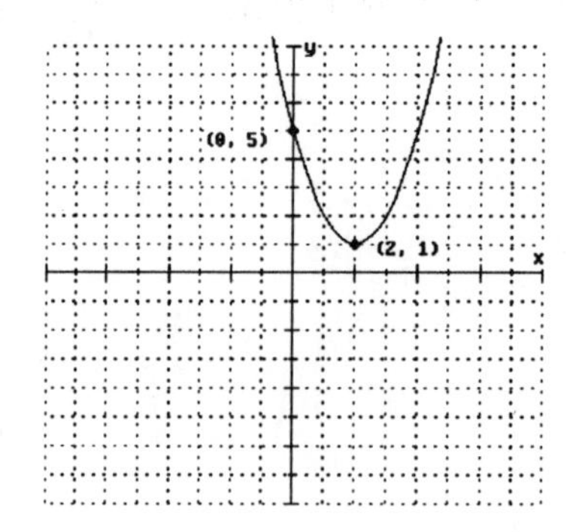

22. $y = x^2 - x - 12$

$y = \left(x^2 - x + \frac{1}{4}\right) - 12 - \frac{1}{4}$

$y = \left(x - \frac{1}{2}\right)^2 - \frac{49}{4}$

Vertex: $\left(\frac{1}{2}, -\frac{49}{4}\right)$

Let $x = 0$: $y = 0^2 - 0 - 12 = -12$
y-intercept: $(0, -12)$

Let $y = 0$: $0 = x^2 - x - 12$
$0 = (x - 4)(x + 3)$
$x - 4 = 0$ or $x + 3 = 0$
$x = 4$ or $x = -3$

x-intercepts: $(4, 0)$, $(-3, 0)$

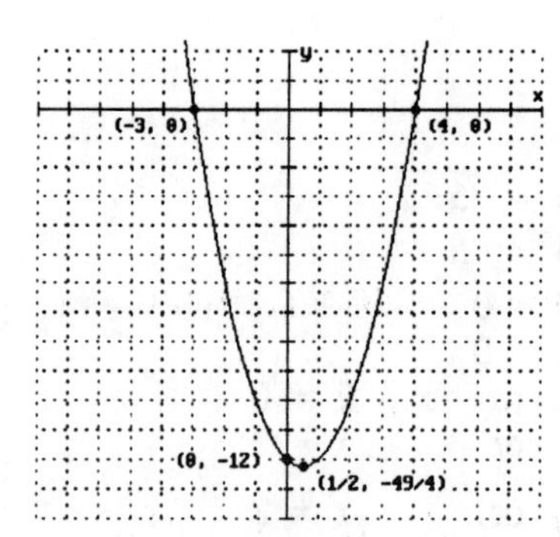

SOLUTIONS TO PRACTICE FINAL EXAMINATION

1. $-17 - (-6) = -17 + 6$
$= -11$

2. $-\frac{3}{8} + \frac{2}{5} = -\frac{15}{40} + \frac{16}{40}$
$= \frac{1}{40}$

3. $6[2 + 7(5 - 9) - 1]$
$= 6[2 + 7(-4) - 1]$
$= 6[2 - 28 - 1]$
$= 6(-27)$
$= -162$

4. $-2(-3)^2 - 24 = -2(9) - 24$
$= -18 - 24$
$= -42$

5. $x^2 - 2yz = (-2)^2 - 2(5)(3)$
$= 4 - 30$
$= -26$

6. $-9 < -5$

7. $|-10| = 10$

8. $|-4| = 4$
$3 - (-2) = 5$
$|-4| < 3 - (-2)$

9. Commutative property for multiplication.

10. Distributive property.

11. $2x + 8 - x - 15 = (2x - x) + (8 - 15)$
$= x - 7$

12. $-7(y - 2) - 3(4 - 6y)$
$= -7y + 14 - 12 + 18y$
$= (-7y + 18y) + (14 - 12)$
$= 11y + 2$

13. $-\frac{4}{7}x = 24$
$-\frac{7}{4} \cdot -\frac{4}{7}x = -\frac{7}{4} \cdot 24$
$x = -42$
$\{-42\}$

14. $2(y - 3) = -(4 - y)$
$2y - 6 = -4 + y$
$2y - 6 - y = -4 + y - y$
$y - 6 = -4$
$y - 6 + 6 = -4 + 6$
$y = 2$
$\{2\}$

15. $\frac{6(a + 1)}{7} = a - 3$
$7 \cdot \frac{6(a + 1)}{7} = 7(a - 3)$
$6(a + 1) = 7(a - 3)$
$6a + 6 = 7a - 21$
$6a + 6 - 6a = 7a - 21 - 6a$
$6 = a - 21$
$6 + 21 = a - 21 + 21$
$27 = a$
$\{27\}$

16. $\frac{1}{2} - b + \frac{5}{2} = b + 3$
$3 - b = b + 3$
$3 - b + b = b + 3 + b$
$3 = 2b + 3$
$3 - 3 = 2b + 3 - 3$
$0 = 2b$
$\frac{0}{2} = \frac{2b}{2}$
$0 = b$
$\{0\}$

17. x = unknown number

$$x + \frac{2}{3}x = 25$$

$$3\left(x + \frac{2}{3}x\right) = 3(25)$$

$$3x + 2x = 75$$
$$5x = 75$$

$$\frac{5x}{5} = \frac{75}{5}$$

$$x = 15$$

The number is 15.

18.

	rate	· time	= distance
train A	80	t	$80t$
train B	92	t	$92t$

train A's distance + train B's distance = 516

$$80t + 92t = 516$$
$$172t = 516$$
$$t = 3$$

It will take 3 hours.

19.

$$6x - 7y = 14$$
$$6x - 7y - 6x = 14 - 6x$$
$$-7y = 14 - 6x$$

$$\frac{-7y}{-7} = \frac{14 - 6x}{-7}$$

$$y = \frac{14 - 6x}{-7} = \frac{-14 + 6x}{7}$$

20.

$$3x - 8 \le 6x + 16$$
$$3x - 8 - 3x \le 6x + 16 - 3x$$
$$-8 \le 3x + 16$$
$$-8 - 16 \le 3x + 16 - 16$$
$$-24 \le 3x$$

$$\frac{-24}{3} \le \frac{3x}{3}$$
$$-8 \le x$$

21.

$$5x - y = -7$$
$$5(-2) - 3 = -7$$
$$-13 = -7$$

False

It is not a solution.

22.

$$10x - 3y = 55$$
$$10(4) - 3y = 55$$
$$40 - 3y = 55$$
$$-3y = 15$$
$$y = -5$$

(4, -5)

23. Starting at the point (0, 1) on the line, you move up 4 and to the right 5 to get to another point on the line, (5, 5). Hence the slope is $\frac{4}{5}$.

24.

$$m = \frac{y_2 - y_1}{x_2 - x_1}$$
$$= \frac{-9 - 12}{7 - 1}$$
$$= \frac{-21}{6} = -\frac{7}{2}$$

25.

$$4x - y = 13$$
$$-y = -4x + 13$$
$$y = 4x - 13$$
$$m = 4$$

26. (3, 6) and (-1, 4)

$$m_1 = \frac{6 - 4}{3 - (-1)} = \frac{2}{4} = \frac{1}{2}$$

(8, 3) and (9, 1)

$$m_2 = \frac{3 - 1}{8 - 9} = \frac{2}{-1} = -2$$

The lines are perpendicular since $m_1 = -\frac{1}{m_2}$.

27. $2x + y = 4$

x	y
0	4
2	0
1	2

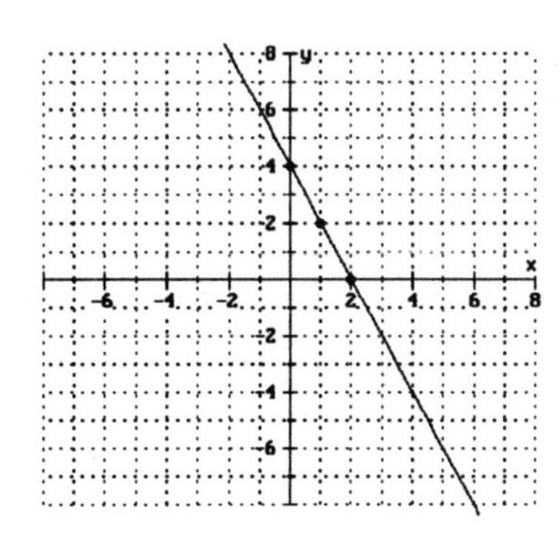

28. $y \le -3x$ (solid line)

$y = -3x$

x	y
-1	3
0	0
1	-3

Test point: (1, 0)

$y \le -3x$
$0 \le -3(1)$
$0 \le -3$
False

Shade the half-plane not containing (1, 0).

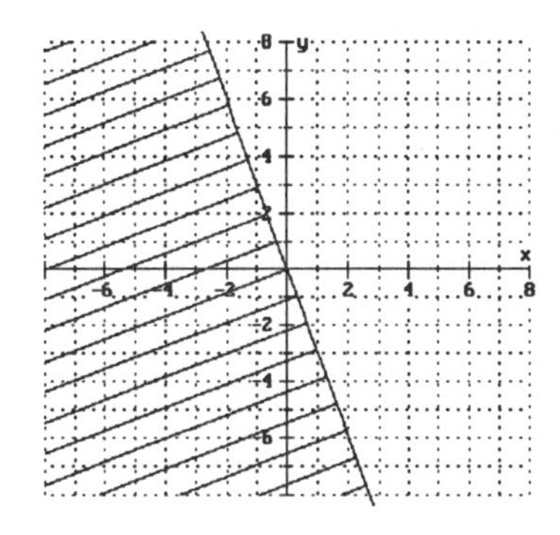

29. $x + 6 = 0$

$x = -6$

Vertical line through (-6, 0).

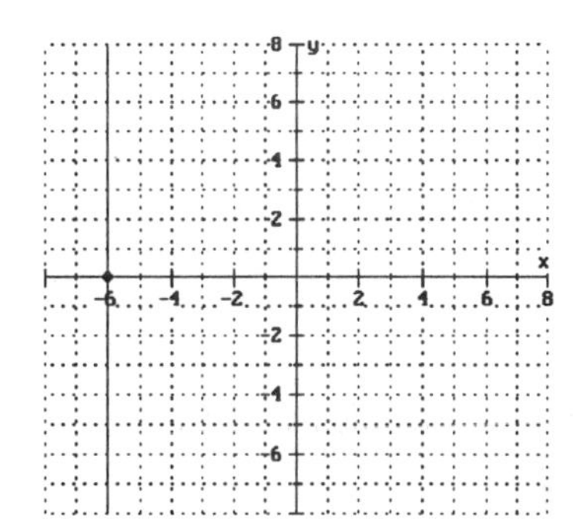

30. $x = 5 + 2y$

x	y
5	0
3	-1
1	-2

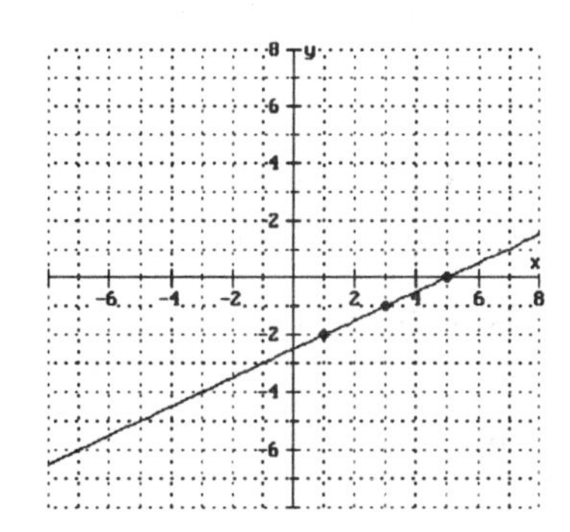

31. $-2^4 = -1 \cdot 2^4$
$= -1 \cdot 16$
$= -16$

32. $\left(\dfrac{8x^6y^2}{14xy^7}\right)^3 = \left(\dfrac{4}{7}x^{6-1}y^{2-7}\right)^3$

$= \left(\dfrac{4}{7}x^5y^{-5}\right)^3$

$= \left(\dfrac{4}{7}\right)^3 x^{(5)(3)}y^{(-5)(3)}$

$= \dfrac{64}{343}x^{15}y^{-15}$

$= \dfrac{64x^{15}}{343y^{15}}$

33. $\dfrac{4^2x^{-10}y^8}{4^5x^{-2}y^{-6}} = 4^{2-5}x^{-10-(-2)}y^{8-(-6)}$

$= 4^{-3}x^{-8}y^{14}$

$= \dfrac{y^{14}}{4^3x^8} = \dfrac{y^{14}}{64x^8}$

34. $4_{\wedge}6500000. = 4.65 \times 10^7$

35. $3.81 \times 10^{-6} = 0_{\wedge}000003.81 \times 10^{-6}$

$= 0.00000381$

36. The degree of each term is:

$2x^4y$: $4 + 1 = 5$
$10x^2y^5$: $2 + 5 = 7$
x^3y^6: $3 + 6 = 9$

The largest of these degrees is the degree of the polynomial. Hence the polynomial is of degree 9.

37. $(2x^4 - 3x^2 + 5x - 11) + (4x^3 + 2x^2 - 6x + 9)$
$= 2x^4 + 4x^3 + (-3x^2 + 2x^2) + (5x - 6x) + (-11 + 9)$
$= 2x^4 + 4x^3 - x^2 - x - 2$

38. $(4x - 1)(3x + 5)$
$= (4x)(3x) + (4x)(5) + (-1)(3x) + (-1)(5)$
$= 12x^2 + 20x - 3x - 5$
$= 12x^2 + 17x - 5$

39. $(2x - 3y)^2 = (2x)^2 - 2(2x)(3y) + (3y)^2$
$= 4x^2 - 12xy + 9y^2$

40.
$$\begin{array}{r} x + 10 \\ x - 2\overline{)x^2 + 8x - 7} \\ \underline{x^2 - 2x} \\ 10x - 7 \\ \underline{10x - 20} \\ 13 \end{array}$$

$x + 10 + \dfrac{13}{x - 2}$

41. $3a^2b - 6a^3b^2 + 9a^5b^3$
$= 3a^2b(1 - 2ab + 3a^3b^2)$

42.

Negative factors of 7	Sum
−1, −7	−8

$x^2 - 8x + 7 = (x - 1)(x - 7)$

43.

Negative factors of 12	Sum
−1, −12	−13
−2, −6	−8
−3, −4	−7

$x^2 - 4x + 12$ is prime since none of the negative factors of 12 have a sum of −4.

44. $3y^2 - 108 = 3(y^2 - 36)$
$= 3(y^2 - 6^2)$
$= 3(y + 6)(y - 6)$

45.

Factors of -15	Sum
1, -15	-14
-1, 15	14
3, -5	-2
-3, 5	2

$x^2 - 2xy - 15y^2 = (x + 3y)(x - 5y)$

46. $8a^3 + 64b^3 = 8(a^3 + 8b^3)$
$= 8[a^3 + (2b)^3]$
$= 8(a + 2b)[a^2 - a(2b) + (2b)^2]$
$= 8(a + 2b)(a^2 - 2ab + 4b^2)$

47. $m^2n + 6m^2 - 4n - 24$
$= m^2(n + 6) - 4(n + 6)$
$= (n + 6)(m^2 - 4)$
$= (n + 6)(m^2 - 2^2)$
$= (n + 6)(m + 2)(m - 2)$

48. $x^2 + 6xy + 9y^2$
$= x^2 + 2(x)(3y) + (3y)^2$
$= (x + 3y)^2$

49. $2x^2 + 19x = -24$
$2x^2 + 19x + 24 = 0$
$(2x + 3)(x + 8) = 0$
$2x + 3 = 0$ or $x + 8 = 0$
$2x = -3$ or $x = -8$
$x = -\frac{3}{2}$

$\left\{-\frac{3}{2}, -8\right\}$

50. length $= x$
width $= x - 5$

Area = length · width
$234 = x(x - 5)$
$234 = x^2 - 5x$
$0 = x^2 - 5x - 234$
$0 = (x - 18)(x + 13)$
$x - 18 = 0$ or $x + 13 = 0$
$x = 18$ or $x = -13$

Since x represents a length, it must be positive.
$x = 18$
$x - 5 = 18 - 5 = 13$

The dimensions are 18 ft by 13 ft.

51. $$\frac{x - 9}{x^2 - 81} = \frac{x - 9}{(x + 9)(x - 9)}$$
$$= \frac{1}{x + 9}$$

52. $$\frac{x^2 - x - 2}{x^2 + 4x + 3} \cdot \frac{x^2 - x - 12}{x^2 + 3x - 10}$$
$$= \frac{(x - 2)(x + 1)}{(x + 3)(x + 1)} \cdot \frac{(x - 4)(x + 3)}{(x + 5)(x - 2)}$$
$$= \frac{(x - 2)(x + 1)(x - 4)(x + 3)}{(x + 3)(x + 1)(x + 5)(x - 2)}$$
$$= \frac{x - 4}{x + 5}$$

53. $$\frac{5}{x - 4} \div \frac{10}{x^2 - 16} = \frac{5}{x - 4} \cdot \frac{x^2 - 16}{10}$$
$$= \frac{5}{x - 4} \cdot \frac{(x + 4)(x - 4)}{10}$$
$$= \frac{5(x + 4)(x - 4)}{10(x - 4)}$$
$$= \frac{x + 4}{2}$$

54. $$\frac{7}{x^2 - 36} + \frac{4}{x + 6}$$
$$= \frac{7}{(x + 6)(x - 6)} + \frac{4}{x + 6}$$
$$= \frac{7}{(x + 6)(x - 6)} + \frac{4(x - 6)}{(x + 6)(x - 6)}$$

$$= \frac{7 + 4(x - 6)}{(x + 6)(x - 6)}$$

$$= \frac{7 + 4x - 24}{(x + 6)(x - 6)}$$

$$= \frac{4x - 17}{(x + 6)(x - 6)}$$

55. $$\frac{3y}{y^2 + 7y + 6} - \frac{4}{y^2 - 2y - 3}$$

$$= \frac{3y}{(y + 6)(y + 1)} - \frac{4}{(y - 3)(y + 1)}$$

$$= \frac{3y(y - 3)}{(y + 6)(y + 1)(y - 3)} - \frac{4(y + 6)}{(y - 3)(y + 1)(y + 6)}$$

$$= \frac{3y(y - 3) - 4(y + 6)}{(y + 6)(y + 1)(y - 3)}$$

$$= \frac{3y^2 - 9y - 4y - 24}{(y + 6)(y + 1)(y - 3)}$$

$$= \frac{3y^2 - 13y - 24}{(y + 6)(y + 1)(y - 3)}$$

56. $$\frac{2}{a} - \frac{3}{4} = \frac{-1}{8}$$

$$8a\left(\frac{2}{a} - \frac{3}{4}\right) = 8a\left(\frac{-1}{8}\right)$$

$$8a\left(\frac{2}{a}\right) - 8a\left(\frac{3}{4}\right) = 8a\left(\frac{-1}{8}\right)$$

$$16 - 6a = -a$$

$$16 = 5a$$

$$\frac{16}{5} = a$$

$$\left\{\frac{16}{5}\right\}$$

57.

$$\frac{4}{x^2 - 9} = \frac{3}{x + 3} + \frac{6}{x - 3}$$

$$\frac{4}{(x + 3)(x - 3)} = \frac{3}{x + 3} + \frac{6}{x - 3}$$

$$(x + 3)(x - 3)\left[\frac{4}{(x + 3)(x - 3)}\right] = (x + 3)(x - 3)\left[\frac{3}{x + 3} + \frac{6}{x - 3}\right]$$

$$(x + 3)(x - 3)\left[\frac{4}{(x + 3)(x - 3)}\right] = (x + 3)(x - 3)\left(\frac{3}{x + 3}\right) + (x + 3)(x - 3)\left(\frac{6}{x - 3}\right)$$

$$4 = 3(x - 3) + 6(x + 3)$$
$$4 = 3x - 9 + 6x + 18$$
$$4 = 9x + 9$$
$$-5 = 9x$$
$$-\frac{5}{9} = x$$

$$\left\{-\frac{5}{9}\right\}$$

58.

$$\frac{9 - \frac{4}{x^2}}{\frac{3}{x} + \frac{2}{x^2}} = \frac{9 - \frac{4}{x^2}}{\frac{3}{x} + \frac{2}{x^2}} \cdot \frac{x^2}{x^2}$$

$$= \frac{9(x^2) - \left(\frac{4}{x^2}\right)(x^2)}{\left(\frac{3}{x}\right)(x^2) + \left(\frac{2}{x^2}\right)(x^2)}$$

$$= \frac{9x^2 - 4}{3x + 2}$$

$$= \frac{(3x + 2)(3x - 2)}{3x + 2}$$

$$= 3x - 2$$

59. x = unknown number

$$x - 14\left(\frac{1}{x}\right) = 5$$

$$x\left(x - \frac{14}{x}\right) = x(5)$$

$$x^2 - 14 = 5x$$
$$x^2 - 5x - 14 = 0$$
$$(x - 7)(x + 2) = 0$$
$$x - 7 = 0 \quad \text{or} \quad x + 2 = 0$$
$$x = 7 \quad \text{or} \quad x = -2$$

The number is 7 or -2.

60.

	time to fill tank	part of tank filled in 1 hr
1st pipe	12	1/12
2nd pipe	18	1/18
together	x	$1/x$

Part of tank filled by 1st pipe in 1 hr. + **part of tank filled by 2nd pipe in 1 hr.** = **part of tank filled by both tanks together in 1 hr.**

$$\frac{1}{12} \quad + \quad \frac{1}{18} \quad = \quad \frac{1}{x}$$

$$\frac{1}{12} + \frac{1}{18} = \frac{1}{x}$$

$$36x\left(\frac{1}{12} + \frac{1}{18}\right) = 36x\left(\frac{1}{x}\right)$$

$$36x\left(\frac{1}{12}\right) + 36x\left(\frac{1}{18}\right) = 36x\left(\frac{1}{x}\right)$$

$$3x + 2x = 36$$
$$5x = 36$$
$$x = \frac{36}{5}$$

It will take $\frac{36}{5}$ hr or $7\frac{1}{5}$ hr with both pipes working together.

61.

$$3x - 7y = 21$$
$$-7y = -3x + 21$$
$$y = \frac{3}{7}x - 3$$

$m = \frac{3}{7}$; y-intercept: $(0, -3)$

62.

$$y - y_1 = m(x - x_1)$$
$$y - 6 = -\frac{2}{3}(x - 4)$$
$$3(y - 6) = 3\left(-\frac{2}{3}\right)(x - 4)$$
$$3y - 18 = -2(x - 4)$$
$$3y - 18 = -2x + 8$$
$$2x + 3y = 26$$

63.
$$m = \frac{y_2 - y_1}{x_2 - x_1}$$
$$= \frac{5 - 4}{-1 - 3} = -\frac{1}{4}$$
$$y - y_1 = m(x - x_1)$$
$$y - 4 = -\frac{1}{4}(x - 3)$$
$$4(y - 4) = 4\left(-\frac{1}{4}\right)(x - 3)$$
$$4y - 16 = -(x - 3)$$
$$4y - 16 = -x + 3$$
$$x + 4y = 19$$

64. A line parallel to $y = 2$ will be horizontal. The horizontal line through (8, −11) has equation $y = -11$.

65. $2x - y = 6$

x	y
0	-6
3	0
1	-4

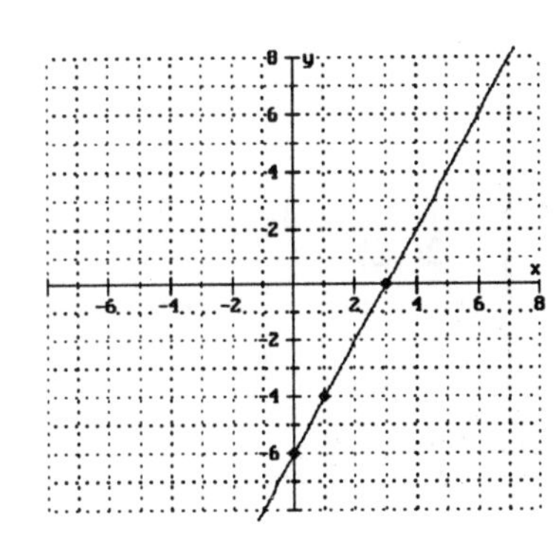

66. $y = |x + 3|$

x	y
-5	2
-4	1
-3	0
-2	1
-1	2

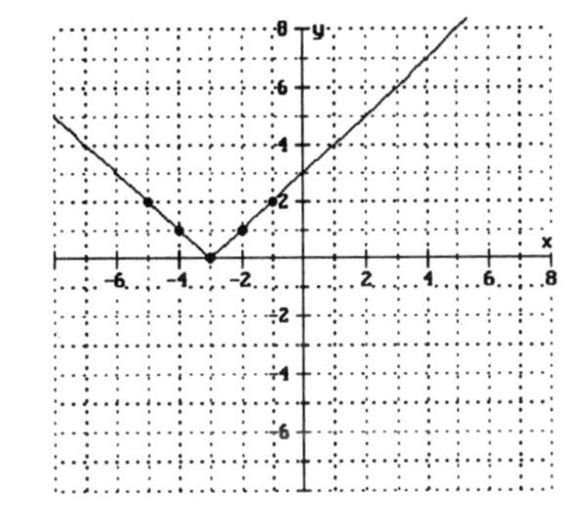

67. $2x - 9y = 4$

x	y
0	-4/9
2	0
1	-2/9

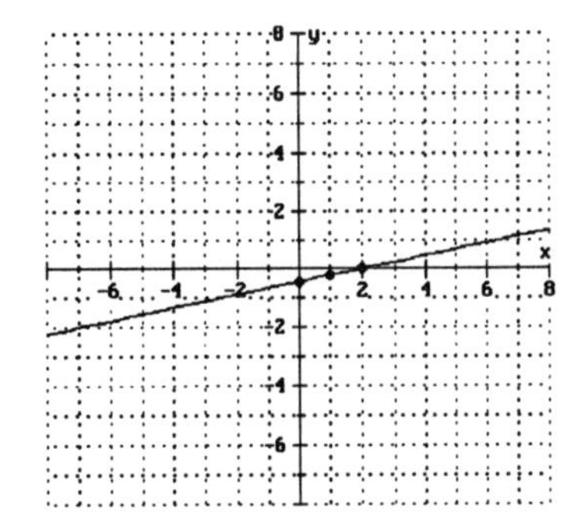

It is a function since the graph passes the Vertical Line Test.

68. $f(x) = -2x + 3$
$f(-4) = -2(-4) + 3 = 11$

69. $g(x) = -2$
$g(0) = -2$

70. The domain will be all real numbers except those values that cause the denominator to equal 0.

$x + 1 = 0$
$x = -1$

Domain: all real numbers except $x = -1$.

71.
$3x - y = 5$
$3(1) - (-2) = 5$
$5 = 5$
True

$x - 4y = -7$
$1 - 4(-2) = -7$
$9 = -7$
False

It is not a solution.

72.
$$\begin{array}{r} x + y = 3 \\ \underline{x - y = 5} \\ 2x \quad = 8 \\ x = 4 \end{array}$$

$4 + y = 3$
$y = -1$

$\{(4, -1)\}$

73.
$$\begin{cases} 2x + 3y = 11 \\ -x + 5y = 14 \end{cases}$$

$$\begin{cases} 2x + 3y = 11 \\ 2(-x + 5y) = 2(14) \end{cases}$$

$$\begin{array}{r} 2x + 3y = 11 \\ \underline{-2x + 10y = 28} \\ 13y = 39 \\ y = 3 \end{array}$$

$2x + 3(3) = 11$
$2x + 9 = 11$
$2x = 2$
$x = 1$

$\{(1, 3)\}$

74. $y = 3x$
Substituting into the 2nd equation,

$x + 15(3x) = -92$
$x + 45x = -92$
$46x = -92$
$x = -2$
$y = 3x = 3(-2) = -6$

$\{(-2, -6)\}$

75.
$$\begin{cases} 8x + 5y = 7 \\ 3x + 2y = 2 \end{cases}$$

$$\begin{cases} -2(8x + 5y) = -2(7) \\ 5(3x + 2y) = 5(2) \end{cases}$$

$$\begin{array}{r} -16x - 10y = -14 \\ \underline{15x + 10y = 10} \\ -x \quad = -4 \\ x = 4 \end{array}$$

$3(4) + 2y = 2$
$12 + 2y = 2$
$2y = -10$
$y = -5$

$\{(4, -5)\}$

76.
$$\begin{cases} \frac{1}{2}x + 5y = 32 \\ \frac{1}{4}x - \frac{1}{6}y = 0 \end{cases}$$

$$\begin{cases} 2\left(\frac{1}{2}x + 5y\right) = 2(32) \\ 12\left(\frac{1}{4}x - \frac{1}{6}y\right) = 12(0) \end{cases}$$

$$\begin{cases} x + 10y = 64 \\ 3x - 2y = 0 \end{cases}$$

$$\begin{cases} x + 10y = 64 \\ 5(3x - 2y) = 5(0) \end{cases}$$

$$\begin{aligned} x + 10y &= 64 \\ 15x - 10y &= 0 \\ \hline 16x &= 64 \\ x &= 4 \end{aligned}$$

$$\begin{aligned} 4 + 10y &= 64 \\ 10y &= 60 \\ y &= 6 \end{aligned}$$

$\{(4, 6)\}$

77.

	number of bills ·	value per bill =	total value
\$5 bills	x	5	$5x$
\$10 bills	y	10	$10y$
together	$x + y$		$5x + 10y$

$$\begin{cases} x + y = 21 \\ 5x + 10y = 150 \end{cases}$$

$$\begin{cases} -5(x + y) = -5(21) \\ 5x + 10y = 150 \end{cases}$$

$$\begin{aligned} -5x - 5y &= -105 \\ 5x + 10y &= 150 \\ \hline 5y &= 45 \\ y &= 9 \end{aligned}$$

$$\begin{aligned} x + 9 &= 21 \\ x &= 12 \end{aligned}$$

He has 12 \$5 bills and 9 \$10 bills.

78.

	amount invested ·	rate of investment =	interest earned
5% acct	x	0.05	$0.05x$
7% acct	y	0.07	$0.07y$
together	$x + y$		$0.05x + 0.07y$

$$\begin{cases} x + y = 9000 \\ 0.05x + 0.07y = 590 \end{cases}$$

$$\begin{cases} -0.05(x + y) = -0.05(9000) \\ 0.05x + 0.07y = 590 \end{cases}$$

$$\begin{aligned} -0.05x - 0.05y &= -450 \\ 0.05x + 0.07y &= 590 \\ \hline 0.02y &= 140 \\ y &= 7000 \end{aligned}$$

$$\begin{aligned} x + 7000 &= 9000 \\ x &= 2000 \end{aligned}$$

She has \$2000 invested at 5% and \$7000 invested at 7%.

79. $y - 2x \leq 4$ (solid line)

$y - 2x = 4$

x	y
0	4
-2	0
1	6

Test point: (0, 0)
$y - 2x \leq 4$
$0 - 2(0) \leq 4$
$0 \leq 4$ True
Shade the half-plane containing (0, 0).

$y \geq 1$ (solid line)
$y = 1$ horizontal line

Test point: (0, 0)
$y \geq 1$
$0 \geq 1$ False
Shade the half-plane not containing (0, 0).

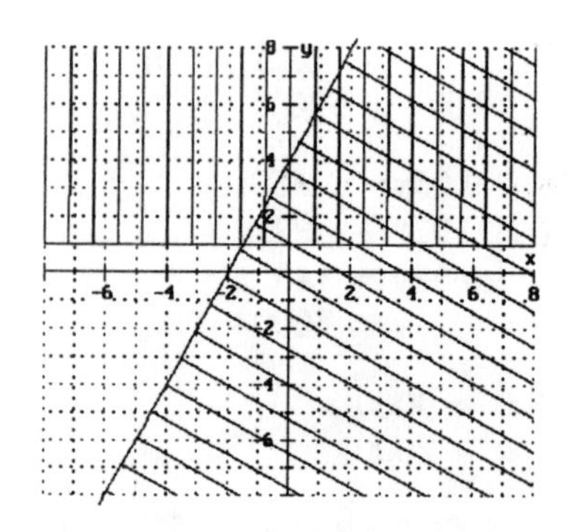

The solution set is where the two shadings overlap.

80. $3x + y > -2$ (dashed line)

$3x + y = -2$

x	y
0	-2
-2/3	0
1	-5

Test point: (0, 0)
$3x + y > -2$
$3(0) + 0 > -2$
$0 > -2$ True
Shade the half-plane containing (0, 0).

$x - y < 4$ (dashed line)

$x - y = 4$

x	y
0	-4
4	0
1	-3

Test point: (0, 0)
$x - y < 4$
$0 - 0 < 4$
$0 < 4$ True
Shade the half-plane containing (0, 0).

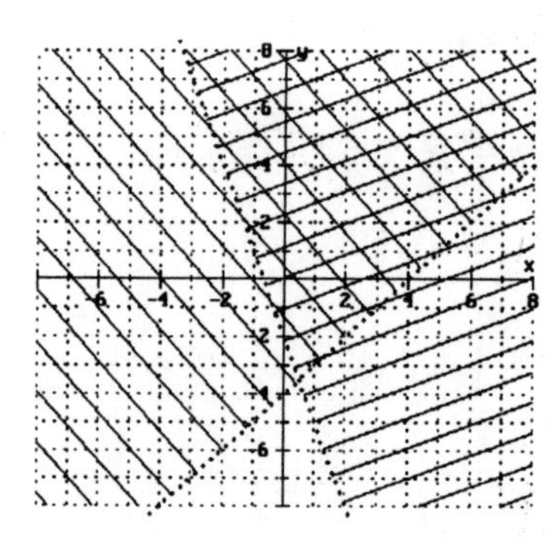

The solution set is where the two shadings overlap.

81. $\sqrt{289} = 17$, because $17^2 = 289$.

82. $-\sqrt[3]{-64} = -(-4) = 4$

83. $81^{3/4} = \left(\sqrt[4]{81}\right)^3 = 3^3 = 27$

84. $25^{-5/2} = \dfrac{1}{25^{5/2}}$

$= \dfrac{1}{\left(\sqrt{25}\right)^5}$

$= \dfrac{1}{5^5}$

$= \dfrac{1}{3125}$

85. $\sqrt{56x^6y^7} = \sqrt{4 \cdot 14x^6 \cdot y^6 \cdot y}$

$= \sqrt{4x^6y^6 \cdot 14y}$

$= \sqrt{4x^6y^6} \cdot \sqrt{14y}$

$= 2x^3y^3\sqrt{14y}$

86. $(x^{4/5})^{10} = x^{(4/5)(10)} = x^8$

87. $\sqrt{45x} - \sqrt{20x} + 9\sqrt{5x^3}$

$= \sqrt{9 \cdot 5x} - \sqrt{4 \cdot 5x} + 9\sqrt{x^2 \cdot 5x}$

$= 3\sqrt{5x} - 2\sqrt{5x} + 9x\sqrt{5x}$

$= \sqrt{5x} + 9x\sqrt{5x}$

88. $\dfrac{3}{\sqrt{5} + 1} = \dfrac{3}{\sqrt{5} + 1} \cdot \dfrac{\sqrt{5} - 1}{\sqrt{5} - 1}$

$= \dfrac{3\left(\sqrt{5} - 1\right)}{\left(\sqrt{5}\right)^2 - 1^2}$

$= \dfrac{3\left(\sqrt{5} - 1\right)}{5 - 1}$

$= \dfrac{3\sqrt{5} - 3}{4}$

89. $\sqrt{3x + 2} = \sqrt{2x - 5}$

$\left(\sqrt{3x + 2}\right)^2 = \left(\sqrt{2x - 5}\right)^2$

$3x + 2 = 2x - 5$
$x + 2 = -5$
$x = -7$

Check: $\sqrt{3(-7) + 2} = \sqrt{2(-7) - 5}$

$\sqrt{-19} = \sqrt{-19}$

Since $\sqrt{-19}$ is not a real number, $x = -7$ is an extraneous solution.

No solution.

90. $d = \sqrt{(x_2 - x_1)^2 + (y_2 - y_1)^2}$

$= \sqrt{(-8 - 3)^2 + (4 - 6)^2}$

$= \sqrt{(-11)^2 + (-2)^2}$

$= \sqrt{121 + 4}$

$= \sqrt{125}$
$= 5\sqrt{5}$

91. $8x^2 - 19x - 15 = 0$
$(8x + 5)(x - 3) = 0$
$8x + 5 = 0$ or $x - 3 = 0$
$8x = -5$ or $x = 3$
$x = -\dfrac{5}{8}$

$\left\{-\dfrac{5}{8}, 3\right\}$

92. $4y^2 = 144$
$y^2 = 36$
$y = \pm\sqrt{36}$
$y = \pm 6$

$\{-6, 6\}$

93. $(5m - 4)^2 = 12$
$5m - 4 = \pm\sqrt{12}$
$5m - 4 = \pm 2\sqrt{3}$
$5m = 4 \pm 2\sqrt{3}$

$m = \dfrac{4 \pm 2\sqrt{3}}{5}$

$\left\{\dfrac{4 + 2\sqrt{3}}{5}, \dfrac{4 - 2\sqrt{3}}{5}\right\}$

94. $2x^2 + 6x - 1 = 0$
$a = 2, b = 6, c = -1$

$x = \dfrac{-b \pm \sqrt{b^2 - 4ac}}{2a}$

$= \dfrac{-6 \pm \sqrt{6^2 - 4(2)(-1)}}{2(2)}$

$= \dfrac{-6 \pm \sqrt{36 + 8}}{4}$

$= \dfrac{-6 \pm \sqrt{44}}{4}$

$$= \frac{-6 \pm 2\sqrt{11}}{4}$$

$$= \frac{2(-3 \pm \sqrt{11})}{4} = \frac{-3 \pm \sqrt{11}}{2}$$

$$\left\{\frac{-3 + \sqrt{11}}{2}, \frac{-3 - \sqrt{11}}{2}\right\}$$

95. $8x^3 = x$

$8x^3 - x = 0$

$x(8x^2 - 1) = 0$

$x = 0$ or $8x^2 - 1 = 0$

$8x^2 = 1$

$$x^2 = \frac{1}{8}$$

$$x = \pm\sqrt{\frac{1}{8}}$$

$$x = \pm\frac{1}{\sqrt{8}}$$

$$x = \pm\frac{1}{2\sqrt{2}}$$

$$x = \pm\frac{1}{2\sqrt{2}} \cdot \frac{\sqrt{2}}{\sqrt{2}} = \pm\frac{\sqrt{2}}{4}$$

$$\left\{0, \frac{\sqrt{2}}{4}, -\frac{\sqrt{2}}{4}\right\}$$

96. $x^2 - 5x + 7 = 0$

$a = 1, b = -5, c = 7$

$$x = \frac{-b \pm \sqrt{b^2 - 4ac}}{2a}$$

$$= \frac{-(-5) \pm \sqrt{(-5)^2 - 4(1)(7)}}{2(1)}$$

$$= \frac{5 \pm \sqrt{25 - 28}}{2}$$

$$= \frac{5 \pm \sqrt{-3}}{2}$$

$$= \frac{5 \pm i\sqrt{3}}{2}$$

$$\left\{\frac{5 + i\sqrt{3}}{2}, \frac{5 - i\sqrt{3}}{2}\right\}$$

97. $\sqrt{-160} = \sqrt{-1 \cdot 16 \cdot 10}$

$= \sqrt{-1} \cdot \sqrt{16} \cdot \sqrt{10}$

$= i \cdot 4 \cdot \sqrt{10}$

$= 4i\sqrt{10}$

98. $(4 + 7i) - (-6 + 2i) = 4 + 7i + 6 - 2i$

$= (4 + 6) + (7i - 2i)$

$= 10 + 5i$

99. $$\frac{2 - i}{3 + 2i} = \frac{2 - i}{3 + 2i} \cdot \frac{3 - 2i}{3 - 2i}$$

$$= \frac{6 - 4i - 3i + 2i^2}{9 - 4i^2}$$

$$= \frac{6 - 7i + 2(-1)}{9 - 4(-1)}$$

$$= \frac{4 - 7i}{13}$$

$$= \frac{4}{13} - \frac{7}{13}i$$

100. $y = x^2 - 8x + 7$

Vertex: $x = \frac{-b}{2a} = \frac{-(-8)}{2(1)} = 4$

$y = 4^2 - 8(4) + 7 = -9$

$(4, -9)$

y-intercept: Let $x = 0$

$y = 0^2 - 8(0) + 7 = 7$

x-intercepts: Let $y = 0$

$0 = x^2 - 8x + 7$

$0 = (x - 7)(x - 1)$

$x - 7 = 0$ or $x - 1 = 0$

$x = 7$ or $x = 1$

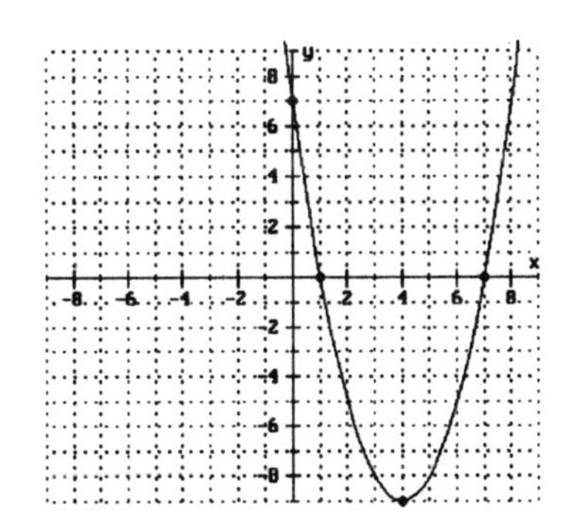